Data Analysis Techniques for High-Energy Physics

Now thoroughly revised and up-dated, this book describes techniques for handling and analysing data obtained from high-energy and nuclear physics experiments.

The observation of particle interactions involves the analysis of large and complex data samples. Beginning with a chapter on real-time data triggering and filtering, the book describes methods of selecting the relevant events from a sometimes huge background. The use of pattern recognition techniques to group the huge number of measurements into physically meaningful objects like particle tracks or showers is then examined, and the track and vertex fitting methods necessary to extract the maximum amount of information from the available measurements are explained. The final chapter describes tools and methods that are useful to the experimenter in the physical interpretation and in the presentation of the results.

This indispensable guide will appeal to graduate students, researchers and computer and electronic engineers involved with experimental physics.

RUDOLF K. BOCK studied and obtained his PhD in Munich, and has been a senior physicist at CERN for more than 30 years. He has extensive experience in most aspects of high-energy physics experiments and, in particular, in the analysis of data both in real time and off-line. He has lectured at many universities and summer schools, and is the author of numerous publications on physics and data analysis methods.

RUDOLF FRÜHWIRTH studied mathematics in Vienna, where he also completed his doctorate. Since 1996 he has held an external readership at the University of Technology, Vienna and he has also worked as a lecturer at the University of Economics, Vienna, as well as at various summer schools. He currently works as a mathematician in the Institute of High Energy Physics in Vienna. He has wide experience in statistical methods, pattern recognition and track reconstruction for high-energy physics experiments, and has published papers and books on data analysis methods.

HANS GROTE obtained his PhD in theoretical physics from the University of Goettingen. In 1967, he joined CERN and worked in experimental data analysis for 20 years. He was responsible for the analysis software of high-energy physics experiments at CERN (SFM, UA2, and CHARM-II). Since 1988, he has worked in accelerator physics and is currently a senior physicist in the optics team of the LHC project.

DIETER NOTZ studied physics at Hamburg University and completed his doctorate at DESY. Between 1974 and 1977 he worked at CERN and is currently a senior physicist at the Deutsches Elektronen-Synchrotron, DESY. He is also the chairman of a European standardisation body for electronics (ESONE).

MEINHARD REGLER received his PhD from the University of Technology in Vienna. He worked at CERN from 1970 to 1975 and then became head of the Experimental Department at the Institute of High Energy Physics, Vienna, where he has been deputy director since 1993. He has been a Professor at the University of Technology, Vienna, since 1989 and has lectured at summer schools throughout the world. He has edited and contributed to several books.

CAMBRIDGE MONOGRAPHS ON PARTICLE PHYSICS, NUCLEAR PHYSICS AND COSMOLOGY

11

General Editors: T. Ericson, P. V. Landshoff

1. K. Winter (ed.): *Neutrino Physics*
2. J. F. Donoghue, E. Golowich and B. R. Holstein: *Dynamics of the Standard Model*
3. E. Leader and E. Predazzi: *An Introduction to Gauge Theories and Modern Particle Physics, Volume 1: Electroweak Interactions, the 'New Particles' and the Parton Model*
4. E. Leader and E. Predazzi: *An Introduction to Gauge Theories and Modern Particle Physics, Volume 2: CP-Violation, QCD and Hard Processes*
5. C. Grupen: *Particle Detectors*
6. H. Grosse and A. Martin: *Particle Physics and the Schrödinger Equation*
7. B. Andersson: *The Lund Model*
8. R. K. Ellis, W. J. Stirling and B. R. Webber: *QCD and Collider Physics*
9. I. I. Bigi and A. I. Sanda: *CP Violation*
10. A. V. Manohar and M. B. Wise: *Heavy Quark Physics*
11. R. Frühwirth, M. Regler, R. K. Bock, H. Grote and D. Notz: *Data Analysis Techniques for High-Energy Physics*, second edition

Data Analysis Techniques for High-Energy Physics

Second edition

R. FRÜHWIRTH, M. REGLER
Austrian Academy of Sciences

R. K. BOCK, H. GROTE
CERN, Geneva

D. NOTZ
DESY, Hamburg

Edited by
M. REGLER and R. FRÜHWIRTH

CAMBRIDGE
UNIVERSITY PRESS

CAMBRIDGE UNIVERSITY PRESS

Cambridge, New York, Melbourne, Madrid, Cape Town, Singapore, São Paulo

Cambridge University Press
The Edinburgh Building, Cambridge CB2 2RU, UK

Published in the United States of America by Cambridge University Press, New York

www.cambridge.org
Information on this title: www.cambridge.org/9780521632195

First published 1990
Second edition 2000

A catalogue record for this publication is available from the British Library

Library of Congress Cataloguing in Publication data

Data analysis techniques for high-energy physics / R. K. Bock. . . . [et al.] – 2nd ed.
p.   cm.
Includes bibliographical references and index.
ISBN 0 521 63219 6 (hardbound)
1. Particles (Nuclear physics) – Experiments – Data processing.
I. Bock, R. K.
QC793.412.D37   2000
539.7′6–dc21   99–20207   CIP

ISBN-13 978-0-521-63219-5 hardback
ISBN-10 0-521-63219-6 hardback

ISBN-13 978-0-521-63548-6 paperback
ISBN-10 0-521-63548-9 paperback

Transferred to digital printing 2006

Contents

Preface to the second edition

If – after a period of ten years – the authors of a book are invited to prepare a new edition, this certainly indicates that the book has enjoyed some success. Naturally, the authors will, in the meantime, have changed their interests and duties, and, therefore, in our case, a new member has joined the team. This has made a sometimes substantial rewriting possible. In addition, R. Frühwirth, who is currently working in the field of track reconstruction, kindly accepted the invitation to make an essential contribution to the overall editing of this revised version.

The first edition was largely based on experience from earlier experiments such as the ones at the ISR or PEP, and was strongly influenced by the lessons learned at the proton–antiproton and electron–positron colliders in Geneva (CERN) and Hamburg (DESY). Another source of inspiration was the research and development work done in preparation for the large electron–positron collider LEP at CERN, the Stanford linear collider SLC, and the electron–proton collider HERA at DESY. Many methods that were new at that time have been field-proven in the meantime, and we are now looking forward to meeting the challenges of LHC, which will come into operation in the year 2006. A similar level of difficulty will be reached soon, probably in 1999, by the experiments at the RHIC heavy ion collider in Brookhaven, with track multiplicities above 1000.

In the field of triggering LHC requires a total change of strategy. With ten to twenty events with high multiplicity superimposed every 25 ns, the first level trigger can be only of a local nature. It requires simple local signatures or combinations thereof, e.g. a muon with large transverse momentum or a large energy deposit in a specific calorimeter window. The bulk of the data has to be pipelined for several microseconds before the first-level trigger can take a decision; the data are then fully digitized and stored in buffers for further processing. The required data reduction rates are extremely large: even with a level-1 reduction of a factor 1000, event rates for subsequent processing are of the order of 100 kHz, and full event collection cannot proceed at more than a few kilohertz. Novel approaches to triggering, for instance the use of neural networks, are described in new examples added in Chapter 1. A gap in the first edition has been closed by adding a section on acceptance and the computation of acceptance corrections. The section on data buses has been extended by a description of the PCI bus.

In the field of pattern recognition, neural networks have been the most important innovation. It seems, however, that they still need to be supplemented by the more conventional methods described in the first edition. Therefore, nearly all of the material in Chapter 2 has been retained. However, several subsections have been added or rewritten. In the area of local track finding methods the Kalman filter has been inserted; the subsection on global methods has been extended by a discussion of the Hough transform, neural networks, elastic arms, and global Kalman filtering. The subsections on compatibility and efficiency have been extended as well.

The sections dealing with pattern recognition in calorimeters and ring imaging Cherenkov counters have been brought up to date. In the field of calorimetry better resolution for higher energies is required. This applies both to current fixed target experiments, such as the high precision experiment at CERN which measures CP violation, and to future collider experiments. In these experiments good identification of jets and precise measurements of missing energy are needed, in conjunction with good time resolution, supported by a more efficient treatment of signals (e.g. initial current method). This in turn leads to a higher performance of data reduction carried out in real time.

Over the past five years, ring imaging Cherenkov counters have been used in large experiments with great success, and new devices are currently under design, including experiments at B-factories and hadron colliders. Pragmatic approaches to the analysis have replaced early attempts at sophisticated maximum-likelihood methods.

Many changes have also been made in Chapter 3. In the sections dealing with track fitting a more compact presentation of the Kalman filter is given, including novel methods of robustification. The same holds for vertex fitting, where the structure of the problem is very well suited to a robustification by the M-estimator. The subsection on material effects (multiple scattering and energy loss) has been rewritten, and a paragraph dealing with the Landau distribution has been added.

Chapters 4 and 5 are probably the ones which were most out of date, because of the rapid change in computing hardware and software technology, and the appearance of ever improving commercial software tools. At the moment the high-energy physics community is at a decisive turning point: both large LHC experiments have decided to switch to the object-oriented programming paradigm, and some collaborations in the USA have already converted a large amount of their software to conform to object-oriented principles. The practical consequences of this momentous step are not yet fully visible; it may be expected, however, that the traditional way of designing, writing, improving and tuning the analysis code will be fundamentally changed. The design, and to a large extent also the implementation, will have to be done by specialists. Similarly, the tuning – a very difficult task because of the extreme load of the detectors – can very likely not be done by a fast intervention of the same physicists who have built the corresponding detector module, but will require intensive communication with the software specialists. On the other hand, the benefits expected from the object-oriented paradigm are manifold: professional design, guaranteed integrity of programs and data at any time, availability of commercial software and database tools, automatic documentation, and others. It is still too early to say how the physics community will adjust to this new working style, and which tools will be at the disposal of the experimenter allowing him or her to access all types of data without becoming involved too deeply with the intricacies of object orientation. We have therefore decided not to rewrite Chapter 5, which would again be out of date very soon in any case; instead the reader will find some hints on new developments in the first two sections of Chapter 4. The rest of Chapter 4 has been extensively revised, guided by the intention to concentrate on the basic issues, not on ephemeral technical solutions.

The references to the literature have been augmented by many recent articles and books; several out-of-date references have been deleted. Some misprints found in the first edition have been corrected.

Finally, words of thanks are due to several people. D. Notz would like to thank M. Dieckvoß for his help in the tedious task of correcting the text in LaTeX format. The editors wish to thank their (former) Ph.D. students G. Fischer, D. Liko and M. Winkler for valuable input from their current research activities, and D. Rakoczy for her scrupulous help with the editing.

M. Regler   Vienna, June 1998
R. Frühwirth
Editors of the second edition

Preface to the first edition

This book brings together for the first time all important data-handling aspects of today's particle physics experiments. For us this was the major reason for writing it. We hope that it will serve our intentions: to present the information which is currently scattered through journal articles, conference proceedings, official and informal laboratory reports, and collaboration notes in a single volume; to facilitate a global view of the problems involved; to show the close connection that exists between the different data-handling fields such as data acquisition, topological and kinematical event reconstruction, and how they are embedded in the framework of hardware and software. Our aim was thus to provide a useful introduction to the field for graduate students, a reference for physicists and engineers working in the field, and a guide to the efficient and successful planning of all data-handling aspects in a particle physics experiment. We hope that the book will, at the same time, prove to be useful for experimenters in fields other than particle physics who encounter similar problems in data acquisition and information extraction.

The first three chapters follow the chronological treatment of the data: real-time data triggering, filtering, and acquisition in the first chapter, recognition of the event topology by assembling tracks and showers from fine-grain raw data in the second chapter, and geometrical and kinematical event reconstruction ('fitting') in the third chapter. Then follow two chapters complementing these methods enumerated: a description of physics analysis methods, such as the principles of data abstraction and selection, and the use of graphics for data interpretation and event display in Chapter 4, and a treatment of the 'managerial' aspects of the software development work, such as the program life cycle and the problems of distributed code writing, in Chapter 5.

Wherever possible, the chapters contain both a presentation and a discussion of the basic principles, and of their application to existing or future experiments. In this way the book should turn out to be more useful than a 'recipe book' containing only detailed descriptions of applications which somehow never quite match the reader's requirements, and which tend to be outdated rather quickly because of the speed of progress in the fields of detectors, electronics, and large computers. We have nevertheless included many real-life examples and a large number of references to original papers for the active researcher in the fields described, since we are aware of the fact that a general description needs to be illustrated by examples, and that in many cases our condensed explanation does not give enough details of the actual implementation of a given method.

It is almost impossible to list all those to whom we are indebted for help, in one form or another, in the long process of the preparation of this book. A number of colleagues were kind enough to read and comment on some portions of the drafts. R. K. Bock expresses his thanks for valuable contributions to A. Clark, T. Hansl-Kozanecka, A. Putzer, B. Schorr, T. P. Shaw, and C. V. Vandoni, D. Notz would like to thank K. Rehlich for many fruitful discussions. He would like to record his special thanks to his wife Elke and his children Katrin, Annika, Dirk, and Wiebke for their patience, forbearance, and encouragement. M. Regler wishes to thank W. Mitaroff for his invaluable help during the preparation of this manuscript; thanks are also due to R. Frühwirth and M. Metcalf for many years of pleasant and stimulating collaboration in the field of track fitting. Several useful suggestions have been made by P. Billoir. H. Grote would like to thank F. Carena, J. C. Lassalle, and M. Metcalf for fruitful discussions. D. Notz and M. Regler would like to thank M. Dieckvoß, E. Ess, U. Kwapil and S. Karsky for their technical assistance in the preparation of the drawings and the text editing. Last but not least we would like to thank the people at Cambridge University Press for their encouragement and cooperation.

Abbreviations

ADC: Analog-to-Digital Converter
ALU: Arithmetic and Logic Unit
AMD: Advanced Micro Devices
AMU: Analog Memory Unit
ANN: Artificial Neural Network
CAM: Content Addressable Memory
CCD: Charge Coupled Device
CERN: European Organization for Nuclear Research
CPU: Central Processing Unit
CMS: Code Management System
DAC: Digital-to-Analog Converter
DD: Data Dictionary
DD: Data Handling Division at CERN
DEC: Digital Equipment Corporation
DESY: Deutsches Elektronen-Synchrotron
DFD: Data Flow Diagram
DSP: Digital Signal Processor
ECL: Emitter Coupled Logic
ERD: Entity Relationship Diagram
FPLA: Field-Programmable Logic Array
HERA: Hadron Elektron Ring Anlage
IBM: International Business Machines
IEC: International Electrotechnical Commission
IEEE: Institute of Electrical and Electronics Engineers
ISR: Intersection Storage Ring
LEAR: Low Energy Antiproton Ring
LEP: Large Electron Positron Ring
LHC: Large Hadron Collider

LSM:	Least-Squares Method
MFA:	Mean-field Annealing
MIMD:	Multiple Instruction Multiple Data
MLM:	Maximum Likelihood Method
MST:	Minimum Spanning Tree
MTBF:	Mean Time Between Failures
MWPC:	Multi Wire Proportional Chamber
NIM:	Nuclear Instrumental Module
PAL:	Programmable Array Logic
PCI:	Peripheral Component Interconnect
PDL:	Process-Description Language
PROM:	Programmable Read-Only Memory
QCD:	Quantum Chromo Dynamics
QED:	Quantum Electro Dynamics
RAM:	Random Access Memory
RICH:	Ring-Imaging Cherenkov Counter
RMS:	Root Mean Square
SA:	Simulated Annealing
SASD:	Structured Analysis Structured Design
SIMD:	Single Instruction Multiple Data
SLAC:	Stanford Linear Accelerator Center
STD:	State-Transition Diagram
TDC:	Time-to-Digital Converter
TOF:	Time-of-Flight
TPC:	Time Projection Chamber
TTL:	Transistor-Transistor coupled Logic

Symbols

$\equiv$	Identical by definition
$\cong$	Approximately equal
$\sim$	Proportional
$\leftarrow$	Is replaced by
(a, b)	Open interval
$[a, b]$	Closed interval
$[a, b)$	Semi-open interval
$\mathbf{x}$	Vector $\mathbf{x}$
$\mathbf{A}$	Matrix $\mathbf{A}$
$\mathbf{AB}$	Matrix multiplication; but $\mathbf{A} \cdot (\mathbf{B} - \mathbf{C})$
$(\mathbf{xx}^{\mathrm{T}})_{ij} = x_i x_j$	
$\mathbf{C}(\mathbf{x})$	Covariance matrix of (random) vector $\mathbf{x}$
$\langle \mathbf{x} \rangle$	Expectation of (random) vector $\mathbf{x}$
$\overset{t}{\mathbf{x}}$	True value of vector $\mathbf{x}$
$\{\mathbf{x}\}$	Region

Introduction

In this book we are concerned with the data-handling aspects of experimental particle physics as it is performed today in laboratories around the world. Particle physics is the science of the fundamental structure of matter: it is a study of the properties of subatomic particles and the way in which they interact. Its ultimate aim is to find a complete description of the elementary constituents of matter and of the forces acting between them, a description which should be as simple as possible. As in all branches of natural sciences, the field is approached from both the theoretical and the experimental points of view. Theory predicts phenomena that can be verified by experiments, and experiments very often provide new insight through unexpected results which, in turn, lead to an improved theoretical description.

Observing phenomena at the subatomic and subnuclear level, on a scale smaller than any other, requires extraordinary instruments. With visible light, objects of a size comparable to the wavelength of this light can be seen in an optical microscope. Smaller distances require quanta of shorter wavelength or, according to de Broglie's principle, quanta of higher energy. Thus electron microscopes operate with accelerated electrons with energies of several thousand electronvolts (keV). Particle accelerators today achieve energies up to 1 TeV (10^{12} eV), which is a thousand times larger than the energy equivalent of the proton mass. This allows the investigation of quarks and gluon quanta deep inside nucleons, at distances down to 10^{-18} m. Accelerators can thus be seen as the 'probe' for observing the ultimately small. The corresponding 'eye' is then the detector, where the incident particles lead to observable effects. The analogy goes even further: in a human being, the final image is the result of distributed brain functions operating on the incoming data. In a particle physics experiment, the final knowledge about a recorded event is the result of an analysis performed by the hardware and software components in the analysis chain. The purpose of our book is to describe precisely these 'brain functions'.

High energies are achieved in particle accelerators where long-lived or stable particles such as protons and electrons acquire the necessary energy in strong electric fields. Since electric fields cannot be made arbitrarily strong, the energy to which a particle can be accelerated in a linear accelerator is limited by the length of the accelerator. For this reason, the circular accelerator – at present – has become the more common type, because here a particle crosses the same accelerating field many times. Such machines are called synchrotrons. The energies achieved today range from several ten thousand million electronvolts (1 GeV = 10^9 eV) to a million million electronvolts (1 TeV = 10^{12} eV). However, for electron–positron circular colliders with $100 + 100$ GeV the ultimate accepted limit of synchroton radiaton losses has been reached. Therefore the next generation of electron–positron colliders will be of the linear type. Tests with very promising results have been done with superconducting cavities, with a field strength of the order of 20 MV/m, as well as with ultra-high-frequency cavities (30 GHz), with a field strength of the order of 100 MV/m.

Particle physics experiments can be divided into two groups: fixed-target experiments, and collider experiments. In the first case, a beam of highly energetic particles is directed at a solid or liquid target, and the resulting secondary particles are observed. The beam particles either come directly out of the accelerator (primary beam), or are created in an intermediate target by the primary beam particles (secondary beam). In this way it becomes possible to provide beams with a great variety of particles, and in particular neutral and short-lived ones, both of which types are unsuitable for direct acceleration. When particles interact at relativistic speeds, only the centre-of-mass energy E_{cm} is available for the interaction proper – to create new particles or to penetrate deeply into the target particles. The fixed-target technique has in this respect a disadvantage, because E_{cm} increases only with the square root of the energy of the incident particle. The remaining energy is used to boost the outgoing particles into the forward direction. Doubling the interaction energy, therefore, requires a quadrupling of the beam energy, and so forth.

This limitation has led to the second mode of experimentation: two counter-circulating beams of highly energetic particles are kept in orbit inside a storage ring and are made to collide at certain 'intersection points' along its circumference. When the two beams consist of packets of particles and their anti-particles, only one set of bending and focussing magnets, and only one beam tube, are needed. Furthermore, since in this case the centre of mass of two colliding particles remains at rest in the laboratory system, the energy E_{cm} available for the interaction itself is now simply equal to the total energy (twice the beam energy in symmetric colliders). Note, however, that the centre of mass of the interacting particles does not remain at rest in the laboratory frame in all colliders. At the CERN 'Intersecting Storage Rings' (ISR), particles of the same charge, namely protons, had been accelerated in opposite directions. The crossing angle between the two beams was 15°. At the 'Hadron-Elektron Ring-Anlage' (HERA) at DESY the kinematics of proton–electron collisions is very asymmetric, so that its observational aspects more closely resemble those of a fixed target experiment. The present large electron–positron colliders (SLC and LEP) are fully symmetric, and so are the detectors. However, the new beauty factory at SLAC will be a strongly asymmetric machine with nominal beam energies of 3.1 and 9 GeV.

The two modes of collision call for two types of experimental set-up: the fixed-target experiments are confronted with energetic particles in the forward direction and, therefore, require long magnetic spectrometers with high-precision tracking detectors and good two-particle resolution, in order to measure the momenta of the outgoing charged particles with sufficient precision from their deflection in a magnetic field. In collider experiments with equal beam momenta and zero crossing angle, no particular direction is privileged by kinematics of point-like particle collisions (although in hadron colliders many secondary particles accumulate around the beam direction; they are fragmentation products of those quarks that do not take part in the collision, the so-called 'spectator quarks'). Accordingly, a compact detector is needed which covers the full solid angle and records a maximum of information about all outgoing particles. Consequently, the lever arm for measuring the deflection of charged particles by a magnetic field is relatively short (the typical scale of a collider detector is 10 m, whereas for fixed-target detectors it may reach 100 m). As a result of the higher interaction energy, the particle multiplicity is on average much higher than in fixed-target experiments. Therefore, to compensate these two effects, high-precision tracking detectors with good two-particle resolution are required here as well.

Apart from the measurement of the momenta of charged particles, the experimental set-up has also to provide clues for their identification. These are given by electromagnetic effects such as Cherenkov radiation, transition radiation, and the relativistic rise of energy loss per unit length, all of which occur when charged particles pass through matter. The third task of the experimental device is to detect neutral particles such as photons, neutrons, or neutral K-mesons. As these particles do not give rise to significant ionization along their path through matter, they must be measured in a destructive way by firing them into a block of matter where they are absorbed and, thereby, create detectable showers of secondary particles. This type of detector is called a 'calorimeter', although the minute amount of heat produced is, of course, not measurable. Instead, one looks for another detectable signal which is in some way proportional to the energy deposited, for instance ultra-violet light emitted by scintillators suitably interspersed with absorbers. Calorimeters serve not only for the detection of neutral particles, but are also used to measure the energy of single charged particles and of highly collimated bundles of particles (so-called 'jets'). For highly energetic particles such as electron–positron pairs from Z^0 decays, the energy resolution of calorimeters is superior to that achieved by curvature measurements in magnetic spectrometers.

In order to provide as much information on secondary particles as possible, huge detector systems are the rule. They frequently consist of a dozen or more subdetectors, each one designed for a specific task. The yoke of the spectrometer magnet in a typical collider experiment has a diameter of 5–10 m and weighs around 1000 tons. The interaction region is surrounded by an arrangement of sophisticated particle detectors with a depth of several metres. Hundreds of thousands of electronic channels perform the analog-to-digital conversion or the time digitization of the many signals coming out of the detector. Complicated electronics is needed for the fast-decision logic that 'triggers' on good events in the presence of a huge background of unwanted ones. High event rates and a high total number of interactions produce a huge amount of data to be processed. This will be dramatic at the 'Large Hadron Collider' (LHC), the forthcoming proton–proton collider with $7 + 7$ TeV beam energy, where some twenty collisions will occur every 26 ns, some of them producing extremely high multiplicities. Furthermore, rare events (which are of course very interesting, but occur at random) must be selected from complex raw data which may contain a million times more events.

So we finally come to the subject of this book, the handling and analysis of these data. This calls in the first stage, during the real-time operation of a detector, upon the most recent technical achievements in the field of electronics and computers, since the analysis begins while an event is still being recorded. Among the hardware components we find: electronic components which allow 'distributed intelligence', i.e. small data processing and storage units at many places in the detector electronics; the latest semiconductor memory chips which can store up to 256 million bits; fast hardwired processors that execute selection algorithms within a few microseconds; microprocessors which allow complex decisions to be taken within a few milliseconds; readout systems and standardized bus interfaces that have been speeded up by at least one order of magnitude during the past decade; distributed parallel processing units that allow sophisticated multilevel triggering, and a considerable reduction of the amount of raw data at an early stage, certainly before they are recorded. Real-time data triggering and filtering is covered in Chapter 1 of this book.

Historically, one used to distinguish between the real-time 'on-line' analysis, and the subsequent steps of the analysis which were performed 'off-line'. The border used to be drawn at the point where the data were recorded on a mass storage device and could thus be 'replayed'. This sharp distinction tends to disappear in modern large experiments where enough real-time computing capacity exists to perform an increasing number of analysis tasks. The first step of this analysis (once an event has been accepted by the real-time filtering) consists normally of 'pattern recognition', where typically the signals belonging to each track are associated, vertices are found, and showers in calorimeters and rings in 'Ring Imaging CHerenkov' (RICH) counters are reconstructed. This task, which is inherently of a combinatorial nature, has in the past mainly been performed on mainframes with large storage systems and high processing speeds. The situation is changing: if enough experience has been gained from the off-line analysis of real data, part of the mass storage for raw data recording can be saved by implementing more and more pattern recognition steps already in the on-line filtering stage. This will, however, normally require from the outset the firm intention to achieve this goal, and a careful planning of the detector in parallel with pattern recognition studies. It goes without saying that data rejected on-line are lost forever, whereas recorded data can be reprocessed should the need arise. If the pattern recognition software has not reached a sufficient state of maturity, this risk should be avoided given today's fast and powerful mass storage devices. Pattern recognition in the context of off-line analysis is discussed in Chapter 2.

The next important task of the data analysis is to extract the ultimate information on charged tracks as provided by the detector's high-precision tracking devices. This then permits a final test on the decisions taken in the course of the pattern recognition stage: an association of tracks to primary and secondary vertices, and the calculation of appropriate input quantities for further physics analysis. This requires the fast solution of the equation of motion for charged particles in a magnetic field, including the efficient storage of the field map, and the flexible handling of the matrix operations involved. A basic understanding of the behaviour of charged particles in matter (multiple scattering, energy loss) is also needed. Track and vertex fitting, including modern filter techniques and their robustification, is covered by Chapter 3.

An important technique in modern data analysis is the application of statistical methods to data abstraction. The development of such methods can be greatly helped by graphical presentation of the data, and by interaction. Graphics also allow the analysis program to be tuned interactively, and may, in addition, help in the recognition of unforeseen relationships between different components of the data. A large number of possible approaches to this problem exist, and the different solutions attempted so far have not yet converged. Broad methods of analysis, including the use of graphics, are treated in Chapter 4.

Large-scale detectors are typically planned, built, and exploited by groups of several hundred physicists and engineers. The fact that many of them take part in the software development imposes certain constraints on the way in which this software is designed, written, and managed. A further complication arises when the analysis programs have to run on different computers (and still have to give the same results), a requirement whose solution is far from obvious. At present, however, the physics community is in a period of upheaval which is tied to the rapid change of hardware and software technology. Some collaborations have already converted their analysis software to conform to object-oriented principles, and many others are likely to follow suit, including the large LHC collaborations. The consequences of this radical reorientation are not yet fully visible. We therefore have decided to refrain from an in-depth discussion of this subject, in accordance with our principle to present to the reader only field-proven methods and applications. The interested reader can, however, find a few hints in Chapter 4.

1

Real-time data triggering and filtering

1.1 Definitions and goals of triggers and filters

The task of a trigger system is to select rare events and to suppress background events as efficiently as possible. To illustrate the trigger problem let us assume that one has to find a friend among the 13 000 000 inhabitants of Mexico City. To find this friend requires a trigger sensitivity of about $1 : 10^7$. With some further knowledge the selection problem can be reduced. He or she is in the city only for a short time and probably lives in a hotel. From 13 000 000 choices one is now down to 10 000 hotel guests. Assuming that he or she lives in a hotel near the centre of the city reduces the search to a group of 400 people. This example demonstrates the various trigger levels for reducing the number of choices.

Data taking is the limiting factor in many experiments. The high rates and the volume of data which must be read out by the data-acquisition system require there to be some time during which no data can be taken. This time is called 'dead time'. In order to reduce dead time one has to improve the quality of the trigger by sophisticated processors which increase the number of good events per time unit.

1.1.1 General properties of particle accelerators

In scattering experiments a high-energy particle beam produced in an accelerator is directed either onto a target where scattering takes place or towards a highly focussed beam coming from the opposite direction. Electrons are liberated from a high voltage triode tube, while protons originate from hydrogen which dissociates into positive ions with the help of oscillating electrons. These charged particles are then accelerated by linear accelerators that have a set of cavities, one behind the other, to accelerate the beam and quadrupoles to focus the beam. They are used

at low energies (400 MeV) to inject particles into circular accelerators or
at high energies to accelerate heavy ions or electrons. At the Stanford
Linear Accelerator Center (SLAC) electrons reach an energy of up to
60 GeV/c.   Linear accelerators are proposed for the future that will
accelerate electrons up to very high energies (500 GeV for the DESY
linear collider (TESLA)) (Brinkmann *et al.* 1997) to avoid the difficulties
when extracting the beam and to avoid energy loss from synchroton
radiation.   The energy loss in a circular accelerator for electrons per
revolution is

$$\Delta E \; [\text{keV}] = 88.5 \cdot E^4 \; [\text{GeV}]/r[\text{m}]. \qquad (1.1)$$

where E is the energy of the electrons in gigaelectronvolts (GeV) and r is
the radius of the accelerator in meters. The energy loss for a machine with
$r = 200$ m and $E = 18$ GeV is then $\Delta E = 46$ MeV. The loss increases to
124 MeV if the beam energy is increased to 23 GeV.

The beams of a linear accelerator are of high quality. They have a good
energy resolution, are well focussed and have very little halo around them.
On the other hand, linear accelerators are long (3.2 km at SLAC), they
need many accelerating elements and deliver short pulses.

One can overcome these disadvantages by using circular machines in
which particles are forced by dipole magnets to stay in a closed orbit.
The cavities are used to accelerate the same particles in each cycle. One
needs less space but the beam quality is not very good: the beam spot
is larger with more halo, the energy resolution is not so good and one
has to compensate for the loss of synchrotron radiation. In the electron
synchrotrons the magnetic field, and therefore the energy, is ramped like a
sine wave. The beam is extracted at the maximum of this wave. For long
extraction times the energy of the extracted beams varies within certain
limits. The typical time taken to accelerate electrons to 7 GeV is 10 ms
and to accelerate protons to 500 GeV is 50 s. The intensity of the extracted
beam is of the order of 10^{13} particles/pulse.

1.1.2 Secondary beams

If one wants to investigate the interactions of neutral particles (e.g. K^0),
unstable particles (e.g. $\pi^{\pm}$, $K^{\pm}$) or antiprotons with matter one has first
to produce these particles. To this end the extracted proton beam hits a
primary target. The particles produced in these collisions are selected and
directed as secondary beams towards the target of the experiment. A beam
transport system containing a set of dipole magnets and quadrupoles is
used to transport charged particles in a certain angle and momentum
interval.

To produce a photon beam, electrons are first directed onto a target of high atomic number A and then deflected by a magnetic field. The energy spectrum of the photons is inversely proportional to the momentum of the photons, thus many low-energy photons are produced. By having lithium hydride in the beam the number of low-energy photons can be reduced. When these photons are directed towards another target they produce e^+e^- pairs. The positrons can then be selected by a magnetic field and a collimator.

1.1.3 Energy balance in scattering experiments

In fixed-target experiments part of the energy of the incoming particles is not available for the interaction but wasted in boosting the particles in the forward direction. Storage rings or *colliders* are used to increase the energy available in an interaction. In a collider two beams from opposite directions interact with each other. The energy W available in an interaction can be expressed in a *Lorentz invariant* form:

$$W^2 = s = (p_1 + p_2)^2 \qquad (1.2)$$

where s is the square of the invariant energy, $p_1 = (m_1, \mathbf{p}_1)$ is the four vector of the incoming particle, $p_2 = (m_2, \mathbf{p}_2)$ is the four vector of the target particle, $(p_2 = (m_2, 0)$ for a fixed target). When a proton of energy $E = 400$ GeV hits a stationary hydrogen target (protons) the square of the invariant mass is

$$\begin{aligned} s &= m_p^2 + m_p^2 + 2Em_p = 754 \text{ GeV} \\ (m_p &= 0.938 \text{ GeV}) \\ W &= \sqrt{s} = 27.5 \text{ GeV} \end{aligned} \qquad (1.3)$$

Subtracting twice the proton mass for the surviving protons (baryon number conservation) leads to the energy available in the reaction

$$M_x = W - 2m_p = 25.6 \text{ GeV} = 2 \times 12.8 \text{ GeV}$$

which means that two colliding protons of 12.8 GeV each release as much energy as a proton of 400 GeV hitting a proton at rest. The equivalent beam energy for a fixed target machine for a given storage ring beam energy E of a storage ring and a particle mass m is then:

$$E_{\text{fixed target}} = 2E_{\text{storage ring}}^2/m \qquad (1.4)$$

Three different types of collider can be distinguished: hadron–hadron machines, lepton–lepton machines and hadron–lepton machines. Several of these machines are listed in Table 1.1.

Table 1.1. *Colliders*

	Particles	Beam energy [GeV]	Luminosity [cm^{-2}s^{-1}]	Crossing [μs]
TEVATRON FERMILAB	$\bar{p}p$	1000	2.5×10^{31}	3.5
CESR CORNELL	e^+e^-	6	6×10^{32}	0.22
PEP II SLAC	e^+e^-	9+3.1	3×10^{32}	2.3
LEP CERN	e^+e^-	100	2.4×10^{31}	11
HERA DESY	e^-p	$27.5e + 920p$	2×10^{31}	0.096
SLC SLAC	e^+e^-	50	0.8×10^{30}	8300
KEK B KEK	e^+e^-	8+3.5	10^{34}	0.002
UNK(II) SERPUKHOV	pp	3000	10^{32}	0.005
LHC CERN	pp	7000	10^{34}	0.025

1.1.4 Luminosity

The number of interactions in a fixed target experiment is proportional to the cross section σ for the type of interaction, the particle flux, and the number of atoms per cubic centimetre in the target multiplied by the length l. The inverse of the last quantity has the dimension of an area and is called the *target constant F*. The number of particles per cubic centimetre is given by the Avogadro constant $N_A \times$ Density ϱ/Atomic weight A.

$$F = A/(N_A \varrho l) \tag{1.5}$$

For a liquid hydrogen target ($\varrho = 0.071$ g/cm^3) with a length of 11 cm the target constant is

$$F = 1 \text{ g}/(6.022 \times 10^{23} \times 0.071 \text{ g/cm}^3 \times 11 \text{ cm})$$
$$= 2.1 \times 10^{-24} \text{ cm}^2 = 2.1 \text{ b}$$

Cross sections are measured in barn (1 b $= 10^{-24}$ cm^2). Assuming a given cross section one can estimate the number of reactions per second by

$$\frac{N_{\text{events}}}{\text{s}} = \sigma \frac{N_{\text{flux}}/\text{s}}{F} = \sigma \times \text{Luminosity} \tag{1.6}$$

The *luminosity* is a measure of sensitivity and gives directly the number of events per second for a cross section of 1 cm^2. For the target described above and a flux of 10^7 particles per second the luminosity is

$$L = 4.8 \times 10^{30} \text{ cm}^{-2} \text{ s}^{-1} = 4.8 \text{ μb}^{-1} \text{ s}^{-1}$$

A cross section of 1 μb would result in 4.8 events per second.

In a storage ring the luminosity depends on several parameters such as the number of particles per bunch N_b, the number of bunches in each beam k_b, the distance between bunches or the revolution frequency f and the beam radii σ of the bunches at the crossing point:

$$L = N_b^2 f k_b/(4\pi\sigma^2) \tag{1.7}$$

All these quantities depend on the energy and on the beam dynamics. A bunch with high density may increase in diameter when interacting with a bunch coming from the opposite direction, with the mirror charge on the beam pipe walls or due to head–tail interactions. It is therefore hard to calculate the luminosity for a given storage ring.

The luminosities for some storage rings are given in Table 1.1. They range from 10^{30} to 10^{34} cm^{-2} s^{-1} which is in the same order of magnitude as the example given above. One would therefore expect that the event rates are of the same order of magnitude.

1.1.5 Time structure of accelerators

1.1.5.1 Time structure at fixed-target accelerators. The particles in an accelerator are accelerated by electric fields which are generated by a radio-frequency system. This defines a timing structure and requires particles that are packed into *bunches*. Because of this timing structure an experiment receives particles for only a fraction of the overall time: this is called the *duty cycle* of the machine. The duty cycle is a measure of the efficiency of an accelerator.

$$\text{duty cycle} = \text{available beam time} / \text{total time} \tag{1.8}$$
$$= \text{duration of a bunch} \times \text{number of bunches per second}$$

The duty cycle at the linear accelerator SLAC is:

$$\text{duty cycle (SLAC)} = 1.2 \ \mu s \times 360/s = 0.04\%$$

At the electron synchrotron DESY there are 50 acceleration cycles per second or 50 spills with a spill length of 1 ms. The duty cycle is then

$$\text{duty cycle (DESY)} = 1 \ ms \times 50/s = 5\%$$

The magnetic field and therefore the energy are ramped like a sine wave. With the help of interferences with higher modes of the accelerating field the shape of the acceleration can be modified in such a way as to produce a flat top which prolongs the spill and can improve the duty cycle by some fraction. At the proton accelerators the cycle time to accelerate particles is of the order of 2–50 s and the burst time is of the order of 300 ms–15 s giving a duty cycle between

$$\text{duty cycle} = 300 \ ms/2 \ s = 15\%$$

and

$$\text{duty cycle} = 15 \ s/50 \ s = 30\%$$

A data-acquisition system at fixed-target accelerators must be organized in such a way that data are collected rapidly during the spill time. Filtering, monitoring and recording of data preferably take place in the time between the spills.

1.1.5.2 Time structure at colliders. In colliders or storage rings both rings are filled with particles which are then accelerated to the nominal energy. If particles collide with their antiparticles the beams can be stored in the same ring. The time to fill a storage ring is of the order of five minutes to two hours for e^+e^- or pp colliders. A $\bar{p}p$ collider can only be filled a few times per day because it takes several hours to collect enough antiprotons.

The time required to accelerate the stored particles ranges from some minutes to about half-an-hour, depending on the mass of the particles and on the nominal energy. After tuning the collider for high luminosity with the help of focussing quadrupoles near the interaction regions, the experiments take data for some hours until the intensity of the beams is so low that a new filling procedure is required.

The time between bunches depends on the circumference of the machine and on the number of bunches. The low energy antiproton ring (LEAR) at CERN (2 GeV $\bar{p}p$) is a direct current (d.c.) machine with a Poisson distributed bunch structure. In the proposed future collider LHC the time between bunch crossings will be 25 ns. At HERA the crossing time is 96 ns. It is not possible to reach a trigger decision within such a short time. All incoming data must be time delayed ('pipelined') for several microseconds while the trigger processors are active (see Subsection 1.6.3). In colliders like LEP (22 µs) or SLC (5500 µs) there is enough time to define a trigger between bunch crossings with hardwired processors.

1.1.6 Event rates at different accelerators

1.1.6.1 Event rates at fixed-target accelerators. The event rates increase proportionally with the cross section, the flux, the target length and density, or, if one combines the last three items, with the luminosity. At energies above 10 GeV the pp cross section is about 40 mb and increases at higher energies. It is easy to produce a high-intensity proton beam because one can extract the proton beam directly from the accelerator. When 10^{13} protons per pulse with a pulse length of 10 s hit a hydrogen target of length 11 cm ($F = 2.1$ b) one gets a peak luminosity of

$$L_{\text{peak}} = 10^{13}/(10 \text{ s} \times 2.1 \text{ b}) = 0.48 \times 10^{12} \text{ b s}^{-1}$$
$$= 0.48 \times 10^{36} \text{ cm}^{-2} \text{ s}^{-1}$$

The number of events per second that must be recorded during the pulse is then 19.2×10^9. This rate is remarkably high. In a typical experiment one measures the differential cross section in which particles are scattered through a certain angle. In these cases the measured cross section drops by up to 10 orders of magnitude, giving 0.2 events per second. Taking the acceleration time of 20–50 s into account the rate decreases to 0.04 events per second and the luminosity averages at $L = 10^{33}$ cm^{-2} s^{-1}.

The total photoproduction cross section γp for energies above 4 GeV is of the order of 120 µb. Photons are produced in secondary beams with an intensity of 10^3 'energy-tagged' photons per 1 ms pulse. The peak luminosity for a hydrogen target 11 cm in length is then

$$L_{\text{peak}} = 10^3/(0.001 \text{ s} \times 2.1\text{b}) = 0.48 \times 10^{30} \text{ cm}^{-2} \text{ s}^{-1}$$

giving an event rate of

$$\text{rate} = 120 \times 0.48 = 58 \text{ events per second}$$

1.1.6.2 Event rates in $\bar{p}p$ or pp colliders. In $\bar{p}p$ or pp colliders the total cross section is 60 mb and above. At $\bar{p}p$ colliders the luminosity is limited by the number of antiprotons, as these cannot be produced in large quantities. A typical value of the luminosity is of the order of 10^{30} cm^{-2} s^{-1}, which results in an event rate of 6×10^4 events per second. For selected reactions like the production of the intermediate vector boson Z^0 the cross section is only 2 nb, which results in a production rate of one Z^0 per 5×10^3 s or 19 Z^0 per day. The planned hadron–hadron collider (LHC) at CERN will consist of two proton rings with magnetic fields in opposite directions. The pp cross section is expected to rise to 135 mb at $\sqrt{s} = 17$ TeV. It consists of three parts: the elastic, the diffractive and the inelastic cross section. Only the inelastic part contributes to the event rate of a general purpose detector. The particles emerging from elastic and diffractive events remain almost exclusively in the beam pipe. With an inelastic cross section of 60% of the total cross section and an expected peak luminosity of $L = 10^{34}$ cm^{-2} s^{-1} the event rate becomes $(6 - 8) \times 10^8$ Hz. With 25 ns between bunch crossings the average number of observed interactions in each crossing is about 15! If one wants to trigger on missing energy which is a signature of reactions with an escaping neutrino one has to limit the luminosity in such a way that one gets only one event per bunch crossing:

$$L_{\text{limit}} = \langle n \rangle / (\text{Time between bunches} \times \sigma_{pp}) \qquad (1.9)$$

Even for $\langle n \rangle = 1$ there is more than one event per crossing in 26% of the interactions. (Poisson distribution $P_1(> 1) = 1 - P_1(0) - P_1(1) = 1 - 2e^{-1} = 0.26$.)

1.1.6.3 Event rates in e^+e^- colliders. In e^+e^- machines on the other hand, the cross section is very small. The single photon exchange cross section is determined by the cross section for $e^+e^- \to \mu^+\mu^-$ which is

$$\sigma_{\mu\mu} = \frac{4\pi\alpha^2}{3s} = \frac{21.9 \text{ nb GeV}^2}{E_{\text{beam}}^2} \qquad (1.10)$$

where α is the fine structure constant. The ratio of hadron production to $\mu^+\mu^-$ production is given by the sum of the squared charges of possible quarks multiplied by 3 for the three colours:

$$R = 3 \sum_{i=1}^{6} Q_i^2 = 5. \qquad (1.11)$$

At LEP energies ($E_{\text{beam}} = 55$ GeV, $L = 1.5 \times 10^{31}$ cm^{-2} s^{-1}) this small cross section gives an event rate of

$$N/s = L \times \sigma = 1.5 \times 10^{31} \text{ cm}^{-2} \text{ s}^{-1} \times 4 \times 21.9 \times 10^{-33} \text{ cm}^2/55^2$$
$$= 0.0005/s$$

or 26 events per day. These rare events must be selected from a background which has mainly two sources: beam–gas interactions and showers in the beam pipe. The rates for these processes are of the order of 10^3–10^4 per second. If one operates LEP at energies necessary to produce the Z^0 the production cross section increases to 50 nb which yields an event rate of 1.8×10^4 events per hour.

1.1.6.4 Event rates at electron–proton colliders. In electron–proton colliders the cross section has two parts: a neutral current (γ, Z^0) and charged current ($W^\pm$) contribution. These contributions can be calculated and depend mostly on the momentum transfer Q^2. Above $Q^2 = 1000$ GeV^2 the cross section for 27.5 GeV electrons and 820 GeV protons is of the order of 150 pb. For a luminosity $L = 10^{31}$ cm^{-2} s^{-1} this results in 4 events per hour.

1.1.7 Background rates

A trigger system should select good events and suppress those reactions which are not interesting. Figure 1.1 summarizes the requirements for a trigger in e^+e^- colliders (Waloschek 1984). At a beam energy of 20 GeV at medium luminosities one expects an event rate of 1 event per 5 minutes. Interactions of the beam with the gas that is left inside the beam tube even under good vacuum conditions or interactions of beam particles with the walls of the beam tube cause high background rates of the order of 10^3–10^4 per second. Particles lose energy due to synchrotron radiation or other instabilities of the accelerator. These particles are no longer kept in a closed orbit at the nominal beam position and hit the collimators or walls of the vacuum system. Background processes of this type do not create tracks coming from the interaction region: the tracks are boosted in the forward direction. A good trigger system must therefore reject tracks that do not have their origin at the interaction point (*vertex detectors*). Cosmic ray events may appear at any time; they are neither correlated with the timing structure of the beam nor with the interaction point.

On the other hand, some background events can be useful for the experiment. *QED* events such as $e^+e^- \rightarrow e^+e^-$ (Bhabha scattering) are used for normalization to measure the luminosity of the collider. They produce two collinear tracks. Another source of background comes from

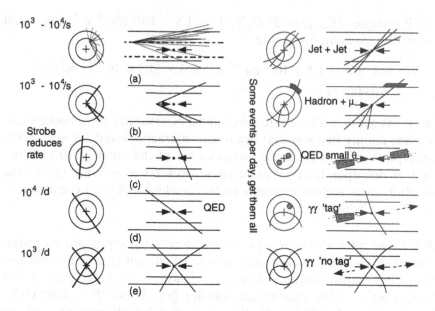

Fig. 1.1. Rates and topologies of e^+e^- background and events. Some selected event topologies are shown at the right part. A trigger must be sensitive to as many event candidates as possible and should discriminate against background as much as possible. (a) Shower originating in the vacuum tube; (b) beam–gas interaction; (c) cosmics ray background; (d) coplanar and collinear beam–beam interaction; (e) multibody beam–beam event.

electrons emitting photons by bremsstrahlung. These photons can react with the photons in the opposite beam: photon–photon- or $\gamma\gamma$-reactions. In high electron beams (1 TeV) the two-photon cross section becomes very large for small two-photon invariant mass.

The main source of background at electron–proton colliders comes from interactions between the proton beam and the gas while the background rates at proton–proton colliders are small compared with the high event rate.

1.2 Trigger schemes

1.2.1 On-line data reduction

A trigger system should select all good event candidates and reject most of the background events. Let us consider a detector with 100 000 detector elements that operates in a storage ring with 0.1 μs time between beam crossings and that generates raw data at a rate of approximately 2×10^{12} bytes per second (one byte contains 8 bits of information). The time to write data onto a storage medium is of the order of 1 Mbyte per second. The on-line data acquisition system must therefore reduce the incoming raw data by at least 6 orders of magnitude. The border line between on-line and off-line data handling is defined here as follows: on-line data acquisition operates with the data from their first appearance in digital form up to writing them onto a permanent storage medium; off-line analysis is any analysis which reads the information from that medium.

In past experiments a simple trigger requiring a coincidence of some electronic signals was used and data were written directly to tape. Nowadays with the advent of microprocessors more sophisticated event filtering can be performed. The speed of the electronic components in an experiment ranges from nanoseconds to seconds and will be briefly discussed here.

The incoming and outgoing particles of a reaction are measured by a detector. One of the most commonly used particle detectors is the *scintillation counter* in which a fraction of energy lost by a charged particle is used to excite atoms in a scintillating medium. Part of the energy in the deexcitation can produce light. The rise time of the light output is of the order of 1 ns and the decay time varies for some plastic scintillators between 1.3 and 4.0 ns.

The scintillator light is converted into an electrical signal and amplified by a *photomultiplier*. Here electrons are emitted in the photocathode by the photoelectric effect and accelerated onto the dynodes of the tubes. The amplification for a 12-stage tube is of the order of 10^7.

Another fast detector is the *Cherenkov counter*. Particles travelling faster than the speed of light in the radiator of a Cherenkov counter produce light at a certain angle. The light is focussed by mirrors onto the photocathode of a photomultiplier.

The spatial resolution of scinillation counters is limited to approximately 1 cm. If better resolution is required one can use *multiwire proportional chambers*, *drift chambers* or *semiconductor detectors* (see Subsection 2.1.1).

The electric signals of a photomultiplier vary both in length and in amplitude. In order to perform logical operations such as AND or OR among several detectors one has to use standard pulses with fixed amplitude, short rise times and small variations in length. *Pulseformers* or *discriminators* are used to generate a standard output signal if the input pulse exceeds a given *threshold*. Typical output pulse lengths and amplitudes are of the order of 20 ns and 0.8 V, respectively (see Subsection 1.6.1). Fast electronic circuits perform logical operations such as AND or OR and operate in the same speed range as discriminators.

When a trigger condition is fulfilled the entire information in the detector is read out into a memory. Simple yes/no information, which just indicates whether there was a particle or not, is stored into a *flip-flop* or a *latch*. The latches are connected to a data-acquisition bus and are read out within approximately 100 ns (see Section 1.8).

Analog information of detectors that measure, for example, particle energy is converted to digital information by Analog-to-Digital Converters (ADCs). An ADC has a reset time of approximately 1 µs. A second event cannot be recorded during that time. If the trigger condition is fulfilled the analog information in the ADC must be converted; this takes from 50 µs to a 1 ms. Many ADCs are then read out by a voltage ramp and a comparator circuit. The trigger system has to reduce the raw input rate to 1 kHz if ADCs of this speed range are used.

If higher rates are required one can use *Flash ADCs* (FADCs), which convert data within 10–100 ns.

The time information needed to measure the speed of particles between two detectors like Time-of-Flight (TOF) counters or to measure the drift time in a drift chamber is digitized by *Time-to-Digital Converters* (TDCs). TDCs need the same amount of time for conversion as ADCs.

If one assumes an event lengh of 100 kbyte and a trigger rate of 1 kHz the transfer rate to a computer is of the order of 100 Mbyte per second, which exceeds the bandwidth of a 'normal' computer bus or data acquisition bus system. Further intermediate stages are needed.

The event is stored *in parallel* by the different detector components called *subdetectors*. With the help of special processors, e.g. *Digital Signal Processors* (DSPs) the input rate may be reduced to 10–100 Hz (see Subsection 1.7.1). At this rate a complete event of 100 kbyte can be transferred to a computer but the data rate is still too high to be recorded onto tape.

Further reduction with filter algorithms is needed. These algorithms can now operate on the entire event and can search for tracks, energy clusters or interaction points and look, for example, for correlations between track elements found in different subdedectors. The computation may take place in programmable processors that can reduce the event rate to 5–10 Hz, leading to a data rate of 0.5–1 Mbyte per second, which matches the tape speed. A good trigger scheme must be designed in such a way that dead time is minimized and the number of good events processed per time unit is increased.

1.2.2 Dead time of electronic components

Each particle detector has a specific resolution time. Particles following each other within a short time interval are not detected as two separate particles. The non-sensitive period of the detector or the electronics is called *dead time*. The loss caused by dead time must be corrected for and must be kept small for reasons of efficiency. In detectors the dead time ranges from several nanoseconds for scintillators up to micro- and

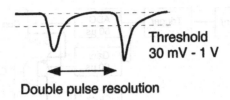

**Threshold
30 mV - 1 V**

Double pulse resolution

Fig. 1.2.   Double pulse resolution for a discriminator. Knowledge of the double
pulse resolution is needed to estimate the maximum rate which can be taken by
a trigger system.

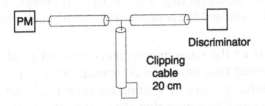

Discriminator

**Clipping
cable
20 cm**

Fig. 1.3.   Clipping cable to make pulses short. For some applications like long
calorimeter signals it might be necessary to shorten pulses with the help of a
clipping cable (PM = photomultiplier).

milliseconds for devices that must be recharged with high voltage like wire
chambers.

The dead time due to electronics and the data-acquisition system covers
a wide range. Fast electronics has a dead time of nanoseconds while the
dead time of a computer can be of the order of minutes if a tape has to
be rewound and a new tape mounted.

Pulseformers or discriminators have a dead time due to the discharging
of capacitors or to the reestablishment of the initial conditions in circuits
with back coupling (see Subsection 1.6.1). For discriminators, instead
of speaking about dead time, one uses the double pulse resolution as
a characteristic quantity. The double pulse resolution is of the order of
10 ns. If the second pulse appears more than 10 ns after the first pulse and
if the first pulse was short (5 ns) then the second pulse will be registered
by the discriminator (Fig. 1.2). To minimize dead time one tries to make
the detector signals short with the help of a *clipping cable* or by pulse
differentiation. For pulse clipping the signal from the scintillator is sent
to the discriminator and to a short coaxial cable which is short-circuited
at the end (Fig. 1.3). Reflections at the end of the cable force the voltage
to zero and the pulse is shorter.

The input signals appear randomly. One must find a way to correct for
losses due to dead time. The entire dead time of a system can be measured
directly if each component has a dead time of more than 20 ns. Either this

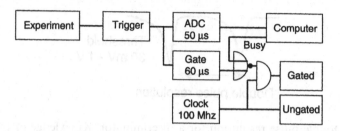

Fig. 1.4. Measurement of dead time in a data taking system. The dead time is measured by comparing the gated and ungated clock. The busy signal of the computer has a variable length (e.g. wait for tape mounting). A gate is used to take the ADC dead time and the computer respond time into account.

dead time is fixed or the components generate a busy signal of variable length. Variable dead time occurs in computers when for example writing data to tape, mounting tapes or waiting for response of an operator. The dead time is measured by comparing the counts of an ungated clock to a clock that is gated with the OR of all busy signals (Fig. 1.4).

To measure the dead time of fast electronic circuits one uses two radioactive sources and compares the single rate of each to the combined rate of both sources. It is assumed that pulses which appear within the dead time do not increase the dead time of the whole system (Stuckenberg 1968). The procedure goes like this. If a scaler has registered n pulses per second it cannot count $n\tau$ pulses for a system with dead time τ. The number of real pulses is then $N = n/(1 - n\tau)$. This number can be approximated for small dead times by $N = n(1 + n\tau)$.

Let A and B be the true rate from each source and Z be the zero rate if no source is present. The zero effect, source A alone, source B alone and A plus B together are then measured and from the equations

$$A + Z = n_a(1 + n_a\tau)$$
$$B + Z = n_b(1 + n_b\tau)$$
$$A + B + Z = n_s(1 + n_s\tau)$$

one can compute the dead time

$$\tau = (n_a + n_b - n_s - Z)/(n_s^2 - n_a^2 - n_b^2) \qquad (1.12)$$

All the other quantities are known and one can therefore compute τ. Care must be taken that both sources have the same geometrical acceptance. With a fast oscilloscope one can measure the dead time directly if the input rate is high. Using the set-up of Fig. 1.5(a) one observes at the oscilloscope the pattern shown in Fig. 1.5(b), which gives the dead time directly.

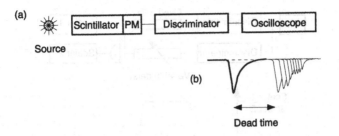

Fig. 1.5.   Measuring the dead time with an oscilloscope (PM = photomultiplier).

1.2.3 *True and wrong coincidences, accidentals*

In a reaction that produces two particles, both particles can be measured with two detectors using a logical AND. If both particles are produced at the same time one has a *true coincidence*. On the other hand, coincidences can be simulated due to the fixed *resolution time* of the detector components. Two independent particles that appear within the resolution time of the detector and the electronics are registered as a coincidence. The rate of these *accidentals* depends on the single rate of each detector and the resolution time. Assume that detector A produces pulses of length τ_1, detector B those of τ_2. A pulse in A that is τ_1 before or τ_2 after a pulse of detector B is registered as a coincidence. For a rate of n_A and n_B in both detectors the number of accidentals can be computed by

$$n_{\text{acc}} = n_A n_B (\tau_1 + \tau_2) = 2 n_A n_B \tau \tag{1.13}$$

For coincidences with several inputs the number of accidentals can be computed by

$$n_{\text{acc}} = k\tau^{k-1} \prod n_k \tag{1.14}$$

The time resolution of a coincidence can be measured with a variable delay line and two scalers (Fig. 1.6).

1.2.4 *Multi-level triggers*

A sophisticated trigger system should be able to reduce the input rate from background processes in an efficient way without losing good events. In order to improve the quality of an experiment using fast processors one has a specific aspect of quality in mind, i.e. the statistical significance or the number of *good* events recorded per time unit.

To reduce dead time and to include complex decisions in the trigger, several trigger levels are required. At each trigger level more information is available to perform better filtering. Assuming that the cross section is so high that the experiment is limited by the tape speed, one can only

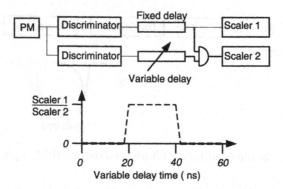

Fig. 1.6. Measurement of resolution time of a coincidence. Knowledge of the resolution time of a coincidence is needed to estimate the maximum rates.

record events with a rate of 7 Hz. In this case one can use three trigger levels.

1. Trigger 1 acts on the prompt information that is available in scintillation counters, proportional chambers or drift chambers with small gaps. Hardwired processors and fast electronics select a rough trigger, which should not run above 10 kHz.

2. The input rate at level 2 is 100 µs per event (10 kHz). Within this time data can be digitized by ADCs. Special processors like digital signal processors (see Subsection 1.7.1) or special processors with *Content Addressable Memories (CAMs)* can do a better track or energy cluster search or vertex fitting. Depending on the readout time of the entire event the output rate should be below 20–100 Hz.

3. At level three the complete event information and digitizations are available. The track filtering can be improved and correlations between various detector elements can be utilized to reduce the data rate to 7 Hz, which matches the tape speed. At this level clusters of fast standard processors can be used to finish their task in 100 ms. If 100 ms are not enough the events can be handed over to several parallel processors. But one must take care when designing a high speed bus. Between each trigger level the events must be buffered. This decouples the different processor speeds and the statistical arrival of the data. The optimal buffer length will be given in Subsection 1.3.1.

With fast processors one can *decrease* the number of recorded background events and at the same time *decrease* the dead time and *increase* the number of processed events per time unit. In order to study the effect of multi-level triggers in a quantitative way (Lütjens 1981) let us assume a given event rate n_e per second. The mean waiting time for an event is then $t_e = 1/n_e$. With a recording time of t_R the number of triggers recorded per second is given by:

$$1 = n_R(t_R + t_e) \qquad (1.15)$$

The fraction of recorded events is then

$$E = n_R/n_e = (1 + t_R/t_e)^{-1}$$

The dead time caused by a long recording time t_R can be reduced by a second stage veto. A fast processor with a good algorithm should be able to detect and reset a trigger caused by a background event within a 'processing time' $t_p < t_R$. With a fast processor the number of triggers aborted per second is Kn_t and the number of triggers recorded per second is n_t. K is called the rejection factor and is a measure of the quality of the algorithm. Equation (1.15) then becomes:

$$\begin{aligned} 1 &= n_t(t_R + t_e) + Kn_t(t_p + t_e) \\ &= \text{Recorded triggers} + \text{Aborted triggers} \qquad (1.16) \\ &= n'_t(t'_R + t_e) \text{ (Processed triggers)} \end{aligned}$$

The fraction of processed events is therefore

$$E' = n'_t/n_e = (1 + t_R/t_e)^{-1} \frac{K + 1}{1 + K(1 + t_p/t_e)/(1 + t_R/t_e)} \qquad (1.17)$$

$$E' = EG$$

G is limited by the fact that the processor itself produces dead time. The limit at large ratio t_R/t_e (high dead time caused by long recording time) is

$$G < G_{max} = \frac{1 + K}{1 + K(t_p/t_R)}$$

A second stage veto reduces the amount of recorded data by a factor of G/K:

$$n_t = (G/K)n_r \qquad (1.18)$$

The number of processed events (recorded plus aborted) is

$$n'_t = (1 + K)n_t$$

The 'reduced' dead time caused by a second stage veto is

$$t'_R = \frac{t_R + Kt_p}{1 + K} \qquad (1.19)$$

Table 1.2. *Influence of fast second stage trigger processors on trigger efficiency, dead time and maximum gain*

Rejection factor K	0	1	3	9	99
Rejection rate $\kappa = K/(K+1)$	0	0.5	0.75	0.90	0.99
Trigger efficiency E [%]	0.99	1.8	3.0	5.0	8.4
Number of recorded triggers n_t/s	9.9	8.9	7.5	5.0	0.84
Number of aborted triggers n_a/s	0	8.9	22.4	45.0	83.2
Number of processed triggers n_p/s	9.9	17.8	29.9	50	84.0
Gain factor G	1	1.80	3.02	50.1	84.9
Reduced dead time t'_R [s] per trigger	0.1	0.055	0.033	0.019	0.011
Maximum gain G_{max}	1	1.82	3.08	5.26	9.17

Figure 1.7 summarizes the effect of a second stage veto for different ratios t_R/t_e and the gain in sensitivity for various rejection rates $\kappa = K/(K+1)$. In this figure we assume that a second stage processor is ten times faster than the recording time: $t_p/t_R = 0.1$, e.g. $t_r = 100$ ms, $t_p = 10$ ms. If the average waiting time of an event is $t_e = 1$ ms one gets the results shown in Table 1.2. The recording time is several orders of magnitudes slower than the time between input triggers at the first stage. With fast processors (fast compared to the recording time) that select good events and reject background events, it is possible with an appropriate algorithm to decrease dead time and to increase the number of events processed per time unit.

1.3 Queuing theory, queuing simulation and reliability

1.3.1 Queuing theory

The results of queuing theory can be used to answer the following questions which are illustrated by Fig. 1.8 (Morse 1958; Allen 1978; Margenau and Murphey 1964). The events occur independently of each other and enter the system with a rate of $\lambda = 5$ events per second. These events are handled by processors with different rates and different buffer lengths.

1. What is the dead time of a system with an input rate λ and a processing rate $\mu = \lambda$?

2. What is the dead time of a system as before with five event buffers in front of the processor?

3. How fast must a processor with a single buffer be to get the same dead time as in (2)?

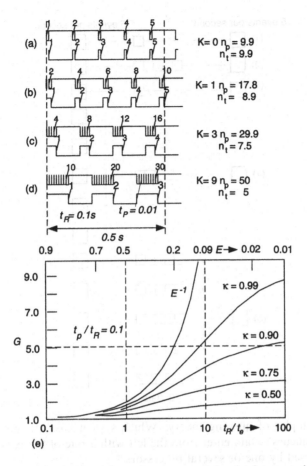

Fig. 1.7. Gain factor obtained by using a second level trigger: t_e is the average waiting time for an event, t_R is the recording time (≈ 0.1 s), t_p is the processing time for the second-stage trigger (here $t_p/t_R = 0.1$, $t_p \approx 0.01$ s). Parts (a)–(d) indicate in which way the number of processed events increases for rejection factors $K = 1$ to 9. In (e) κ is the rejection rate of the second stage trigger. If events occur at a rate of $t_e = 0.001$ s ($t_R/t_e = 100$) and if 90% of the triggers are rejected by the second level processor the gain factor is 5.

4. What is the average queue length of system (4)?

5. What is the average queue length of system (5)?

The answers to these questions will be given at the end of this subsection. The time a computer or a variable flow trigger processor needs to handle an event depends on the complexity of the event, like the number of tracks, the amount of energy deposited in clusters, the number of background hits etc. If one measures the time each event needs and arranges this sequence in order of decreasing length one can plot the number of events

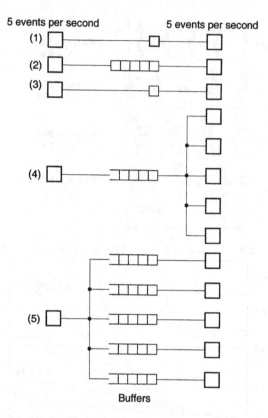

Fig. 1.8. Examples of queuing theory. Which system has little dead time and short waiting queues? Data enter from the left with a rate of 5 events per second and are processed by one or several processors.

that need longer than a given time t. By dividing by the total number of cases one gets the probability $S_0(t)$ that the computation will take longer than time t (Fig. 1.9). The derivative

$$s(t) = -\mathrm{d}S_0(t)/\mathrm{d}t \tag{1.20}$$

is the *probability density function*. It is a rate since its dimensions are probability divided by time. The probability that an event is completed in the interval $[t, t + \mathrm{d}t]$ is given by $s(t)\,\mathrm{d}t$. The average computing time is

$$T = \int_0^\infty S_0(t)\,\mathrm{d}t = \int_0^\infty ts(t)\,\mathrm{d}t \tag{1.21}$$

Irregular arrivals may be described in a manner quite analogous to service times. One measures the times between arrivals, and from these constructs a curve of the probability $A_0(t)$ that the next arrival comes later

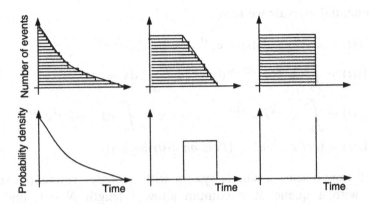

Fig. 1.9. Distributions of service times (top row). The right hand distribution represents a system with constant service time; the left-hand one represents a system with exponential service times, while the middle one needs fixed minimum and maximum service times. The lower diagrams show probability densities that an operation is completed at time t.

than a time t after the previous arrival. Similarly one defines the mean rate λ of arrivals as the reciprocal of the mean time between arrivals:

$$T_a = 1/\lambda = \int_0^\infty A_0(t)\,\mathrm{d}t \tag{1.22}$$

The probability density function $a(t)$ gives the probability that the next arrival comes between t and $t + \mathrm{d}t$ after the previous one:

$$a(t) = -\mathrm{d}A_0(t)/\mathrm{d}t \tag{1.23}$$

In the case where the probability of occurrence of the next arrival is independent of the time since the last arrival, the probability is given by

$$A_0(t) = \mathrm{e}^{-\lambda t}$$

and the probability density function $a(t)$ by:

$$a(t) = \lambda \mathrm{e}^{-\lambda t} \tag{1.24}$$

This is called the *exponential distribution*.

We sometimes wish to know the probability that n arrivals occur within an interval of duration t. This probability is equal to

$$A_n(t) = \int_0^t a(x)A_{n-1}(t - x)\,\mathrm{d}x \tag{1.25}$$

For exponential arrivals we have

$$a(t) = \lambda e^{-\lambda t}, \quad A_0(t) = e^{-\lambda t}$$

$$A_1(t) = \int_0^t \lambda e^{-\lambda x} e^{-\lambda(t-x)} dx = \lambda e^{-\lambda t} \int_0^t dx = \lambda t e^{-\lambda t} \tag{1.26}$$

$$A_2(t) = \int_0^t \lambda e^{-\lambda x} \lambda x e^{-\lambda(t-x)} dx = \lambda^2 e^{-\lambda t} \int_0^t x dx = \lambda^2 t^2 e^{-\lambda t}/2 \tag{1.27}$$

$$A_n(t) = (\lambda t)^n e^{-\lambda t}/n! \quad \text{(Poisson distribution)} \tag{1.28}$$

We will now discuss a simple system with a single exponential service channel, with a queue of maximum allowed length $N - 1$, and with exponential arrivals. The mean arrival rate should be $\lambda = 1/\tau_a$ and the mean service rate $\mu = 1/\tau_s$. The various states of this system can be characterized by the total number of units in the system, the number in service plus the number in the queue. We can expect that the system will settle down to a statistical steady state so that, for example, the numbers of units in the queue is independent of time. To show the principle how one can solve this problem we write the conditions that are required to find one event in the queue in the time interval $[t, t + dt]$ (Fig. 1.10).

1. At time t there was one event in the system. No new event came in, no event left the system in the time interval dt. The probability is

$$(1 - \lambda\, dt)(1 - \mu\, dt)P_1(t)$$

2. At time t there was one event in the system. One event entered and one event left the system. The probability is

$$\lambda\, dt\, \mu\, dt\, P_1(t)$$

3. At time t there was no event in the system. One event entered, no event left the system. The probability is

$$\lambda\, dt(1 - \mu\, dt)P_0(t)$$

4. At time t there were two events in the system. One event left. The probability is

$$(1 - \lambda\, dt)\mu\, dt\, P_2(t)$$

The probability $P_1(t + dt)$ is the sum of all probabilities given above:

$$P_1(t + dt) = P_1(t) + dP_1 = P_1(t) - (\lambda + \mu)dt\, P_1(t)$$
$$+ \lambda\, dt\, P_0(t) + \mu\, dt\, P_2(t) + \text{high order terms} \tag{1.29}$$

This leads to

$$dP_n = [\lambda P_{n-1} + \mu P_{n+1} - (\lambda + \mu)P_n]dt \qquad (1.30)$$

If this is equal to zero then P_n will be independent of time. This gives

$$\mu P_{n+1} + \lambda P_{n-1} - (\lambda + \mu)P_n = 0, \quad \text{for } n > 1 \qquad (1.31)$$

For $n = 0$ this has a special form because there is no P_{-1}. Condition (1) needs to be modified and condition (3) is not possible. This leads to

$$\lambda P_0 - \mu P_1 = 0 \qquad (1.32)$$

This set of equations can easily be solved. If we express all Ps in terms of P_0 we have

$$P_n = (\lambda/\mu)^n P_0 = \rho^n P_0 \qquad (1.33)$$

$$\rho = \lambda/\mu = \tau_s/\tau_a \qquad (1.34)$$

We now want to compute the dead time of a system with N buffers or an upper limit of the queue. The equations hold for n from 0 to $N-1$. For $n = N$ one gets

$$\lambda P_{N-1} - \mu P_N = 0 \qquad (1.35)$$

and the solution $P_n = \rho^n P_0$ holds for $0 \leqslant n \leqslant N$. We can derive P_0 by adding all P_n and requiring that the sum of all Ps is unity:

$$1 = \sum_{n=0}^{N} P_n = P_0(1 + \rho + \ldots + \rho^N) \qquad (1.36)$$

Using

$$(1 + \rho + \rho^2 + \ldots + \rho^N)(1 - \rho) = 1 - \rho^{N+1}$$

gives

$$P_0 = \frac{1 - \rho}{1 - \rho^{N+1}}$$

$$P_n = \frac{(1 - \rho)\rho^n}{1 - \rho^{N+1}} \qquad (1.37)$$

The system cannot accept more events and will produce dead time if the queue is full. The dead time is therefore

$$\tau = \frac{(1 - \rho)\rho^N}{(1 - \rho^{N+1})}, \qquad \rho = \frac{\lambda}{\mu} \neq 1, \quad N = \text{number of buffers}$$

$$\tau = \frac{1}{N + 1}, \qquad \rho = 1 \qquad (1.38)$$

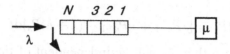

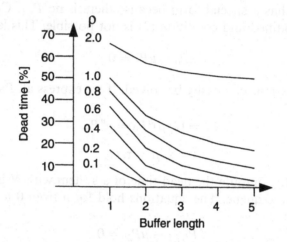

Fig. 1.10. Queue for a system with finite queue length. If the queue is full the system will cause dead time.

Fig. 1.11. Dead time as a function of buffer length and processor speed. ρ is the ratio of service rate to arrival rate. $\rho = 1$ means average arrival time equals average service time. In this case the dead time drops from 50% to 16.6% if five buffers are used.

The dead time for different buffer lengths and processor speeds is given in Fig. 1.11. If one waits long enough there will be a steady-state situation with an average *queue length* in the system of

$$L = \sum_{n=0}^{N} nP_n = \rho \frac{1 - (N+1)\rho^N + N\rho^{N+1}}{(1-\rho)(1-\rho^{N+1})} \tag{1.39}$$

which reduces to

$$L = \begin{cases} \rho + \rho^2 & \rho \ll 1 \\ N/2 + N(n+2)(\rho-1)/12 & \rho \to 1 \\ N - (1/\rho) & \rho \gg 1 \end{cases} \tag{1.40}$$

For $\rho \geqslant 1$ the solutions are not stable. The queue is increasing to infinity. From the queue length one can compute the average waiting time which is $W = L/\lambda$.

We will now discuss systems with an infinite number of buffers (infinite queue length). Each incoming event will enter the queue, there is no dead

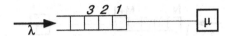

Fig. 1.12. Infinite queues. Each event enters the queue.

time (Fig. 1.12). For an infinite queue the steady-state solutions are

$$P_n = (1 - \rho)\rho^n \tag{1.41}$$

$$L = \rho/(1 - \rho)$$

A system with $\rho = 1$, i.e. arrival rate equal to service rate, will not give a stable solution. The queue length diverges to infinity.

Let us now turn to systems with one queue and several processors, say M processors (Fig. 1.13). Condition (4) now changes. If there are two events in the system and one event leaves we get the probability

$$(1 - \lambda)\mathrm{d}t\, 2\mu\, \mathrm{d}t\, P_2(t)$$

because two events are handled by two processors (if $M \geqslant 2$). The equations of detailed balance for steady-state operation are therefore

$$\mu P_1 - \lambda P_0 = 0$$

$$(n + 1)\mu P_{n+1} + \lambda P_{n-1} - (n\mu + \lambda)P_n = 0, \quad 0 < n < M \tag{1.42}$$

$$M\mu P_{n+1} + \lambda P_{n-1} - (M\mu + \lambda)P_n = 0, \quad M \leqslant n$$

For a system with maximum queue length N the equation for $n = N$ is

$$\lambda P_{N-1} - (M\mu + \lambda)P_N = 0 \tag{1.43}$$

The solution of this system is

$$P_n = (M\rho)^n P_0/n! \quad 0 \leqslant n < M \tag{1.44}$$
$$P_n = M^M \rho P_0/M! \quad M \leqslant n \leqslant N \tag{1.45}$$
$$\rho = \lambda/(M\mu) \tag{1.46}$$

An experiment with five processors of processing rate μ and no extra buffers ($M = N = 5$) will then produce the following dead time:

$$P_1 = M\rho P_0$$
$$P_2 = (M\rho)^2 P_0/2!$$
$$P_3 = (M\rho)^3 P_0/3!$$
$$P_4 = (M\rho)^4 P_0/4!$$
$$P_5 = (M\rho)^5 P_0/5!$$

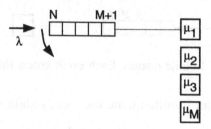

Fig. 1.13. Several servers are working on the queue. Queuing theory should answer the question whether several slow processors are better than a single fast processor.

Normalizing

$$1 = \sum P_n$$
$$= P_0[120 + 120M\rho + 60(M\rho)^2 + 20(M\rho)^3 + 5(M\rho)^4 + (M\rho)^5]/120$$

leads to dead time $P_5 = 5^5\rho^5 P_0/120$. For $\rho = 0.6$ this gives 8.05%. For a system with infinite queue length P_0 is given by

$$P_0 = \frac{1}{\sum\limits_{n=0}^{M-1} [(M\rho)^n/n! + (M\rho)^M/M!(1-\rho)]} \tag{1.47}$$

The queue lengths for different systems are shown in Fig. 1.14.

We can now answer the questions from the beginning of this subsection.

1. For a system with one buffer and an arrival time that equals processing time, i.e. $N = 1$ and $\rho = \lambda$, the dead time is

$$\tau = 1/(N+1) = 50\%$$

2. With five buffers the dead time drops to

$$\tau = 1/(N+1) = 1/6 = 16.6\%$$

By using just derandomization of the arrival times the dead time drops by a factor of 3!

3. A fast processor with a single buffer should produce 16.6% dead time.

$$\tau = 16.6\% = (1-\rho)\rho/(1-\rho^2) = \rho/(1+\rho) \implies \rho = 0.2$$

The processor must be five times faster than a processor in case (2). But buffers are much cheaper than fast processors.

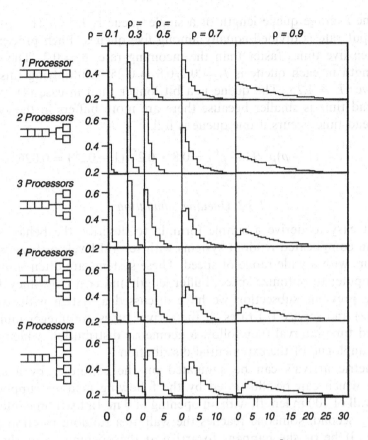

Fig. 1.14. Queue lengths for different systems. One to five processors are used to process data. From left to right the processors are slower or the input rate gets higher. One can imagine that the queue length goes to infinity if the service rate is equal to the arrival rate.

4. The average queue length is $L = \sum nP_n$. In our example we have five processors with a processing power that is as high as the incoming rate, $\rho = 0.2$. Applying the formula mentioned above for P_0 and P_n with $M = 5$ gives

$$L = 0.367\,82 + 0.183\,91 \times 2 + 0.061\,30 \times 3 + 0.015\,33 \times 4 = 0.980\,84$$

For most of the time only one buffer is occupied. A system with five buffers will generate a dead time

$$\tau = 1^5 \times P_0/5! = 0.31\%$$

with

$$P_0 = 120/(120 + 120 \times 1 + 60 \times 1^2 + 20 \times 1^3 + 5 \times 1^4 + 1^5) = 0.368$$

5. The average queue length of a single queue is $L = \rho/(1 - \rho)$. The input rate is divided equally among five queues. Each processor is then five times faster than the incoming rate, $\rho = 0.2$. The queue length in each queue is $L = 0.2/0.8 = 0.25$. Multiplying this by 5 gives $L = 1.25$. The queue is a bit longer than in case (4) but the dead time is smaller because there are more buffers in the system. Dead time occurs if one queue is full.

$$\tau = (1 - \rho)\rho^5/(1 - \rho^6) = 0.8 \times 0.2^5/(1 - 0.2^6) = 0.026\%$$

1.3.2 Queuing simulation

It is not easy to derive a simple formula to describe the behaviour of queues in complicated readout systems with several levels of triggers and processors with a wide range of speed. These systems are often simulated in a computer to optimize speed, buffer length and cost (Dewdney 1985).

In the previous subsection we have discussed a system with average arrival (λ) and service (μ) rates. If the arrival times are integer multiples of a fixed time interval they follow a geometric distribution, which is the discrete analogue of the exponential distribution.

Geometric arrivals can be simulated on the computer by a simple method, which can be illustrated by the following picture: suppose we have a wall of 100 m length with an opening of 5 m. In fixed time intervals, say every second, someone reaches the wall at a random position along the wall. If he or she happens to arrive at the opening he or she may pass through. The time lapse between arrivals then follows a geometric distribution and depends on the ratio w of the width of the opening to the length of the wall (Fig. 1.15). The following FORTRAN subroutine returns a random number distributed according to the geometric distribution with parameter w. It calls the function RNDM which generates a uniform random number in the interval $[0, 1]$.

```
      INTEGER FUNCTION ITDIST(W)
      ITDIST=0
  2   ITDIST=ITDIST+1
      IF(RNDM(ITDIST).GT.W) GOTO 2
      RETURN
      END
```

Noninteger exponential arrivals can be generated by a fast method, which is based on the following general principle. Suppose that $f(t)$ is

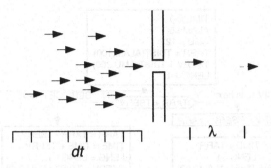

Fig. 1.15. Simulation of geometric arrivals. People arrive in fixed time intervals at a wall with an opening *w*. The time difference between people passing through the opening follows a geometric distribution.

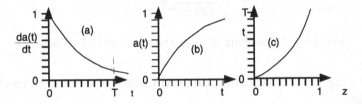

Fig. 1.16. Generation of exponential arrival times in the interval $[0, T]$. The left hand function is the probabability density function. The function must be integrated and than inverted. With the correct constant of integration a uniformly distributed random number between 0 and 1 will generate events that follow the desired distribution.

a probability density function and that $F(t)$ is its associated cumulative distribution function:

$$F(t) = \int_{-\infty}^{t} f(u)\,du \qquad (1.48)$$

If z is a uniform random number in the interval $[0, 1]$, then $t = F^{-1}(z)$ is distributed according to $f(t)$. The principle is illustrated in Fig. 1.16.

We now apply this principle to the generation of exponential arrival times in the interval $[0, T]$. The probability density function and its cumulative distribution function are given by

$$f(t) = \frac{\lambda e^{-\lambda t}}{1 - e^{-\lambda T}}, \quad F(t) = \frac{1 - e^{-\lambda t}}{1 - e^{-\lambda T}} \qquad (1.49)$$

Solving the relation $z = F(t)$ for t gives $t = F^{-1}(z)$:

$$t = -\frac{\ln(z(e^{-\lambda T} - 1) + 1)}{\lambda} \qquad (1.50)$$

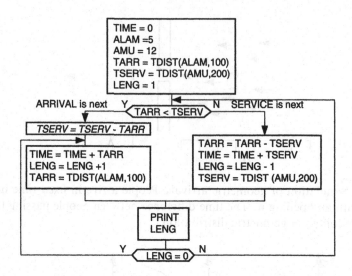

Fig. 1.17. Flowchart of a program to simulate a queue.

The following FORTRAN subroutine will return the desired exponential arrival time:

```
FUNCTION TDIST(ALAM,TMAX)
C=EXP(-ALAM*TMAX)-1
TDIST=-ALOG(Z*C+1)/ALAM
RETURN
END
```

With this routine we can now construct and simulate a queue. TIME is the running time starting at TIME = 0. TARR is the arrival time and TSERV is the service time of the running process. LENG is the length of the queue. The flowchart of a program with $\lambda = 5$ and $\mu = 10$ that simulates a system with one input, one output and an infinite number of buffers is shown in Fig. 1.17.

With similar programs one can simulate readout systems with several buffers, processors with different speeds and trigger systems with several levels.

1.3.3 Reliability theory

The results from *reliability theory* can be used to find an answer to the following questions (Schorr 1974; Lala 1986).

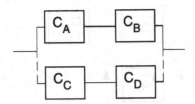

Fig. 1.18. Example of reliability theory. How many components must run in a *series-to-parallel interconnection* to get the same reliability as that of a single component?

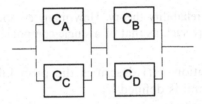

Fig. 1.19. Example of reliability theory. How many components must run in a *parallel-to-series interconnection* to get the same reliability as that of a single component?

1. We have components with given *mean times between failure* (MTBF). For a given task it is necessary to operate two components in series, which will reduce the MTBF or decrease the reliability. How many components must be placed in parallel so that the entire system is as reliable as a single component (Fig. 1.18)?

2. How many components must be used under the conditions shown in Fig. 1.19? Here the system can operate if (AB) or (AD) or (CB) or (CD) are operational.

3. How many components must be used under the conditions shown in Fig. 1.20? Here the stand-by components are only switched on if one of the operating components fails. We assume that the switching is done in zero time and that stand-by components do not age.

The answers to these questions will be given at the end of this subsection.

But what is reliability? The probability $R(t)$ that an unrepairable system performs a specific function without failure under certain conditions for a specific time of length t is called the reliability or reliability function of the system. Unrepairable means that failures of the system during operation lead, in practice, to unrepairable consequences. The probability that a system operates correctly shortly after switching on is nearly 1 and that it still works after an infinite amount of time is 0. This leads to $R(0) = 1$ and $R(\infty) = 0$. An example for $R(t)$ could be $R(t) = \mathrm{e}^{-\lambda t}$.

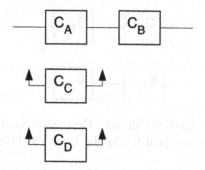

Fig. 1.20.   Example of reliability theory. How many components must *stand by* to get the same reliability such as that of a single component?

The failure distribution $F(t)$ describes the probability of failure of a system before time t and is defined by

$$F(t) = 1 - R(t) \tag{1.51}$$

For an exponential reliability function, $F(t)$ is equal to $F(t) = 1 - e^{-\lambda t}$. The failure density function $f(t) = F'(t)$ describes the lifetime of a system. The failure rate function

$$\lambda(t) = f(t)/R(t) \tag{1.52}$$

gives the probability $\lambda(t)\,dt$ that the system, having reached the age t, will fail during the interval $[t, t + dt]$. For an exponential reliability function $\lambda(t)$ is a constant.

Very often one does not know the reliability function but one can say something about the average lifetime of the system. One determines the MTBF, which is defined as the expected lifetime of the system:

$$\text{MTBF} = \int_0^\infty t\,dF(t) = \int_0^\infty tf(t)dt = \int_0^\infty R(t)dt \tag{1.53}$$

We will now discuss how the reliability changes if two components operate logically in series (Fig. 1.21). This means that if one component fails the whole system goes down.

T_i is the lifetime of the component C_i. If T is the lifetime of the system then $R(t)$ is the probability that $T > t$ or

$$R(t) = P(T > t)$$

For a serial connection we require that

$$R(t) = P(T_1 > t, T_2 > t)$$

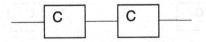

Fig. 1.21. Components operate in series. If one component goes down the whole system is down. How does the reliability change?

If the individual lifetimes are independent it follows that

$$R(t) = \prod_{i=1}^{2} P(T_i > t) = \prod_{i=1}^{2} R_i(t) \tag{1.54}$$

The failure rate function of the series connection is then equal to

$$\lambda(t) = \sum_{i=1}^{2} \lambda_i(t)$$

For constant component failure rates the failure rate of the system sums as

$$\lambda = \sum \lambda_i$$

If both components have the same reliability we get

$$R(t_i, n\lambda_0) = e^{-n\lambda_0 t}$$

and

$$\text{MTBF} = 1/\sum \lambda_i = (1/n)\text{MTBF}_{\text{component}} \tag{1.55}$$

Two components in series with the same MTBF will have only half the MTBF of a single component.

In systems with parallel components one may distinguish *hot parallel systems* in which all components are in the operating state, and *cold parallel connections* or stand-by components, which are available to replace an operating component (Fig. 1.22). Let T be the lifetime of the system. Then we have

$$R(t) = 1 - P(T < t) = 1 - P(T_1 < t, \ldots, T_n < t)$$

The system fails if all components have failed.

$$R(t) = 1 - \prod_{i=1}^{n} P(T_i < t) = 1 - \prod_{i=1}^{n} [1 - P(T_i > t)] \tag{1.56}$$

A system of components $C_1, \ldots, C_n$ in hot parallel connection has the reliability function

$$R(t) = 1 - \prod_{i=1}^{n} (1 - R_i(t))$$

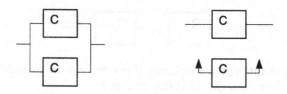

Fig. 1.22. Components operate in parallel. One can distinguish cold and hot redundancy. In cold redundancy a stand-by component takes over if the operating component goes down.

For a system with components of the same reliability the entire reliability improves to

$$R(t) = R(T; n\lambda_0) = 1 - (1 - e^{-\lambda t})^n$$

The failure rate $\lambda(t)$ is no longer a constant but

$$\lambda(t, n\lambda_0) = \frac{n\lambda_0}{\sum_{j=1}^{n-1}(1 - e^{-\lambda_0 t})^{-j}} \tag{1.57}$$

This function increases with time and converges to 1 for $t \to \infty$. The MTBF changes to

$$\mathrm{MTBF} = (\sum 1/k)\mathrm{MTBF}_{\mathrm{component}}$$

Two parallel components increase the MTBF to

$$\mathrm{MTBF}_2 = 1.5 \times \mathrm{MTBF}_{\mathrm{component}}$$

or only by 50 %. However, the time interval at 90% reliability increases by a factor of 4 (Fig. 1.23).

We will now discuss the situation of stand-by components (Fig. 1.22) with cold connection. The components $C_1, \ldots, C_n$ are called cold redundant or stand-by components if they are not in operating state but are available to replace failing components. We assume that components do not age as long as they are stand-by components, which is true for mechanical devices but not necessarily for electronics. Further we assume that an ideal switch can replace the failed component by a stand-by component immediately after failure.

The system consists of just one operating component C_0 for which all the stand-by components are spare parts. If C_0 fails C_1 will replace it and so on. The system fails if all components have failed. This leads to the following mean time between failures:

$$\mathrm{MTBF} = \sum \mathrm{MTBF}_{\mathrm{component}} \tag{1.58}$$

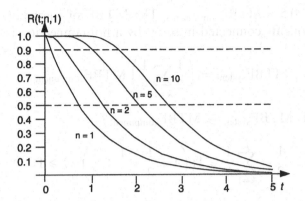

Fig. 1.23. Reliability of coupled systems with hot connection. The components have equal failure rates. On average one does not gain very much for long periods of time, but the time with a reliability at 90% increases by a large factor.

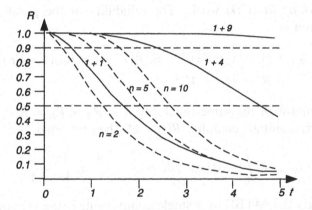

Fig. 1.24. Reliability of systems with cold connections. The reliability of systems with cold connections or stand-by components is higher than for systems with hot connections, assuming that stand-by components do not age. The dashed line shows the corresponding systems with hot connections.

In a system of m operating components in a series connection and n components as stand-by components, we get a failure after $n + 1$ components have failed. For identical components with constant failure rate one gets

$$\text{MTBF} = \frac{n+1}{m} \text{MTBF}_{\text{component}} \qquad (1.59)$$

Generally speaking, a system with cold redundancies is much more reliable than one with hot redundant connections (Fig. 1.24). This comes from the fact that stand-by components do not age.

We can now answer the questions asked at the beginning of this subsection. In the first example the MTBF for two serial components is

$\text{MTBF}_{\text{serial}} = 0.5 \times \text{MTBF}_{\text{component}}$. The MTBF of the whole system of several components connected in series by a hot connection is

$$\text{MTBF}_{\text{system}} = \left(\frac{1}{2} \sum_{k=1}^{n} \frac{1}{k} \right) \text{MTBF}_{\text{component}} \tag{1.60}$$

For which n is $\text{MTBF}_{\text{system}} \geqslant \text{MTBF}_{\text{component}}$?

$$\frac{1}{2} = \sum_{k=1}^{n} \frac{1}{k} \geqslant 1; \left(1 + \frac{1}{2} + \frac{1}{3} + \frac{1}{4} \right) / 2 \geqslant 1$$

We need eight components (four parallel branches with two components in each branch) to get an MTBF for the whole system that is at least as high as that for a single component.

In the second example a four component system can operate if (AB) or (AD) or (CB) or (CD) work. The reliability of the parallel-to-series interconnection is

$$R_{\text{system}} = [1 - (1 - R_A)(1 - R_C)][1 - (1 - R_B)(1 - R_D)]$$
$$= R^4 - 4R^3 + 4R^2$$

for components with the same reliability $R = R_A = R_B = R_C = R_D$. The MTBF for exponential reliability $R(t) = e^{\lambda t}$ then becomes

$$\text{MTBF}_{\text{system}} = \int_0^\infty R(t)_{\text{system}} \, dt = \frac{4}{2\lambda} - \frac{4}{3\lambda} + \frac{1}{4\lambda} = \frac{11}{12\lambda} \tag{1.61}$$

which is nearly the MTBF of a single component. Using six components, i.e. two series of three components, the MTBF of the system is equal to the MTBF of a component times 73/60, or higher by about 20%.

In the third example, for a system with two components in series, we need only one spare component to get the same reliability compared to a single component ($n = 1$, $m = 2$):

$$\text{MTBF}_{\text{system}} = (2/2)\text{MTBF}_{\text{component}} = \text{MTBF}_{\text{component}} \tag{1.62}$$

We need only three components in the case of cold redundant connection compared to eight components for hot redundant connection and six components for parallel-to-series connection.

1.4 Classifications of triggers

1.4.0.1 Classification of triggers based on physical processes. When designing a trigger for an experiment one has to search for physical criteria to select event candidates from background processes. After having defined the trigger conditions one has to check systematically with Monte Carlo studies the way in which a trigger can influence the results of an experiment.

In accelerators or storage rings the beams are very small in diameter and have a fixed timing structure. In an accelerator beam particles are extracted from the accelerator at the end of the acceleration cycle. The trigger electronics should therefore only accept events if they appear within a short time window ('spill gate'). At colliders the beam signal is generated by a pick-up coil at each bunch crossing.

The various final states of an interaction define which class of trigger may be sufficient. Some of these classes are as follows.

1. Track multiplicity of the event.

2. Direction of particles.

3. Deflection or curvature of particles to measure momentum.

4. Coplanarity of the event.

5. Type of particle.

6. Deposited energy in all or part of the detector to measure total or transverse energy.

7. Missing energy.

8. Invariant mass.

9. Interaction point of event ('vertex') or secondary interaction ('kinks', 'V^0').

1.4.0.2 Classification of triggers based on realization in electronics. In the classical spectrometer experiment at fixed target machines simple coincidences are used to count the number of scattered particles and to define the particle flux in the beam. Their ratio multiplied by the target constants and correction factors determines the cross section. The trigger decision is available within 100 ns. Data are then either recorded by a computer, or scalars are incremented to measure total rates. Thanks to technological advances in the field of microelectronics new and powerful tools are available to upgrade the triggering system. The advent of microprocessors and the increasing availability of special integrated circuits with a better cost/performance ratio has allowed the implementation of sophisticated multi-level trigger schemes in the event selection. This leads to a wide range in time from 100 ns to 10 ms and in complexity from two coincidences to programs with 1000 statements and 15 000 cycles. Also the prices range from 1000 Swiss francs for a coincidence or a microprocessor to 100 000 Swiss francs for special hardware devices filling an entire rack of electronics.

Because of this variety one tries to group trigger systems into various classes (Conetti 1984).

1. Fixed-flow triggers.

2. Variable-flow triggers.

3. Logical triggers.

4. Arithmetical triggers.

5. Program-driven processors.

6. Data-driven processors.

As a first step one can distinguish *fixed-flow* and *variable-flow triggers*. Fixed-flow triggers produce their trigger signal at a given time independent of the complexity of the event. Variable-flow triggers contain counters, loops and programs. Only a minimum and maximum time are known. Fixed-flow processors can be grouped into *logical triggers* dealing with logic operations like AND and OR and *arithmetic triggers* asking for a certain pulse height above threshold or a certain number of counters. Variable-flow processors are either *data driven* or *program driven*. In a *data driven processor* an operation is executed when all data needed for the execution are available. This is implemented by sending tokens or data ready signals. A *program-* or a *demand-driven processor* demands an execution that requires evaluation of its arguments.

1.4.1 Trigger on event topology

1.4.1.1 Trigger on track multiplicity. The *track multiplicity* of an event depends on the available energy, the physical process and the average energy of the emerging particles. In an experiment investigating hadronic final states, in general, the production of hadrons can be described by a multistring model leading to multiplicities of 30 charged particles at energy $\sqrt{s} = 500\,\mathrm{GeV}$ and about 70 charged particles at $\sqrt{s} = 20\,\mathrm{TeV}$ (Kunszt 1987). To trigger on these events one can surround the interaction region by several scintillation counters and then trigger on the number of scintillation counters that give a hit.

New phenomena or new particles can result in events with low multiplicities. Examples of new particles are the production of the Higgs boson H or supersymmetric particles like sleptons $\tilde{l}$ or winos $\tilde{W}$. These events show only a few tracks and cannot be detected by a multiplicity trigger because background events often also show high rates of low multiplicities.

1.4.1.2 Trigger on direction of particles. The *direction* of particles together with their momenta can be used to trigger on events with high momentum transfer. In fixed target experiments or in experiments at storage rings, most events are produced at low momentum transfer with particles going in the forward direction. When probing matter at small distances one is interested in events with high transverse momentum p_T. To trigger on these events one has to detect particles emerging from the interaction region perpendicular to the incoming beam particle and having high energy. This is preferably done by calorimeters.

1.4.1.3 Trigger on momentum of particles. If, for example, the inelastic scattering of electrons is to be investigated in a fixed target experiment, the scattered electron is detected at a given angle and a given energy. This then defines a fixed momentum transfer. The detectors used are called spectrometers. They can be rotated perpendicular to the beam around the interaction region. The momentum of the electron is measured by its *deflection* in a magnetic field. In front of and behind the magnet are multiwire chambers to reconstruct the tracks, and behind the last track are scintillation counters to generate a fast trigger.

1.4.1.4 Trigger on coplanarity. *Coplanar* events with a well-defined simple kinematics and known cross section are used for measuring the incoming flux in a separate reaction. In e^+e^- storage rings one uses the coplanar and collinear reaction $e^+e^- \rightarrow e^+e^-$ (Bhabha scattering), the cross section of which is known, to measure the luminosity.

Three jet events in e^+e^- reactions are coplanar if one jet is the fragment of a gluon that is radiated by quark–gluon bremsstrahlung. The fact that in this case the jets are coplanar does not mean that there are no particles outside the 'plane'.

1.4.2 Trigger on type of particle

For many experiments a topological trigger is not sufficient. Often one wants to trigger on a certain type of particle. In the search for the Higgs boson one could use the decay mode $H \rightarrow ZZ \rightarrow llll$, where l is a lepton: $l = e, \mu$ or τ. Another decay of the Higgs boson could be $H \rightarrow WW \rightarrow l\nu l\nu$, resulting in two leptons and two escaping neutrinos. In charge exchange reactions such as $K^-p \rightarrow \bar{K}^0 n$ one would trigger on the $\bar{K}^0$. Triggering on types of particles is summarized in Table 1.3. A

Table 1.3.　*Triggers for type of particle*

Particle	Identification method
γ, π^0	photons and neutral pions, γ and π^0
	electromagnetic showers in lead-scintillator, NaI
	bismuth–germanium–oxide (BGO), lead glass,
	lead–liquid argon
e	electromagnetic showers like γ
	Cherenkov identification
	transition radiation
μ	penetration through magnetized iron
$\pi^\pm$	Cherenkov identification
$K^\pm$	Cherenkov identification
K^0	multiplicity increase $n \rightarrow n + 2$
	tracks not from interaction point
	charged particle veto
$p, \bar{p}$ (fast)	Cherenkov identification
p (recoil)	range
	solid state detector
$n, \bar{n}$	plastic scintillator
	liquid scintillator
	^{3}He-filled proportional wire chambers
	n–p elastic scattering
	charged particle veto
$\Lambda\bar{\Lambda}$	multiplicity increase $n \rightarrow n + 2$
	fast p

good overview can be found in Fernow (1986). In addition, one can use TOF measurements or record dE/dx ionization losses (see Subsection 3.3.1.5).

1.4.2.1 Photons and Neutral Pions, γ and π^0. When high energy photons interact with dense matter they create e^+e^- pairs. The electrons then radiate photons again by bremsstrahlung. A single photon thus creates many electromagnetic particles. A collection of all these particles is called an electromagnetic shower. Thin lead sheets can be used to convert the photons. The electrons created are then detected in scintillators or liquid noble gases. This is called a sandwich calorimeter. Another possibility for detecting photons is the production of light in heavy transparent material, as in lead-glass Cherenkov counters or in NaI scintillation counters. More detailed information can be found in Section 2.5. Neutral pions are reconstructed in detectors with fine granularity by measuring the two photons and reconstructing the invariant mass.

1.4.2.2 Electrons, e. Electrons, like photons, are detected by their electromagnetic showers. To distinguish them from photons a scintillation counter in front of the shower counter indicates the appearance of a charged particle. In certain momentum ranges one can use Cherenkov counters.

Another method for detecting electrons uses the phenomenon of transition radiation. A moving particle emits radiation when crossing different dielectrics. The transition radiation is emitted into a small cone around the particle's direction with typical angle of $\Theta = 1/\lambda$, $\lambda = E/m$. The local intensity increases linearly with λ. Here λ denotes the *Lorentz factor*. To achieve sufficient intensity particles must cross a few hundred radiator boundaries.

1.4.2.3 Muons, μ. Muons have the ability to penetrate a considerable depth of matter before being absorbed. A muon detector consists of a hadron absorber (for example 1 m of iron) and a detector for charged particles, such as a multiwire chamber or a hodoscope of scintillation counters.

If one wants to measure the momentum of the muons one uses segmented magnetized iron. The major background comes from hadron *'punch through'* or *'shower leakage'*. It is possible for a small fraction of hadrons to reach the hodoscopes.

1.4.2.4 Charged pions, π. In hadronic events with high multiplicity it is more likely for any track to be a $\pi^{\pm}$ than anything else. One therefore assumes that the observed tracks are pions. If one has to discriminate pions from kaons or protons one uses Cherenkov counters or one measures TOF or ionization loss.

Cherenkov counters have the disadvantage that they can mainly be used only for particles travelling on the optical axis. To overcome this limitation *Ring Imaging Cherenkov Counters (RICH)* are used for particles diverging from an interaction point. A spherical mirror focusses the cone of the Cherenkov light onto a 'ring image' on a spherical detector (see Section 2.6).

1.4.2.5 Kaons, K. To discriminate fast charged kaons from pions one uses two threshold Cherenkov counters. One counter sets a veto for pions while the second counter gives a positive signal for kaons above a certain

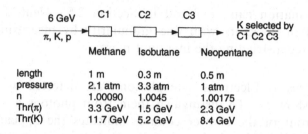

	Methane	Isobutane	Neopentane
length	1 m	0.3 m	0.5 m
pressure	2.1 atm	3.3 atm	1 atm
n	1.00090	1.0045	1.00175
Thr(π)	3.3 GeV	1.5 GeV	2.3 GeV
Thr(K)	11.7 GeV	5.2 GeV	8.4 GeV

Fig. 1.25. Threshold Cherenkov counters to select K-mesons. The beam at 6 GeV contains π, K and p. The experiment studies reactions with K-mesons, which are selected by Cherenkov counters. The threshold is adjusted in such a way that the contamination of $\bar{p}$ for a K^- signal is below 1%; for a $\bar{p}$ signal the K^- contamination is 20% (n = refractive index, Thr = threshold).

threshold. This second counter also gives a signal for pions. Therefore one needs the first counter to flag the pion. Neutral kaons K_s^0 decay after a short distance (of the order of some centimetres) into charged pions. The multiplicity increases then by two. The background produced by Λ production is reduced in the offline analysis in which the invariant mass of the $\pi^+\pi^-$ or the $p\pi^-$ system is computed.

1.4.2.6 Protons, p. Fast protons are identified with Cherenkov counters. Slow protons can be measured by TOF or ionization loss. Figure 1.25 shows a set-up of several Cherenkov counters in a hadron beam to select π, K or p.

1.4.3 Trigger on deposited energy

The resolution of wire chambers in terms of track energy decreases with energy while that of calorimeters increases. Therefore, at high-energy accelerators with energies above 100 GeV calorimetry takes a prominent place. A particle hitting a shower counter deposits energy there. The light output of the scintillator or, generally speaking, pulse height is a measure of the energy of the incoming particle or jet.

In a calorimeter that surrounds the interaction region hermetically one can simply trigger on total energy or transverse energy being above a certain threshold. This threshold can be rather high, of the order of half the total centre-of-mass energy. In calorimeters with high granularity different types of triggers can be set up, such as the global sum for the entire detector, local sums if part of the detector gets high energy, or sums that depend on the angle or transverse momentum p_T.

1.4.4 Trigger on missing energy

A trigger on missing energy is used in experiments that search for reactions with escaping neutrinos. In *ep* colliders, for example, one distinguishes neutral-current and charged-current events. In neutral-current events the electron is scattered and a photon or a Z^0 is exchanged. In charged-current events the electron couples to a neutrino and a W^-. The neutrino is not detected, resulting in a large amount of missing energy. To distinguish these two processes one can either detect the electron or one can build for a *hermetic* detector a trigger for charged current events that have a large fraction of missing momentum and a small amount of total energy.

1.4.5 Trigger on invariant mass

The gauge boson Z^0 can decay into an e^+e^- pair. e^+e^- pairs are also produced at high rates by photons in electromagnetic interactions. Triggers on invariant mass are important for spectrometer experiments looking for di-muon events with high mass. These events are produced at low rates. Therefore spectrometers with a large aperture are used (Greenhalgh 1984). From the trigger point of view one has to find tracks in the chambers behind the magnet and compute the invariant mass:

$$M^2 = (P_1 + P_2)^2 = (E_1 + E_2, \mathbf{p_1} + \mathbf{p_2})^2$$
$$= E_1^2 + 2E_1E_2 + E_2^2 - \mathbf{p_1^2} - \mathbf{p_2^2} - 2\mathbf{p_1p_2}(1 - \Theta^2/2) \qquad (1.63)$$
$$(\cos \Theta \approx 1 - \Theta^2/2)$$

Neglecting the particle masses at high energies leads to $M^2 \approx \mathbf{p_1p_2}\Theta^2$. The invariant mass of two particles at high energies is thus approximately given by $M = \Theta(\mathbf{p_1p_2})^{\frac{1}{2}}$, where $\mathbf{p_1}$ and $\mathbf{p_2}$ are the momenta of the two particles and Θ their opening angle at production.

1.4.6 Trigger on interaction point (vertex)

In storage rings the position of the interaction point ('vertex') is determined by the beam crossing. Transverse to the beam direction it is determined by the beam size, usually to better than 1 mm; in the longitudinal direction it is known within several centimetres. In e^+e^-colliders many background processes result from beam–gas interactions. A good vertex trigger can be used to discriminate against this sort of background (see Section 3.4). In some storage ring experiments the charged particles are detected by chambers with wires aligned parallel to the beam line. The only way to determine the longitudinal position of outgoing tracks along these wires is with the help of charge division (see Subsection 1.5.10).

In fixed target experiments a vertex trigger is necessary to study short lived particles like the *F* or the A_c. These particles contain the *charm quark* and are produced at currently accessible energies at a rate that is 1/1000 of that of 'normal' hadrons with light quarks. The main difficulty consists of finding a very selective signature to separate the signal from the combinatorial background in exclusive decay channel mass plots. A distinct signature is the finite lifetime in the range of 10^{-13}–10^{-11} s. This implies that tracks of decaying particles, when extrapolated back to the primary vertex, have an average impact parameter of 30–3000 µm. The requirements for detectors to resolve secondary vertices from the decay of short living particles are as follows.

1. Very good spatial resolution.

2. Very good two particle separation, since they operate close to the interaction point.

3. High rate capability.

1.4.7 Acceptance

In order to study the reactions of elementary particles, one normally directs accelerated particles on a solid heavy target or lets them collide in a storage ring with particles travelling in the opposite direction. To measure the outgoing fragments of a reaction the interaction point is surrounded with a set of detectors of various kinds. A possible physical reaction can only be registered by the detectors if the outgoing particles meet some criteria.

- The particles must have enough energy to reach the detector at all and for charged particles, to traverse sufficiently many layers of position-sensitive detectors.

- In a storage-ring experiment, particles must not escape via the beam pipe. In fixed-target experiments the detectors covering the interaction region exclude the entrance and the exit of the beam, and particles must not escape through these holes. In general, they must not remain invisible because of cracks in the detector.

- Charged particles can only be reconstructed in tracking chambers if they cross sufficiently many layers of position-sensitive detectors. A minimum of three space points is required to safely recognize a track without information from other detectors or from the vertex position.

- An event is not registered if a particle hits a veto counter. For instance, particles not travelling within a few millimetres of the beam line can be rejected by a veto counter with a hole for the beam. Sometimes cosmic ray particles may be rejected by special veto counters on top of an experiment.

When designing an experiment it is important to understand which fraction of a physics reaction is (partially or entirely) registered by the detector in general, and by the trigger. One difficulty is that one does not know in advance how the (unknown) reaction will behave. One has some assumptions, which may be correct or not.

In order to keep the discussion simple we will study here only the geometrical acceptance of an experiment, neglecting inefficiencies of the individual detector modules. This means that for single particles the acceptance is one if the detector is hit; otherwise it is zero.

As a first example, the simple case of reactions at a collider with a homogenuous angular distribution is studied. The idea is to determine the total cross section by an experiment as shown in Fig. 1.26. This requires that the measured cross section is corrected by the acceptance function. The beam particle enters the target through a hole in the beam counter C and either interacts or exits the target through a hole in counter G. C is a veto counter used to disable the trigger if the incoming beam particle is not on the beam axis. The trigger accepts an event if some of the counters D–K have received a signal. To measure the cross section in a perfect detector, we count the number of events, divide that number by the number of beam particles hitting the target, and multiply it by the target constant (see Equation (1.6)).

If we require that at least four scintillation counters have to give a signal we must first know how many events are produced with a particle multiplicity of four or more. In addition, the particles must be well separated such that at least four particles hit different scintillators. This creates inefficiencies simply by the geometry of the detector for all those events where only three scintillators are hit, although four or more charged particles have been created.

A naive approach to get the overall efficiency would be to integrate the *n*-dimensional angular distribution over the angular ranges that are physically covered by the detectors in order to evaluate the acceptance correction factor. In practice many other effects have to be considered: inefficiencies of the detectors, loss of low energy particles, multiple scattering etc. The usual way out is to replace the analytic integration by a Monte Carlo simulation. However, this is often not possible without prior assumptions on the event topology. In the Monte Carlo program events are generated, and the tracks are followed through the detector,

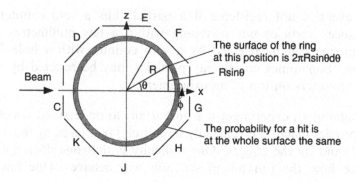

Fig. 1.26. Acceptance for particles beeing produced in all directions θ and ϕ so that each surface element of a sphere is hit with equal probability. The counters E or J from $\theta = 3\pi/8$ to $5\pi/8$ cover 38%, F or H from $\theta = \pi/8$ to $3\pi/8$ cover 27%, D or K cover the same as F or H. Neglecting the hole, counter C rejects 3.8% of the events.

simulating the particle's record as closely as possible to reality. Then one sums up which fraction of events have been seen by the trigger. Large experiments spend very large amounts of computing time in simulations of this type. The technique of Monte Carlo simulation is a topic in its own right, and one which is, however, beyond the scope of this book.

In our simple experiment one needs some prior information from other experiments about the particle multiplicity and about their angular distribution. For easier understanding we come back to the purely geometric approach. We will first discuss in more detail the acceptance for events with a *single* outgoing particle, which is model independent for each specific angle. The Monte Carlo approach would in this case result in a multidimensional look-up table for single particle acceptance.

We define a coordinate system with its origin at the interaction point, x pointing in the direction of the incoming particle, z in the vertical direction, and y perpendicular to both axes (Fig. 1.26). We use polar coordinates in a somewhat unusual way, x being the polar axis:

$$x = r \cos \theta$$
$$y = r \sin \theta \cos \phi \qquad (1.64)$$
$$z = r \sin \theta \sin \phi$$

Uniform angular distribution corresponds to uniform distribution in ϕ and $\cos \theta$. In this case each element of equal surface of a sphere surrounding the interaction point is hit with the same probability. Integration over the

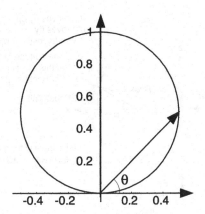

Fig. 1.27. Polar diagram to show the probability (not normalized to one) that a particle is produced in a ring of opening angle $d\theta$. The maximum number of particles is produced at $\theta = 90°$. Note that $|d\cos\theta| = |\sin\theta\,d\theta|$ and that the length of the arrow is $\sin\theta$.

whole spere gives the well-known expression $R^2 \times 4\pi$ for the area:

$$\int_0^\pi \int_0^{2\pi} R^2 \, d\phi \sin\theta \, d\theta = R^2 \times 4\pi \qquad (1.65)$$

If one wants to see directly how many events are produced with a particle in a ring with opening angle $d\theta$ one can use a polar diagram as shown in Fig. 1.27. The length of the arrow is proportional to the probability that particles are produced at the polar angle of the arrow.

We now discuss the case in which a beam particle is as heavy as the target at rest (proton on proton). In this case no particle can go in the backward direction. The (non-relativistic) kinematics is shown in Fig. 1.28. The origin of the coordinate system moves to the surface of the sphere. A movement of the angle β by $d\beta$ results in a movement depending on the distance to the sphere by $2R\cos\beta$. But the direction is perpendicular to c and not perpendicular to the radius of the sphere. Therefore one has to project this direction to the sphere which changes the infinitesimal distance to $d\beta/df = \cos\beta$ or $df = d\beta/\cos\beta$. For the entire sphere we get:

$$\int_0^{\pi/2} \int_0^{2\pi} R^2 2\cos\beta \, 2\sin\beta \, d\phi \, d\beta / \cos\beta = R^2 \times 4\pi \qquad (1.66)$$

This gives again $R^2 \times 4\pi$ but we have integrated only from 0 to $\pi/2$. If we integrate from 0 to 30°, from 30° to 60° or from 60° to 90° we get π, 2π and π which is twice as much as in the example of Fig. 1.26. The polar diagram for this process is shown in Fig. 1.29. No particles are produced at 0° or 90°.

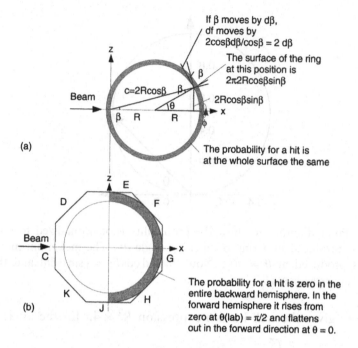

Fig. 1.28. For beam particles as heavy as the target the kinematics changes depending on the momentum of the beam particle. It can easily be seen that no particle can travel in the backward direction. In this simple case where we again assume homogenuous scattering in the rest frame of the reaction (a), we get within an angle from 0 to θ twice as many hits compared to the case shown in Fig. 1.26. The maximum angle in the laboratory frame (b) for θ is 90°. The detectors C, D or K do not count anything. Detector F or H counts about twice as much and detector G about four times as much as in Fig. 1.26.

As a last example we will discuss the situation of electron–proton scattering at high energies. In most cases the electron goes very close to the forward direction. Scattering by an angle of more than a few degrees is very rare. The angular distribution can be approximated by the following expression:

$$d\sigma/d\theta \approx \cot(\theta/2) \tag{1.67}$$

This function has a singularity at 0°. If we integrate from 1° to π we get a finite value:

$$\int_{1°}^{\pi} \int_{0}^{2\pi} \cot(\theta/2) d\phi \, d\theta = 2\pi 2 \ln(\sin(\theta/2))|_{1°}^{\pi} = 59.58 \tag{1.68}$$

If we would like to know how many events have a track between 30° and π we change the limits at the integral and get 17.0 or 28%. In Fig. 1.26 the counter G with an opening angle of 22.5° and a hole of 1° opening

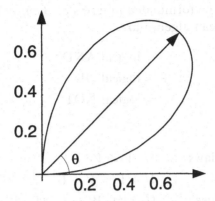

Fig. 1.29. Polar diagram for a process as shown in Fig. 1.28. The maximum number of particles is produced at $\theta = 45°$.

angle would see 66% of the events. With a hole of 0.25°, G would detect 73%.

1.5 Examples of triggers

1.5.1 Fixed-flow triggers

1.5.1.1 A simple fixed-flow trigger to measure angular distributions. Fixed-flow triggers operate in a time range of 50 ns to 1 μs. Their result appears at a fixed time after the operation has been initiated. The synchronization of several trigger systems and the gating for analog signals is tedious but in principle easy. Delay lines can have fixed length. Fixed-flow triggers are used at the first level of a trigger system.

Figure 1.30 gives a typical example of such a trigger. The number of particles hitting the target is defined by a coincidence $A \cdot B$ of the beam counters and a veto of a counter C with a hole to discriminate against particles not coming along the beam. The scattered particles are detected by the counters D and E. The cross section can then be computed by

$$\sigma = \text{Number of events} \times F/\text{Flux}$$

where F is the target constant. In this example one would like to measure the flux of the incoming and scattered particles.

This type of trigger is a typical logical trigger. One writes down the conditions under which a trigger should appear. The flux is counted if the flux counters are set and if the particles travel along the beam line. An event occurs if a particle is scattered through a certain angle. The trigger condition is then

$$\text{Event} = A \cdot B \cdot \overline{C} \cdot D \cdot E$$
$$\text{Flux} = A \cdot B \cdot \overline{C}$$

A logical trigger can be formulated in terms of *Boolean algebra*. The basic operations in a Boolean algebra are

$$\cdot = \text{logical AND}$$
$$+ = \text{logical OR}$$
$$\overline{} = \text{logical NOT}$$

The main rules are:

Commutative laws	$U \cdot V = V \cdot U$	
	$U + V = V + U$	(1.69)
Associative laws	$(U \cdot V) \cdot W = U \cdot (V \cdot W)$	
	$(U + V) + W = U + (V + W)$	(1.70)
Distributive laws	$(U + V) \cdot W = U \cdot W + V \cdot W$	
	$(U \cdot V) + W = (U + W) \cdot (V + W)$	(1.71)

A further important rule is *de Morgan's law*. If the signals are inverted one has to interchange an AND with an OR operation:

$$\overline{A + B} = \overline{A} \cdot \overline{B}, \quad \overline{A \cdot B} = \overline{A} + \overline{B} \tag{1.72}$$

As has already been shown, one also has to take the timing and the pulse lengths into account to get correct results.

Particles produce a light pulse in the scintillator that is transformed to an electrical signal by a photocathode and amplified by a photomultiplier. The scintillator signals are sent to a discriminator, which produces a pulse of -0.8 V (NIM standard) and a variable length between twenty and several hundred nanoseconds. These standard pulses enter a coincidence to establish an AND operation to count the beam particles and another coincidence to detect the scattered particles. Detected events can then be recorded by a computer. The scintillator pulse height must be digitized by an ADC.

1.5.1.2 A fixed-flow trigger to find curved tracks. Another example of a fixed-flow trigger is a processor that finds curved tracks in a cylindrical drift chamber operating at a storage ring. Neighbouring wires are grouped together. More than 5000 coincidences check in parallel for possible tracks. This processor needs 100 000 wrapped wires and occupies a rack of electronics. The trigger information is available 350 ns after the drift time of the chamber (Fig. 1.31).

To avoid having these many wires and gates one can use variable-flow processors. Such a processor solves the track-finding problem of Fig. 1.31 in the following way. The chamber is split into segments and the segments

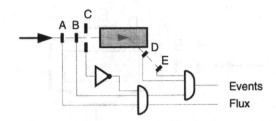

Fig. 1.30. Typical experiment to measure angular distributions. The scattered particles are counted in D and E. The incoming flux is measured by $A \cdot B \cdot \overline{C}$. The ratio of events and flux times the target constant and some correction factors determines the cross section.

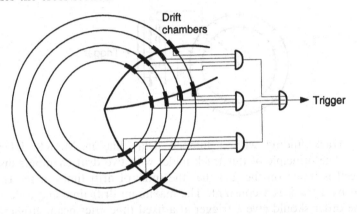

Fig. 1.31. Trigger processor for drift chambers. To find many curved tracks in a cylindrical wire chamber many coincidences (1000–5000) are required.

are connected to a track-finding device. If there is no track in the segment the next sector will be tested. If there are some hits, a track search will start for this segment. This sort of processor needs more time than a hardwired device but it is cheaper and more compact. Another example of a variable-flow trigger are microprocessors, which are used to search for tracks and to compute invariant masses. All variable-flow processors need provisions for time-out to avoid long waiting times.

1.5.1.3 A simple arithmetic trigger to measure total cross section. Arithmetic triggers decide whether a value lies above a certain threshold or within two limits. A simple application is the use of a window discriminator to find within nanoseconds whether the number of counters is, for example, at least four (Fig. 1.32). The trigger condition for an event is then: $P = $ TRUE, if four counters of D, E, ..., K have recorded a hit:

$$\text{Event} = A \cdot B \cdot \overline{C} \cdot P$$

This condition combines a logical and an arithmetic trigger.

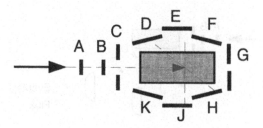

Fig. 1.32. Experiment to measure total cross section. Whenever four or more counters have fired the trigger condition should be fulfilled.

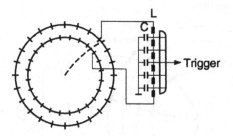

Fig. 1.33. Track finding with a lumped delay line in MARK III (Lankford 1984a). The principle of this track finder with two drift chambers displaced by half a cell is based on the fact that the sums of drift times in the two cells transversed by a track is a constant. The maximum drift time and a delay time of the same order should give a trigger at a fixed time after beam crossing.

1.5.2 Track finding with a lumped delay line

The MARK III detector at SLAC uses two layers of drift chambers with 1 cm drift space to trigger on tracks in 200 ns (Lankford 1984a). The drift cells of the two layers are displaced by half a cell. The trigger decision makes use of the fact that the sum of the drift times in the two overlapping layers is a constant for high energy (straight) tracks originating from the interaction point. It uses a *chronotron* composed of a *'lumped element delay'* line with ten *'taps'* to define a hit (Fig. 1.33 shows only six taps).

1.5.3 Track finding with memory look-up tables

In chambers with more layers that are operating in a magnetic field, one defines masks corresponding to possible tracks. The number of masks can be very high because the position and the curvature of a track can vary. For chambers with a few layers which are separated by more than a few centimetres, high-energy tracks can be handled like straight tracks. This reduces the number of masks considerably. For a circular chamber with

500 wires per layer and four layers one has to test 500 positions, and for each position three directions for positive, negative and straight tracks. This gives 1500 masks. A trigger processor can be built by connecting 1500 AND circuits to all the wires in the correct combination (Fig. 1.31).

Such a processor has the following three shortcomings.

1. It is very bulky: it needs many wire connections.

2. It is not flexible: if a wire does not work one cannot reconfigure the processor.

3. The processor does not allow for inefficiencies: at high rates a wire has some dead time, and a single inefficient wire can inhibit the finding of tracks even if the other three layers show a clean hit.

To gain flexibility and to allow for inefficiencies, one can use memories that are initialized in such a way that they output a valid trigger for each allowed input, including inefficiencies. In the TASSO experiment at DESY, a special memory, the RAM C 10115, was used for the central proportional chamber (Jaroslawsky 1977; Synertek 1976; Platner 1976). In this memory one does not specify a single address to get the contents of a cell. The address lines are organized like a 7×5 matrix and the output is the OR of all addressed cells. Two wire layers can then be connected to the columns and rows of the address inputs. Possible tracks are marked by a 1 in the diagonal. The output gives a track candidate for two layers. This output is then combined with possible track elements for two other layers. In this way one can take wire inefficiencies into account and can fake a signal if the corresponding wire is broken (Fig. 1.34).

Processors described in this subsection can also be used for drift chambers at storage rings with a time between beam crossings longer than 1.5 µs or as a second level trigger. The processors need 500 ns to reach a trigger decision and 1 µs is needed to reset the electronics (ADCs,TDCs) if no trigger has occurred. The ADCs and TDCs are open at each beam crossing. Either they are reset or they start conversion to digitize the event for the computer. The time limit of 500 ns is not given by the processors but by the drift time of the drift chamber. This requires short cells or a cell structure as shown in Fig. 1.35, which guarantees that each emerging particle from the interaction region passes close to at least one wire. The trigger processors rely on the fact that the particles must have their origin in a small vertex region and that high-energy particles can be approximated by straight lines across small detector devices.

The vertex detector of the TASSO experiment has eight layers, which can be split into two parts (Rehlich 1980; Notz 1984). Track elements are then searched in each part separately by defining masks. The two parts

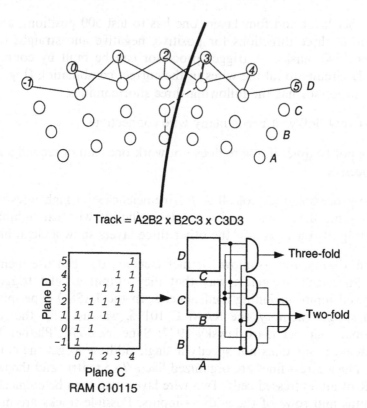

Track = A2B2 x B2C3 x C3D3

Fig. 1.34. Proportional chamber processor with RAM C 10115. The processor should find tracks in four cylindrical layers of a wire chamber. Possible tracks and road widths can be generated by software.

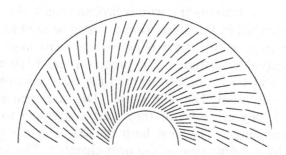

Fig. 1.35. Cell structure for short drift times. To avoid a trigger delay due to long drift times a special cell structure can guarantee that high-energy tracks from the interaction pass close to a wire.

are then split into sectors of ten wires. Ten wires can produce 1024 linear combinations of which a few are real track candidates. An ideal track without inefficiency is found if the address lines 0, 2, 5, 7 are set (Fig. 1.36). This address cell must therefore contain a one. This technique is called

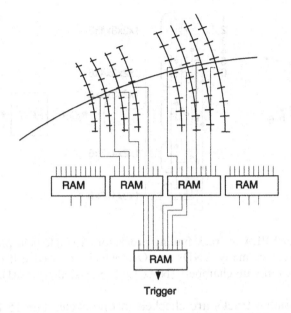

Fig. 1.36. TASSO track chamber processor with three RAMs per sector. Chambers with more than six layers would require large memories to define all masks and coincidences. In these cases one can decompose the chamber into subchambers and look first for track segments, which are then linked together (Rehlich 1980; Notz 1984).

memory look-up (see Subsection 1.6.1). To allow for inefficiencies a track should also be generated if lines 0, 2, 5 or 0, 2, 7 or 2, 5, 7 are set. In this way one can get in one memory cycle track segments for four layers.

1.5.4 Trigger on tracks with field-programmable arrays

The TASSO processor can be used without pretriggers at storage rings with a bunch crossing rate above 2 µs or as a second level trigger (Notz 1981; Stuckenberg 1981). In this case, 1 µs is used for the trigger decision and 1 µs is required to reset the electronics. The processor gets its input from six drift chamber layers with 72–216 wires, from 48 TOF scintillation counters and from 48 bits of track information from a faster proportional chamber processor. The processor should produce a trigger if there are more than three, four or five tracks with a minimum transverse momentum of at least 200–300 MeV, but it should also give a trigger for QED events as $e^+e^- \rightarrow e^+e^-$ or $\mu^+\mu^-$. These events are characterized by two high-energy particles emerging in opposite directions from the interaction point.

To perform at this high speed many operations run in parallel. In a first step one has to find tracks. For each of the 72 wires of the innermost

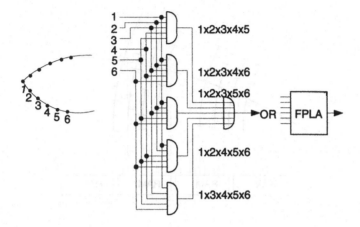

Fig. 1.37. Use of FPLA in track finding processor. A single field programmable logic array can replace many ANDs and ORs which are needed if one wants to allow inefficiencies in wire chambers (Notz 1981; Stuckenberg 1981).

chamber 15 possible tracks are checked in one cycle. The 15 tracks have positive or negative signs and different momenta; 72 reference wires × 15 track candidates gives 1080 possible masks. These possible masks are hardwired. Within each mask one has to combine several neighbouring wires by a wired OR to form a track point for the trigger masks.

From these six possible track points one has to decide whether there is a valid track or not. This is achieved in a very simple manner by a six-fold coincidence. If one also wants to take also inefficiencies into account, one has to use several five-fold coincidences and combine their results by an OR. This requires a lot of hardware. Another possibility is the use of *Programmable Read Only Memories (PROMs)* or *RAMs* as discussed in the previous subsection. For each valid address the generated output represents a trigger. If one wants to gain speed one has to use gates instead of memories (Fig. 1.37). One can benefit from the speed of gates and use nevertheless compact electronics with the help of *Field-Programmable Logic Arrays (FPLAs)* or *Programmed Array Logic (PALs)* (see Subsection 1.6.1). The basic logic of a FPLA consists of a programmable AND array whose outputs feed a programmable OR array. The designer decides how to configure the AND and OR connections. With a special program device the FPLAs are programmed by the customer by burning the fuses. Another approach are PALs which operate in a similar way as FPLAs, but the OR combinations are fixed (see Subsection 1.6.1).

At the output of the possible tracks one gets 1080 bits which must be further reduced to a yes/no trigger (Fig. 1.38). The 1080 tracks are now compared to the 48 tracks of the proportional chamber. Only those tracks in the processor that also have a hit in the proportional chamber are allowed. In order to cut on transverse momentum a mask selects only high-momentum tracks. The final tracks are then combined with information from scintillation counters surrounding the drift chamber. Forty-eight possible bits remain which are then sent to a majority logic to discriminate against low-multiplicity events. In a separate branch, high-energy tracks emerging from the opposite direction are combined and marked as QED events. Typical rates at 15 GeV beam energy are of the order of some triggers per second.

Beam crossing rate	260 kHz
QED rate e^+e^-	1 Hz
Two prong events	7.5 Hz
Three prong events	1.3 Hz
Four prong events	0.5 Hz

1.5.5 Track finders in the trigger with variable-flow data-driven processors

Hardwired parallel processors can compute the number of tracks in a few cycles by comparing the wire chamber hits with predefined masks. These processors are large devices with several thousands of wrapped wires and electronic circuits. Variable flow processors are small because they compute sequentially the possible tracks from all combinations. They need more time (5 μs–2 ms) to find tracks, but the tracks are more precise because these processors act on single wires and not on wire clusters. Figure 1.39 shows one view of a set of wire chambers – drift chambers or proportional chambers. In the case of drift chambers only the information that a wire was hit is used, not the drift time.

A typical track-finding algorithm works as follows. Combine each hit from chamber 1 with those from chamber 3, fit a straight line and predict the position of a hit in chamber 2. If there is a hit within a certain road width a track is formed. For N hits per plane this algorithm needs N^3 operations. The algorithm to find a hit in chamber 2 must be very fast. A track is given by a straight line $y = Ax + B$. The coordinates of points in chamber 1 are $(x_{11}, y_1), (x_{12}, y_1), (x_{13}, y_1), \ldots$ and in chamber 3 $(x_{31}, y_3), \ldots$, respectively. First one has to find the parameters A and B of the track, then one can predict a point in chamber 2:

$$y_1 = Ax_{11} + B$$
$$y_3 = Ax_{31} + B$$
$$A = \frac{y_1 - y_3}{x_{11} - x_{31}}$$
$$B = y_1 - x_{11}\frac{y_1 - y_3}{x_{11} - x_{31}}$$

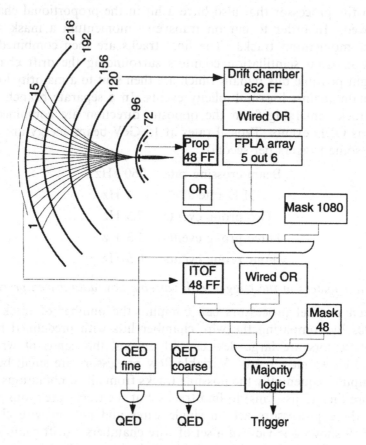

Fig. 1.38. TASSO track chamber processor with FPLAs. For each of the 72 inner wires 15 masks are generated and compared to the data. The 1080 masks are implemented by hardwired field-programmable logic arrays (FPLAs) to avoid these many ANDs and ORs. A trigger is generated for hadronic events (majority $\geqslant$ 3 tracks) or QED events as e^+e^- or $\mu^+\mu^-$ pairs (Notz 1981; Stuckenberg 1981).

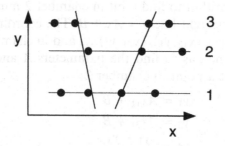

Fig. 1.39. Wire chambers in a spectrometer experiment. The beam is coming from below. The coordinate system is the one used in the algorithm.

The predicted value x_{21} in chamber 2 is then

$$x_{21} = (y_2 - B)/A \tag{1.73}$$

$$= \frac{(x_{11}y_1 - x_{11}y_3)/(x_{11} - x_{31}) + y_2 - y_1}{(y_1 - y_3)/(x_{11} - x_{31})} \tag{1.74}$$

$$= x_{11}\frac{y_2 - y_3}{y_1 - y_3} + x_{31}\frac{y_1 - y_2}{y_1 - y_3} \tag{1.75}$$

$$= x_{11}a + x_{31}b \tag{1.76}$$

The chambers are in fixed positions y_1, y_2, y_3. Therefore a and b are constants and can be computed at the beginning of the experiment. In a von-Neumann computer one would have to write a program with three nested loops to find the tracks. In a special purpose processor one can save computing time by getting rid of the innermost loop with the help of a Content Addressable or Associative Memory (CAM).

Let us take the following example to explain the function of a CAM. A teacher would like to know whether somebody in the class is 10 years old. He or she can ask each pupil if he or she is 10 years old. This is a sequential process like the innermost loop. Alternatively, he or she can ask that those who are 10 years old should raise their hand. This indicates a match in the operation of a CAM. In our case the information of chamber 2 is not stored in an ordinary memory but in a CAM. After prediction of a value x_2 one adds some bits for the road width and checks in one cycle whether the CAM contains that value. In this case a track is found (see Subsection 1.6.1).

We can now decompose the algorithm into several steps and try to produce a modular system for track finding.

1. Store all x coordinates in separate buffers for each chamber.

2. For the loops store data in lists and then count by an index generator.

3. Take a hit from chamber 1 and multiply it by a. Take a hit from chamber 3 and multiply it by b.

4. Add the two results and predict a hit in chamber 2.

5. Match the predicted position with a possible hit.

6. Count the number of tracks.

These steps can be realized with electronic modules, which were developed by LeCroy (Levit and Vincelli 1985) and by Nevis Laboratories and which are in use at Fermilab (Kostarakis *et al.* 1981). The algorithm can be performed by the following system (Fig. 1.40).

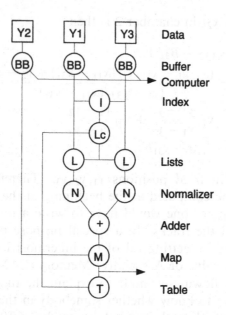

Fig. 1.40. Variable flow data driven track finding processor. The processor consists of separate building blocks, which are connected by a data and a control path. An operation is executed whenever data are ready (Kaplan 1984).

The modules are data driven. As soon as input data are ready the operation is executed. Each module implements a simple operation like adding two 16-bit numbers, comparing a quantity with upper and lower limits, or computing a function with a look-up table. The design cycle time is 25 ns.

1.5.6 A microprogrammed track processor with CAM and look-up tables

The processors discussed so far are fast because many steps are done in parallel by a large amount of hardware. The microprogrammed processor for the drift chamber of the TASSO experiment is a microprocessor that finds tracks within 1 ms (Schildt, Stuckenberg and Wermes 1980). In the processors described up to this point, only the hit information of the drift chambers is used. This approach is sufficient if one wants to trigger on more than three tracks. If one would like to trigger on two tracks, which are not required to be coplanar, this coarse method gives many fake triggers because many wire combinations can generate a two-track trigger.

For low multiplicity events one has to use both the wire and the drift time information to find cleaner tracks. The processor reads the digitized drift times and then searches for tracks. The track reconstruction is

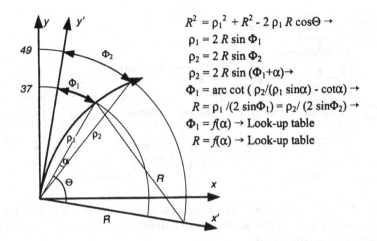

$$R^2 = \rho_1{}^2 + R^2 - 2\rho_1 R \cos\Theta \rightarrow$$
$$\rho_1 = 2R \sin\Phi_1$$
$$\rho_2 = 2R \sin\Phi_2$$
$$\rho_2 = 2R \sin(\Phi_1+\alpha) \rightarrow$$
$$\Phi_1 = \text{arc cot}\,(\rho_2/(\rho_1 \sin\alpha) - \cot\alpha) \rightarrow$$
$$R = \rho_1/(2\sin\Phi_1) = \rho_2/(2\sin\Phi_2) \rightarrow$$
$$\Phi_1 = f(\alpha) \rightarrow \text{Look-up table}$$
$$R = f(\alpha) \rightarrow \text{Look-up table}$$

Fig. 1.41. The track parametrization for a fast-track processor. With this track parametrization one gets from two hits in two layers the value α. It is then easy to compute Φ and R with the help of look-up tables and predict the next hit.

performed in a plane perpendicular to the beam axis ($R\Phi$ plane). As the magnetic field is assumed to be homogeneous all charged particle tracks are segments of circles in the projection on this plane. In order to reach a fast-tracking algorithm the circles are parametrized as shown in Fig. 1.41. Two hits in the first and second cylinder and the interaction point are used to define a circle. The radius and the tangent of the circle are determined by the radii of the cylinders (ρ_1, ρ_2) and the angle between the two hits:

$$R = 0.5\rho_1/\sin\Phi_1 = 0.5\rho_2/\sin\Phi_2 \tag{1.77}$$

$$\Phi_1 = \text{arc cot}[\rho_2/(\rho_1 \sin\alpha) - \cot\alpha] \tag{1.78}$$

Since ρ_1 and ρ_2 are constants the equations only depend on the angle α, which is the angle between the two vectors pointing from the interaction point to the hits in the first and second cylinders. With a given α one can compute the starting values R_1, Φ_1 and R_2, Φ_2. Radius and tangent are calculated at most three times. After that the rest of the track following is just a hit search. A track is accepted if there are at least six hits in nine layers. The angle α is used only between layers $1 + 2$, $1 + 3$, $1 + 4$, $2 + 3$ and $2 + 4$. The maximum value of α does not exceed 0.255. Taking into account the resolution $\Delta\alpha = 0.000\,38$ due to drift time digitization only 665 different values of α are possible for high-energy tracks. The α values are stored in five look-up tables with maximum storage size of 1024×16 bits. The maximum value for Φ is 0.66 or 1722 possible values, and for R it is 8191 cm or 8192 values for a resolution of 1 cm. Within a few memory cycles one gets the parameters of a possible track from

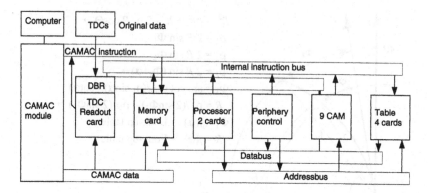

Fig. 1.42. A '*non-von-Neumann*' track processor. This processor uses look-up tables and content addressable memories to compute tracks within 1 ms (Schildt, Stuckenberg and Wermes 1980).

the first three layers. The task is then to check whether there are at least six hits corresponding to that predicted track. This is done by the use of CAMs. A sequential search is too slow, as we have seen in subsection 1.4.1. *Association* and *matching* are done by the memory itself. It is also possible to mask off part of the associative word. Therefore one can search around the predicted point in the next chamber.

The track search is controlled by an ECL *bit-slice processor* (Motorola 1977) with a 72-bit-long *microinstruction word*. The processing unit can manipulate the registers within one cycle while an access to the look-up tables takes place simultaneously. The program is stored in a $1k \times 72$-bit memory. The original drift chamber data contain wire addresses and digitized drift times. This information is transformed into an octal integer representation from 0–$177\,777_8$, corresponding to the interval 0–2π. Up to 40 bits per layer can be stored into the CAMs. The architecture of the processor is shown in Fig. 1.42. A typical two-prong event needs approximately 5000 microinstructions. A clean Bhabha event $e^+e^- \rightarrow e^+e^-$ requires 0.6 ms, and a hadronic event requires 4.5 ms. It is probably a weak point of this processor that the first three layers must fire. If the first layers are inefficient no track parameters can be set up to start a track search.

1.5.7 Examples of triggers on energy

1.5.7.1 Example of analog trigger in the ASP experiment. We will now discuss the ASP detector at SLAC that surrounds the interaction region by four walls of five layers of lead glass separated by proportional chambers (Lankford 1984b). The 640 photomultipliers are divided between digitizers

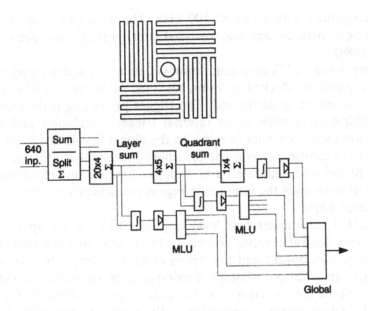

Fig. 1.43. Energy trigger for the ASP experiment with lead glass walls. The top part shows the experiment with its beam pipe and the lead-glass counters. Several energy triggers are generated from the total energy sum and quadrant sums (Lankford 1984b).

and the trigger. The experiment and the trigger are sketched in Fig. 1.43. The trigger system forms first 80 sums of eight multipliers and then another 20 sums for each layer. These 20 layer sums each go to integration circuits and are then discriminated to define hit layers. The hit layers address a memory look-up, which, in turn, defines allowed combinations of hits; for example, the first two layers of a quadrant but not the last two. The layer sums are also summed up into few quadrant sums, which are integrated and then each discriminated by three levels defining three energy thresholds. The resulting 12 signals address a memory look-up that counts and defines combinations of quadrant hits. The quadrant sums are also summed to form a total energy sum, which is also integrated and discriminated to three thresholds.

1.5.7.2 Example of energy trigger in the ZEUS calorimeter. The trigger system of the ZEUS experiment at the HERA electron proton storage ring has to deal with 96 ns intervals between bunch crossings (ZEUS 1986). The number of genuine physics events is small compared to the large background originating from beam–gas and halo interactions along the 70 m straight section in front of the detector. The trigger system has to select the interesting physics events with a rate of 1 Hz while rejecting a large background with a rate of 100 kHz. The maximum rate of events that can be written to tape and analysed later on is of the order of 5 Hz (ZEUS 1989).

All data in the ZEUS data acquisition system are stored in either analog or digital pipelines, clocked at 96 ns, for 5 μs, while the first-level trigger calculations are being performed. The trigger processing is deadtimeless. Each subdetector completes its internal trigger calculation and passes information for a particular crossing to the global first-level trigger.

The ZEUS calorimeter consists of depleted uranium-scintillator towers of $20 \times 20 \text{ cm}^2$ with separate parts to detect hadrons and electrons. The calorimeter surrounds the interaction region hermetically, covering 99.8% of the stereo angle.

The ZEUS calorimeter first level trigger (CFLT) is set up to detect charged and neutral current processes. In these events the current jets (from the proton quark) and the lepton emerge on opposite sides of the beam axis. In addition, photoproduction and exotic physics events are identified. Some typical triggers of the calorimeter are shown in Fig. 1.44. The CFLT passes trigger information to the global trigger within 2 μs.

Mounted at the detector are front end cards (FEC), which pass 5% of the photomultiplier pulses to the trigger sum cards (TCS). Here the data are integrated, summed for two trigger towers, and sent as a differential signal down a 60-m-long shielded twisted pair cable to the electronics house. The gains are adjusted in such a way that $E_{\max} = 400 \text{ GeV}$ in the forward calorimeter and $E_{\max} = 100 \text{ GeV}$ in the barrel and rear calorimeter correspond to 2 V. Each calorimeter tower is read out by two wavelength shifters. To take inefficiencies or failing electronics into account the wavelength shifters of each tower are connected to different parts of the electronics (Fig. 1.45).

Fifty-six calorimeter trigger towers are connected to one trigger crate containing 14 trigger encoder cards (TEC) and two adder cards. Each TEC is connected to four hadronic and four electromagnetic compartments of four trigger towers (Fig. 1.46). The TEC receives analog signals. The signals are amplified with low and high gain and digitized by 8-bit flash ADCs. The overflow of the high gain ADC enables a multiplexer which decides which signal to choose. For each of the four towers the TEC digitizes and linearizes the electromagnetic and hadronic energy. The linearized energies are multiplied by geometric factors to calculate the total energy E_{total}, the transverse energy E_T, E_x and E_y using programmable look-up tables. The TEC sums the electromagnetic and hadronic energies for all four counters together.

In parallel with the energy sums, the TEC also performs tests for quiet towers, electromagnetic showers and minimum ionizing particles. The TEC also tests against six different energy thresholds. These bits are used

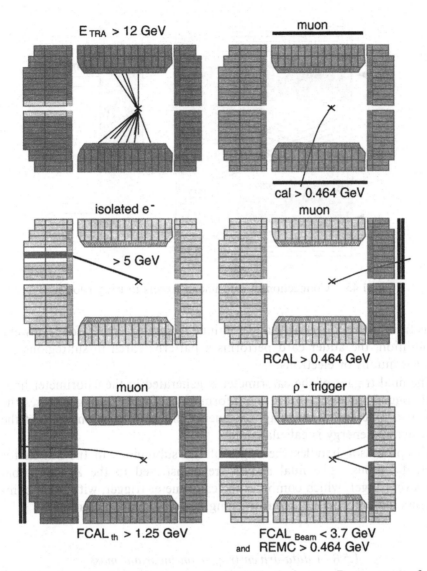

Fig. 1.44. Some typical triggers for the ZEUS calorimeter. Protons enter from the right, electrons from the left. The inner part of the calorimeter with fine granularity measures electromagnetic showers while the outer parts measure hadrons. The left part of the calorimeter is called 'forward', the right part is called 'rear' (FCAL,RCAL).

to remove some parts of the calorimeter in the beam region from certain triggers.

Seven TEC cards in each half of the crate send their information via a special backplane to the adder card in the middle of each crate. This is done in parallel with a 12 ns clock rate. The adder cards receive partial

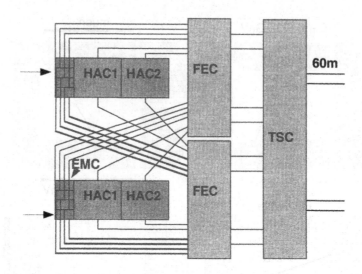

Fig. 1.45.   Connection of calorimeter towers to trigger logic.

sums from the encoder cards and continue to sum up energy for 56 towers. In addition, the adder card performs a pattern search in subregions for isolated muons or electrons.

The final trigger of the calorimeter is generated in the calorimeter first-level trigger processor. From the information of the 16 trigger crates the total hadronic and electromagnetic energy, the transverse energy and the total missing energy is calculated.

The processor handles the edges of the subregions in the search for isolated leptons. The final results are transferred to the ZEUS global first-level trigger, which combines the calorimeter trigger, with other components like information from tracking devices or muon chambers.

1.5.8 A data-driven trigger on invariant mass

Triggers on invariant mass are important for spectrometer experiments looking for di-muon events with high mass. These events are produced at low rates. Therefore spectrometers with a large aperture are used (Greenhalgh 1984). From the trigger point of view one has to find tracks in the chambers behind the magnet and compute the invariant mass (Fig. 1.47).

$$M^2 = (P_1 + P_2)^2 = (E_1 + E_2, \mathbf{p}_1 + \mathbf{p}_2)^2$$
$$= E_1^2 + 2E_1E_2 + E_2^2 - \mathbf{p}_1^2 - \mathbf{p}_2^2 - 2\mathbf{p}_1\mathbf{p}_2(1 - \Theta^2/2) \qquad (1.79)$$
$$(\cos\Theta \approx 1 - \Theta^2/2)$$

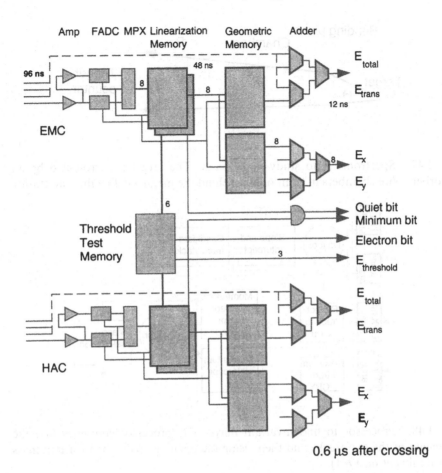

Fig. 1.46. Trigger encoder card to compute E_{total}, E_T E_x, E_y. Analog signals are amplified by low and high gain. EMC = electromagnetic calorimeter, HAC = hadronic calorimeter, FADC = flash analog digital converter, MPX = multiplexer.

Neglecting the particle masses at high energies leads to $M^2 \approx \mathbf{p}_1\mathbf{p}_2\Theta^2$. The invariant mass of two particles at high energies is thus approximately given by $M = \Theta(\mathbf{p}_1\mathbf{p}_2)^{\frac{1}{2}}$, where $\mathbf{p}_1$ and $\mathbf{p}_2$ are the momenta of the two particles and Θ their opening angle at production. The processor has to find the tracks and compute their slopes and intercepts at the centre of the magnet. This information allows the calculation of the bending angle, making use of the bend plane and point target approximation. The bending angle is inversely proportional to the momentum of the particle. The two intercepts at the bend plane determine the opening angle Θ in the xz plane. The Θ_y angle is only computed with the help of the y hodoscope behind the iron wall. The computing time needed to find tracks depends on the number N_i of hits per chamber. Testing all combinations

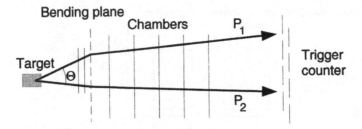

Fig. 1.47. Spectrometer to study $\mu^+\mu^-$ pairs. The target is surrounded by an absorber. Wire chambers in front of and behind the magnet define the trajectories.

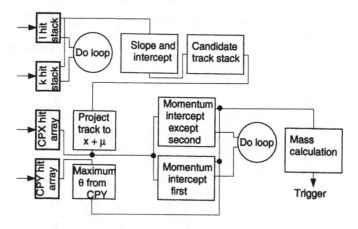

Fig. 1.48. Processor to find invariant mass. The processor computes first the momenta of the two tracks and then, using the opening angle, the invariant mass (Kostarakis *et al.* 1981).

of two chambers needs $N_1 N_2$ tests. One can save time by defining roads and checking only the combinations within these roads. In the case of the di-muon experiment the hodoscope counters in the iron wall define a rough direction of the incoming tracks.

The quantities recorded are the inverse momentum of the track, its intercept at the magnet centre, a two-bit number for vertical position and a sequential road counter (Fig. 1.48). Two stacks are used for the track information in two roads. After having found all the tracks in one road and the first track of the second road, the final stage of the processor can start combining the track information and computing the invariant masses. The square root of the product of the momenta, the three-dimensional opening angle and the di-muon mass are calculated. The processor gives a trigger if the invariant mass is above a certain threshold. The processor needs 5–10 µs to compute the invariant mass and reduces the primary trigger rate by a factor of 10.

1.5.9 Triggering on neutral pions with neural networks

The neutral pion π^0 decays with 98.8% probability into two photons. The opening angle of the two photons depends on the energy of the pion and is of the order of 10° for a pion of 1 GeV and 1° for a pion of 10 GeV. Shower detectors with lead scintillator segmentation and read-out by photomultipliers have a typical size of 20×20 cm^2. With such a device the energy of two well separated photons can only be determined if the decaying pion has an energy below 1 GeV. At higher energies finer segmentation is required. This is possible by using silicon strip detectors. With these detectors, which have a typical size of 3×3 cm^2, one can separate the photons and measure their energy if the pion has an energy of up to 10 GeV.

We describe here the detection system for neutral particles of the ZEUS detector. It consists of a shower counter for electromagnetic particles with a segmentation size of 10×20 cm^2, a shower counter for hadrons (20×20 cm^2) and a silicon strip detector inside the electromagnetic shower counter (see Fig. 1.44, Fig. 1.49 and ZEUS 1986). The two photons generate the following signals in the detector.

- A signal with the maximum energy deposition in one tower for hadronic showers plus signals with less energy in the surrounding eight towers. For two photons this gives 18 signals.

- Eighteen signals from the electromagnetic shower counters giving 36 signals for two photons.

- One measurement of the opening angle.

- Nine signals of the silicon diodes for each 'island' (cluster) gives 18 signals for both photons.

- Distance between the island measured by the silicon diodes.

- Number of silicon diodes for both islands.

- Two impact angles for the photons.

The total amount of information for position, energy, size of shower or island adds up to 77 analog quantities. The question is how this large amount of information can be used to construct a trigger (Fig. 1.49).

One possibility, which is of recent origin, is to use a *neural network*. The development of neural nets was inspired by the human brain with its high degree of parallelism, which enables it to recognize patterns in very short time. A neural net is a mathematical abstraction of some basic aspects of biological nervous systems.

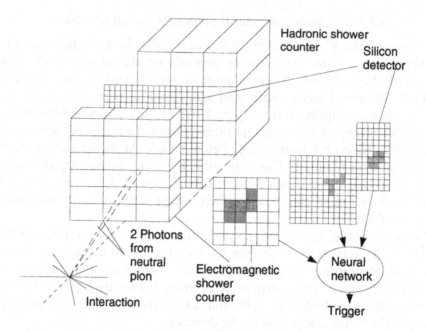

Fig. 1.49. Neutral pion identification by measuring the invariant mass of the two photons. A neural net takes 77 analog quantities as input and determines the possibility of a pion mass (Fricke 1996).

Neural nets consist of *nodes*, which exchange information. These nodes are called *artificial neurons*. Each node i receives input signals x_j from other nodes; each input x_j is multiplied by a weight w_{ij}. The total input of the node is the sum of all weighted input signals minus a threshold α_i. Applying a *transfer function* $g(x)$ to the total input gives the output signal y_i of node i:

$$y_i = g\left(\sum_{j=1}^{n} w_{ij}x_j - \alpha_i\right)$$

In its simplest form the transfer function is a step function that gives an output of 1 if the weighted sum of the inputs is above the threshold, and -1 otherwise. Frequently one uses continuous transfer functions like $g(x) = \tanh(x)$ which change smoothly from -1 to 1 in the vicinity of $x = 0$. The output of a node is often connected to other nodes. If a node is connected neither to the sensory input nor to the output it is called a hidden node. If the nodes are connected only in a non-recursive way one speaks about a feed-forward network (Fig. 1.50). It is a remarkable fact that a feed-forward network with a single hidden layer is capable of approximating nearly all mathematical functions to an arbitrary precision, provided that the hidden layer consists of sufficiently many

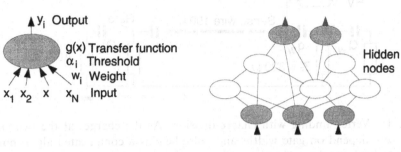

Fig. 1.50. Functional description of a node within a neural network. The input x_i is multiplied by the weight w_i and summed up. By applying a threshold α_i and a transfer function $g(x)$ the output is computed. In the right-hand part one can see a general neural net with input, hidden and output nodes (Fricke 1996).

nodes (Hornik 1991). Such a network is described by many parameters for the weights and thresholds. Finding the best weights or thresholds for the approximation of a particular function is called *learning* or *training* of the network. In order to achieve this goal one needs a training sample, consisting of a set of input vectors and their corresponding function values or output vectors (see also Subsection 2.2.2).

Initially the weights and thresholds are taken randomly from a given interval avoiding that two parameters are identical. Learning then proceeds iteratively. In each iteration one presents to the network a given input vector and the desired output. The difference between the desired output and the actual output is then used to modify the weights and thresholds such that the discrepancy is minimized. Mathematically speaking, one looks for the global minimum of an error function. Since the error functions of even moderately large networks have numerous local minima this is a non-trivial task. Therefore a huge number of minimization methods has been proposed and discussed in the literature. After successful training the parameters learned by the network should be close to the optimum ones.

In the ZEUS experiment a feed-forward neural net was used to identify neutral pions or electrons. The network was realized by a computer program and not in hardware. The patterns of the training samples were generated by a Monte Carlo program. Based on that type of 'trigger' a program was written which tries to find electrons (Abramowicz, Caldwell and Sinkus 1995). In subsection 1.6.2 we will describe a chip which implements a neural net in hardware.

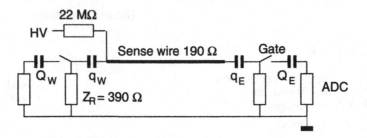

Fig. 1.51. Vertex finding with charge division. As the charges at the two ends of the wire depend on gate widths and pulse heights a complicated algorithm is needed to evaluate the correct vertex position.

1.5.10 Examples of triggers on interaction point

1.5.10.1 Trigger on charge division. In storage rings the position of the interaction point (vertex) is determined by the beam crossing. Transverse to the beam direction it is determined by the beam size, usually to better than a millimetre, and in the longitudinal direction it is known within several centimetres. In many storage ring experiments, as for example in TASSO, the charged particles are detected with wire chambers aligned parallel to the beam line. The position of the outgoing tracks along the beam direction can be determined with the help of charge division.

The charge of the ionizing particle is divided and appears at the two ends of the sense wire at the capacitors q_E and q_W in (Fig. 1.51). Let ρ be the resistance of the signal wire ($\rho = 190\ \Omega$) and $Z_R = 390\ \Omega$ the internal resistors needed to keep the charge; then the position z along a wire of length L is given by

$$z = [(Q_E - Q_W)/(Q_E + Q_W)](L/2)[\rho/(\rho + Z_R)]F(t). \qquad (1.80)$$

Q_E is the charge of the capacitor in front of the ADC, which is loaded from the small capacitor during the gating time. The charges in the small capacitors try to equalize via Z_R. Therefore many corrections are required to determine the correct position of the vertex. $F(t)$ is the correction function, which depends mainly on the gating time and on the charges in the capacitors. The following steps are performed to find the correct z-position.

1. Subtract pedestals in all ADCs.

2. Correct for gains. This is done with look-up tables. Instead of computing $g_E Q_E - g_W Q_W$ one computes $G Q_E - Q_W$; $G = g_E/g_W$.

3. A track passing close to a wire gives a large pulse, which needs a long time before compensation (Fig. 1.52). The correction for these pulses are small and one would like to use a short gate.

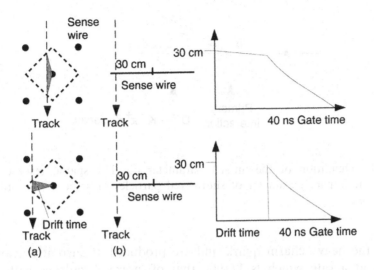

Fig. 1.52. Corrections to determine the vertex. Tracks at different distances from the wires produce different charges at different times at the small capacitors. These charges tend to equalize via the internal network. This compensation depends on the gating time and must be corrected in several steps.

4. Particles far away from a sense wire induce small pulses. They do not appear before the end of the drift time and tend to equalize rapidly. For these pulses the gate must be at least as long as the drift time plus some charge time for the capacitors.

After having found a hit along a single wire one has to compute the vertex. Four layers are used in the vertex calculation. Hits in two neighbouring layers are averaged and the final vertex is computed by

$$z_V = (R_1 Z_2 - R_2 Z_1)/(R_1 - R_2) \tag{1.81}$$

where R_1 and R_2 are the radii of the chambers. As these quantities are fixed one can use look-up tables with six bits for each Z_1, Z_2 to determine the final vertex.

The time needed to find a vertex is of the order of 10 ms. This rather long time is still sufficient to stop the readout of the experiment into the main computer if this requires more than 10 ms (see Subsection 1.2.4).

1.5.11 A trigger on interaction point for short-lived particles with a microstrip detector

In the NA32 experiment at CERN, a vertex trigger was necessary to study short-lived particles such as the F or the Λ_c. These rare particles

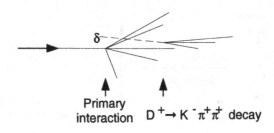

Fig. 1.53. Definition of the impact parameter δ. The spatial resolution to measure the impact parameter of short-lived particles must be of the order of micrometres.

contain the heavy charm quark and are produced at currently available energies at a rate which is 1/1000 that of 'normal' hadrons with light quarks. The main difficulty consists of finding a very selective signature to separate the signal from the combinatorial background in exclusive decay channel mass plots. A distinct signature is the finite lifetime in the range of 10^{-13} s to 10^{-11} s. This implies that tracks of decaying particles, when extrapolated each to the primary vertex, have an average impact parameter of 30–3000 μm (Fig. 1.53).

One possible choice of detector is a silicon microstrip detector, which has the following advantages. The charge deposition is very localized and there is a large amount of charge produced by minimum ionizing particles. The charge transport to the readout electrodes preserves localization and the charge collection is very fast (< 10 ns). Another possible detector is a charged coupled device (CCD). The great advantage of CCDs is the fact that they are two-dimensional devices. The pixel size is of the order of 23 μm × 23 μm, the effective detector thickness is 15 μm. The schematic principle of a CCD is shown in Fig. 1.54. MOS capacitors deposited over a thin space charge layer on a n^+-p-p^+ structure generate a series of potential wells, which store the electric charges created by the incoming particles (Poenaru and Greiner 1997). Fifty thousand pixels are read out in 16 ms. The spatial resolution is $\sigma_x = \sigma_y = 5$ μm, and the double track resolution is 40 μm.

The experimental set-up at the target region of a big spectrometer experiment (Weilhammer 1986) is shown in Fig. 1.55. Seven microstrip detectors in the beam telescope define the incoming particle. The outgoing tracks are measured by two CCDs and eight microstrip detectors. Their spatial resolution is $\sigma \approx 2.6$ μm. Depending on the decay topology the background rejection is in the range of 1 : 4 000 to 1 : 40 000.

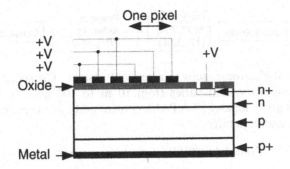

Fig. 1.54. Charged coupled device (CCD). A CCD is used to measure tracks in two dimensions. The pixel size is of the order of 23 µm × 23 µm.

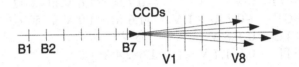

Fig. 1.55. Schematic view of the NA32 vertex detector. B1 to B7 are microstrip detectors in the beam telescope. V1 to V8 are microstrip detectors in the forward direction. For good reconstruction, two-dimensional readout is performed by two CCDs (Weilhammer 1986).

1.6 Implementation of triggers

1.6.1 Electronic components

The readout of a detector device proceeds through several steps. A simple example showing the readout of a photomultiplier channel is sketched in Fig. 1.56. The signal output of the photomultiplier (PM) is connected to a discriminator, which produces a standard signal if the PM signal is above a certain threshold. This first decision is made by a module built in the *Nuclear Instrumental Module* standard (NIM). It operates at a speed of < 10 ns. Only simple logical operations are possible at this level. More complex logic operations require a higher level of integration and have to be implemented by *Emitter Coupled Logic* (ECL) circuits, or *Transistor–Transistor coupled Logic* (TTL) circuits. In this category we will discuss RAMs, FPLAs, CAMs and PALs. These integrated circuits (ICs) operate at a speed of 5–100 ns. Fast special processors like Digital Signal Processors (DSPs) make use of a special architecture. These have a data and an instruction bus and contain link ports that allow connection to other DSPs. They can particularly be used to connect various detector components with each other. DSPs can be programmed in assembler or high-level languages (C). At the end of the readout chain data enter a computer via electronic buses (VME, PCI) or via networks (SCI, ATM),

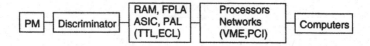

Fig. 1.56. Several electronic steps in a readout system. The speed range of the elements in a readout system varies from 10 ns to 100 ms. Typical elements are programmed array logic (FPLA,PAL), content addressable memories (CAM), microprocessors and computers.

Table 1.4. *Table of electronic standards*

	logic 0	logic 1
NIM	0 V	−0.7 to −1.6 V at 50 Ω
ECL	−0.7 to −1 V	−1.6 to −1.9 V at 50 Ω
TTL	0–0.8 V	2–5 V
ETL	0.6–1.4 V	1.6–2.4 V

and are stored on disks or on tape. These buses and networks tend to use low voltages like *Enhanced Transceiver Logic* (ETL) or *Low Voltage Differential Signalling* (LVDS). The typical signal levels of the various electronic standards are given in Table 1.4.

1.6.1.1 Pulseformers, discriminators. The output signal of a detector varies in length and in amplitude depending on the characteristics of the detector and the energy of the particle. These signals cannot be used to perform logical operations like AND or OR. A pulseformer has to transform the input signal into an output signal of well defined length and amplitude. The input signal has all the vagaries that random rates, shapes, amplitudes and cable techniques can produce. The signals of a photomultiplier can vary between 0.1 V and 10 V in amplitude and between 2 ns and 50 ns in length. All these signals should produce a constant output pulse of −0.8 V and a pulse length of about 5 ns or more than 20 ns. The output signals should appear soon after the input signal has passed over a given threshold. This propagation time is of the order of 7 ns. To allow for good time measurement in TOF counters, the time jitter of the variation of the propagation time must be below 0.3 ns. The input sensitivity of a discriminator can be changed typically between −30 mV and −1 V. It is a challenge to the manufacturer to keep the threshold at a constant value, independent of temperature and input rates. The speed of a discriminator is given by the double-pulse resolution, which is of the order of < 10 ns. A pulse that appears within the double-pulse resolution will not be registered.

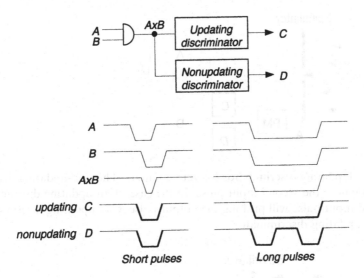

Fig. 1.57. Pulseforms of updating and non-updating discriminators. For short pulses both discriminators behave in the same way. For long pulses the non-updating discriminator will generate several pulses.

When two pulses appear at the input of a discriminator with a time difference larger than the double-pulse resolution but shorter than the output pulse length an *updating discriminator* prolongs the output pulse while the *non-updating discriminator* will ignore the second pulse. Both types of discriminator are used depending on the application.

1. Discriminators are used within fast electronic logic to produce a standard pulse after a coincidence. The two discriminator types will then show the behaviours indicated in Fig. 1.57. To avoid oscillations for long pulses an updating discriminator must be used.

2. A veto counter to suppress the beam halo should be equipped with an updating discriminator to be sure that the veto works efficiently (Fig. 1.58).

3. In experiments that count beam particles at a high rate a non-updating discriminator is advisable. Although the dead time is higher than for an updating discriminator the counting rates do not drop so much at high rates (Fig. 1.59).

A discriminator contains the following building blocks (LeCroy 1985).

1. A *Schmitt trigger* produces a pulse of variable length but constant amplitude. It goes to logic 1 when the input pulse is above threshold and changes back to logic 0 if the input pulse decreases below 50% of the threshold.

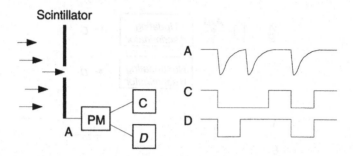

Fig. 1.58.　Updating discriminator for veto counters. The non-updating discriminator will ignore the second input pulse. In the case of an updating discriminator the second input pulse will prolong the output pulse. C = updating discriminator, D = non-updating discriminator.

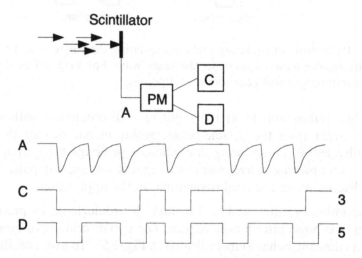

Fig. 1.59.　Non-updating discriminator for high rates. If high rates are registered a non-updating discriminator counts more pulses than an updating discriminator. But we also see that the rate is too high for both discriminators and needs to be corrected. C = updating discriminator, D = non-updating discriminator.

2. A *differentiator* produces a small spike of 2 ns length at the leading edge of the Schmitt trigger output.

3. A *multivibrator* or pulseformer produces a pulse of a predetermined length.

4. An *amplifier* produces a signal of given amplitude (−800 mV) and cable impedance (50 Ω) with short rise and fall time (2 ns).

The main building blocks of a discriminator are shown in Fig. 1.60.

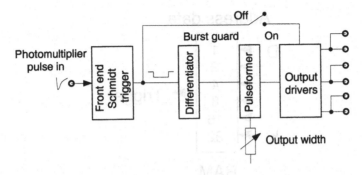

Fig. 1.60. Building blocks of a discriminator (LeCroy 1985). The Schmitt trigger limits the pulse and gives an output as long as the input is above 50% of the threshold. The differentiator is used to trigger only on the leading edge and to be independent of the pulse length.

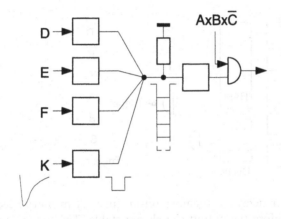

Fig. 1.61. Possible system to trigger on four counters. The output signals (16mA) of the pulseformers are added by a resistor. A window discriminator with thresholds between −2.8 V and −3.6 V selects triggers with four hits.

1.6.1.2 Window discriminators. Window discriminators are used if the input signal is to be between predefined limits. An application of the use of a window discriminator is shown in Fig. 1.61. We assume here an experiment in which four scintillation counters should give a hit to form a multiplicity trigger as shown in Fig. 1.32.

1.6.1.3 Look-up tables. Another possibility to 'compute' the number of counters having a hit is by means of a memory look-up table with seven input address lines and one output line (Fig. 1.62). The memory is set to zero in all cells apart from addresses 15 (= $D \cdot E \cdot F \cdot G$), 23 (= $D \cdot E \cdot F \cdot H$), 39 (= $D \cdot E \cdot F \cdot J$), 71 (= $D \cdot E \cdot F \cdot K$), 30 (= $E \cdot F \cdot G \cdot H$), and so on.

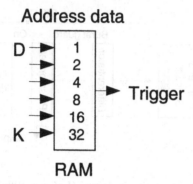

Fig. 1.62. Memory look-up table to trigger on four hits or more. Scintillation counters are connected to the address lines of a memory. Those memory cells that correspond to addresses with four address lines like 15 ($= 1 + 2 + 4 + 8$) contain a 1.

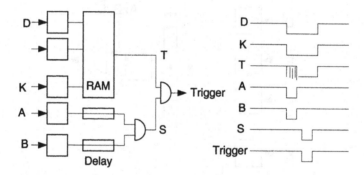

Fig. 1.63. Trigger delay for memory set-up time. A memory needs some time (5 ns – 200 ns) before the output signals are stable. The trigger signals must be therefore delayed by this time.

To avoid trigger spikes during memory set-up time (5–20 ns) the trigger must be clocked or the final decision must be delayed (Fig. 1.63).

Memory look-up tables are widely used to avoid the computations of complex expressions at each trigger. Instead of computing the expression or the function one stores the results for all possible arguments into a memory. This method works well if the range for an argument is not very large and if the function is smooth. Memories are also used to select a certain trigger combination from the trigger signals. This method does not work for a large number of input channels because each additional channel doubles the size of the memory. Very often it is possible to split the experiment into several small and independent sections and use cascaded memories. The access time for memories varies between 0.5 ns for GaAs memory (Bursky 1985; Mourou, Bloom and Lee 1986), 5 ns for

ECL and 10–100 ns for static TTL memory. Memories are cheap and therefore used in many applications.

1.6.1.4 FPLAs. We have seen several examples in which a field programmable logic array (FPLA) can be used to replace many AND and OR integrated circuits. If, for example, one wants to allow one inefficiency in a possible track road through six chambers, the trigger condition is

$$\text{trigger} = 1 \cdot 2 \cdot 3 \cdot 4 \cdot 5 + 1 \cdot 2 \cdot 3 \cdot 4 \cdot 6 + 1 \cdot 2 \cdot 3 \cdot 5 \cdot 6 + 1 \cdot 2 \cdot 4 \cdot 5 \cdot 6$$
$$+ 1 \cdot 3 \cdot 4 \cdot 5 \cdot 6 + 2 \cdot 3 \cdot 4 \cdot 5 \cdot 6$$

Precisely this function can be performed by an FPLA or a PAL. A FPLA consists of a programmable AND array and a programmable OR array (Fig. 1.64). The customer writes down the desired logical equations and derives from these how to burn the fuses in the FPLA by a PROM (Programmable Read Only Memory) burn device.

1.6.1.5 PALs. A PAL is constructed in a similar way. The PAL has a fixed OR array instead of a programmable one. In the example given in the previous subsection one cannot use a PAL, which has only four lines ORed together because six ORs are needed. In this case one has to use a PAL that is configured in a different way with six lines ORed together. In principle PROMs, FPLAs and PALs contain the same basic elements as an AND array and an OR array (Monolithic Memories 1985). In PROMs the AND array is fixed and represents the address lines, while the OR array is programmable and contains the contents of each cell. The AND array is organized in such a way that only one memory cell is enabled for all address combinations (Fig 1.64). An FPLA has a programmable AND and OR field allowing many logical combinations. For a PAL the original OR field of a PROM is fixed and the AND field allows flexibility to perform logic equations. PALs can be programmed by a standard PROM programmer. The PAL appears to the PAL programmer like a PROM. (A PAL programmer is not a person but a hardware device to burn the fuses and to implement the program into the integrated circuit.) During programming some outputs are selected for programming and the inputs are used for addressing. Special PAL assemblers are available that process the logic equations and produce a binary table to burn the right fuses.

1.6.1.6 CAMs. CAMs are widely used in track finding processors to test whether a predicted track position matches the hits in a chamber. Within a memory cycle (12 ns) a CAM can deliver yes/no information if there are hits near the predicted position. Using conventional memories or programs one has to process all hits of a chamber.

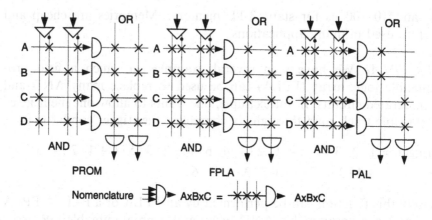

Fig. 1.64. PROM, FPLA and PAL. A PROM has a fixed AND array to perform the address decoding. An FPLA has a programmable AND and OR field, while a PAL has a programmable AND field. All devices are programmed by the customer with a PROM device to burn the fuses.

A CAM can be used as a normal memory to write or read information to or from certain memory locations. The operation *associate* reports on the output whether there was a match and on which address the corresponding value can be found. CAMs do not contain as many storage cells as ordinary memories. An 18-pin chip may contain eight words of two bits.

Figure 1.65 shows the schematics of a CAM. I_1 contains the information that should be stored into the memory or that should be used as a mask to interrogate the memory. If the mask matches the content of a memory cell the corresponding address line goes down, indicating the address of the matching word.

1.6.2 A chip for neural networks

In Subsection 1.5.9 we described a trigger based on neural nets. This trigger was realized in software only. If the trigger is to be applied in real time it has to be implemented in hardware. We describe here the SAND/1 (Simply-Applicable Neural Device) chip, which is optimized for neural nets. It is based on the principle of a systolic array, which is a set of connected processors that operate in a synchronous way and exchange information after each processing cycle. Four parallel processor elements form the heart of the array (Gemmeke 1997). Each processor element has a multiplier and two adders. A post-processing module allows the determination of the largest and smallest output activity. The chip runs at 50 MHz clock frequency and reaches a performance of 200 million

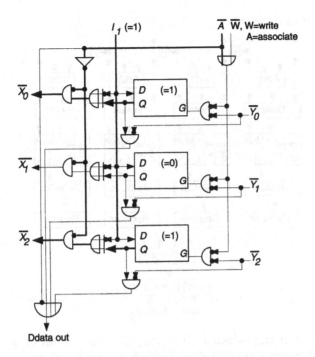

Fig. 1.65. Organization of a content-addressable memory. The CAM can be used as an ordinary memory. But in addition it can indicate if one of its memory cells contains a specific value. In this case the corresponding address line Y goes down. In this example negative logic is used. $Y_0 = 0$ means that something should be stored into address 0.

connections per second. Input data and weights are 16-bit wide. The internal precision is 40-bit. An auto cut module reduces the word width from 40 to 16 bits. The block schematics of the chip are shown in Fig. 1.66.

1.6.3 Pipelines

In an experiment one has to get all information such as pulse heights, timings etc. from the detector to a computer for further analysis. On the other hand the output of a device is also needed to generate a trigger – a yes/no decision to take or not to take the event. The output is therefore split into two branches (Fig. 1.67). One branch takes a known fraction of the output signal to an ADC converter, the other part is sent to the trigger. In the trigger several logical operations are performed, which requires some time, in the range 50 ns to several microseconds. When the trigger appears at the gate of the ADC the original signal is already lost. One therefore has to delay the original information and store it in a pipeline.

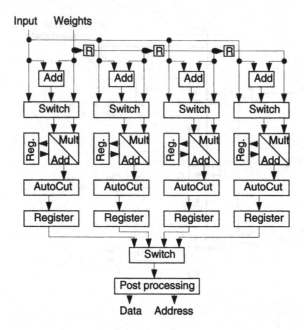

Fig. 1.66. A neural chip which realizes the typical behaviour of a neural net. Input data and weights are added or multiplied. Input and output values are 16-bit wide and normalized from −1 to 1 (Gemmeke 1997).

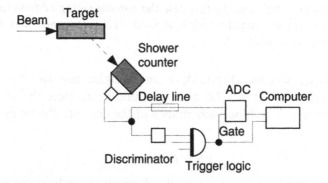

Fig. 1.67. The necessity of a pipeline. The trigger needs some time for its decision. During this time part of the analog signal is stored into a pipeline and is available at the entrance to the ADC if the trigger processing has finished. This method requires a fixed time for the trigger.

1.6.3.1 Delay lines. Delay lines of 50 m length can be used if the trigger does not need more than 250 ns. The signals from the photomultiplier get attenuated on the cable. This attenuation depends on the frequency. As all photomultiplier signals have the same shape or frequency the attenuation is constant. A long cable can contain the data of several events, one

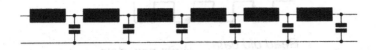

Fig. 1.68. Lumped delay line. The delay time is given by $n\sqrt{LC}$, where n is the number of LC elements, L is the inductivity and C is the capacity of each element.

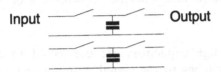

Fig. 1.69. Principle of the Analog-Memory Unit (AMU). Several capacitors are charged if there are data available and discharged to an ADC if there was a trigger.

behind the other. For this reason such an arrangement is called a pipeline. A cable is an approximation of coupled inductances and capacitors. Using discrete elements one gets a *lumped delay line*, which can be used for delay times of 1 µs (Fig. 1.68). A disadvantage of such delay lines is the distortion of the pulse shape, but the charge of the pulse at the end of a delay line is proportional to the charge at the beginning. Delay lines need a good resistor adaption at the ends to avoid reflexions on the cable.

1.6.3.2 AMU (Analog-Memory Unit). An analog cable pipeline can take information at any time and deliver an attenuated signal at the output after a fixed time. It can only be used to store information for a short time (< 1 µs). If one wants to store analog information for a longer time one can use capacitors. This technique is used in *sample-and-hold ADCs*. The input signal charges a capacitor. At the arrival of a trigger some microseconds later, the capacitor will be discharged and the signal will be digitized. If one wants to use this technique as a pipeline one has to use several capacitors that are charged by the raw data and discharged some time later when there has been a trigger. This has lead to the development of an *Analog-Memory Unit (AMU)* at the Stanford electronics centre (Freytag and Walker 1985). The device contains 256 analog storage cells consisting of switching transistors, a storage capacitor and a differential readout buffer. The principle of this device can be seen in Fig. 1.69.

To avoid the need for many input/output connection switches to the outside of the integrated circuit, one uses for each cell two input and two output switches arranged in rows and columns of a 16 × 16 matrix.

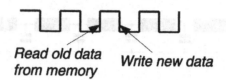

Fig. 1.70. A memory used as pipeline. On one edge of the clock data are read from the memory and on the other edge new data are stored into the memory. If data arrive with 100 ns time difference a memory with 50 ns access time is needed.

In practice the storage capacitors are connected through Field Effect Transistor (FET) switches to the signal input. The time constant of the input combination is 1 ns resulting in a bandwidth above 100 MHz. The wide dynamic range of more than 11 bits is basically a result of the large stored charge of 10^7 electrons at full scale. Readout of the analog information proceeds through a matched pair of source followers with a reference voltage accessible from the outside to inject correction levels.

1.6.3.3 Digital pipeline. To avoid the thermal effects within electronic pipelines one can use a digital pipeline if the input pulse can be digitized in the order of 10–50 ns (depending on the application). Flash ADCs reach this high speed. They compare the incoming pulse with a chain of comparators and present digital information within 10 ns. This information can then be transferred to a shift register or to a memory and can be kept for as long as needed. The memory needs twice the speed of the input rate if it should operate without dead time (Fig. 1.70). At the rising edge of the clock pulse one takes the information out of the memory to the event buffer. On the falling edge new information is fed into the memory. The disadvantage of all pipelines apart from the delay line is that one cannot use pipelines to measure exact timing information, which is required for TOF measurements.

1.7 Multiprogramming

1.7.1 Digital Signal Processors (DSPs)

In recent years *Digital Signal Processors* (DSPs) have undergone an enormous development, mainly in response to demands from the fields of speech and pattern recognition, fast Fourier transforms, and digital filtering. DSPs are essentially special purpose *'Reduced Instruction Set Computers'* (RISC), optimized for fast multiplication. With the commonly used *Harvard architecture*, i.e. separate data and instruction paths, DSPs

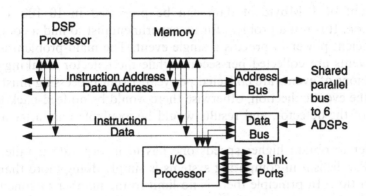

Fig. 1.71. The ADSP-21062 digital signal processor with separate instruction and data bus. The shared parallel bus can connect directly to up to six ADSPs. The six link ports can exchange information with other ADSPs.

can reach very high speeds. In its simplest configuration the memory allows separate access, one for data and one for instructions.

For example, the ADSP-21060 processor integrates on a single chip a fast general purpose floating-point processor with a dual-ported four megabit RAM, DMA controllers and six point-to-point link ports (Super Harvard Architecture Computer, SHARC) (ADSP 1995, SHARC 1997). The peak performance reaches 120 million floating-point operations per second. The six external communication links can be used to connect the processors with each other.

In addition up to six processors can communicate via a 240 MByte/s 32-bit address and 48-bit data bus. This bus maps the processors together in a unified address space (Fig. 1.71). The HERA-B experiment will use the ADSP processors with six DSPs on one VME board for its second-level trigger (Lüdemann, Ressing and Wurth 1996). Via the link ports it is possible to connect a processor from one VME board to another one.

The large memory integrated on the chip allows synchronous access without wait states. No time is wasted to drive large off-chip drivers and external buses.

1.7.2 Parallel processing

1.7.2.1 Needs for computing power and parallel processing. An experimentalist working at a large 4π detector needs a huge amount of computing power for event processing, Monte Carlo generation, and physics analysis. In addition, accelerator physicists and theorists need a lot of computing power to simulate accelerators or to do model calculations, for example in lattice gauge theory.

An event of 1 Mbyte of data can be processed in 10–100 s on a workstation. It is not a problem for an experimentalist to find a computer with sufficient power to process a single event. The main problem is that several events are collected per second while the detector is taking data. Thus millions of events are collected per year. Event processing must keep up with the event collection, otherwise there would be no feed-back to the running of the detector, and results would be available only after a long delay.

In order to obtain higher speeds one has to incorporate parallel processing. Parallelism in computing systems is simply doing more than one thing at a time. In principle there is no limit to the number of concurrent actions offering improvement in computing speed.

1.7.2.2 Classification of parallel processing. Following the classification of Zacharov (1982) one can distinguish the following categories.

1. Parallelism within *functional units*. Arithmetic, logical, and other operations can be implemented in parallel bit-serial mode by bit groups (bit slice processors). This category does not affect the way in which a problem is formulated.

2. Parallelism within *processing elements*. Different operations are executed in parallel on different operands, for example. While a floating point operation takes place some integer or index operations are executed. Another kind of concurrency is shown, for example, by a multiplier that can process a stream of operands in an overlapped or pipelined fashion. Pipelining is very powerful and can give significant improvements in execution speed.

3. Parallelism within *uniprocessing* computers. An example for a single processor is concurrent I/O or DMA (direct memory access).

4. Parallelism in *many processor systems*. An obvious form of concurrency is a computer system containing several processors. The processors may or may not share the main memory, and communicate with each other. In most cases the separate processors are used to execute independent jobs. This gives higher throughput but the concurrency is not used to speed up the execution of individual jobs.

Another classification of parallel processing comes from the field of *vector processors*.

1. *Single Instruction Multiple Data* (SIMD) architecture is used in vector processors to operate on data fields.

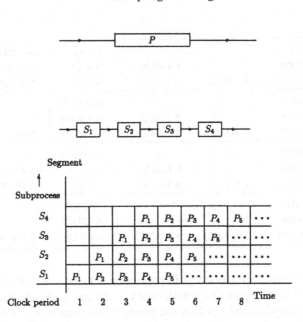

Fig. 1.72. The effect of pipelining in a computer to speed up. The instructions are decomposed into small segments for fetch, decode, normalize, multiply and address calculation. All segments run independent of each other and can be overlapped. In this example five instructions are done in eight cycles instead of 20 cycles.

2. *Multiple Instruction Multiple Data* (MIMD) vector processors can handle several instructions concurrently on data fields.

1.7.2.3 Pipelining for execution. In many cases the execution of an instruction can be decomposed into several steps such as: fetch instruction, decode instruction, calculate the operand address, fetch the operands and execute the instruction. Another example is: fetch a floating point number, normalize, multiply the number, normalize the result and store it back. In a pipelined machine the instructions are decomposed into separate segments S_1, S_2, S_3, S_4, which can operate independently of each other. The advantage of pipelining can be seen in Fig. 1.72 where five instructions are performed in eight instead of 20 machine cycles.

There are, however, some problems with pipelining. The benefits of pipelining can be lost in the following cases. A computer fetches instructions one after another from the memory and then places them in a pipeline for further decoding. If a GOTO statement occurs in the program, or if there is a conditional branching, then the pipeline must be cleared. A way out of this problem is the existence of two pipes; one is used if the condition for a branch is fulfilled and the other if it is not. Another

EXEC			L 1,4(2,3)	A 4,2(4,3)
ADDR.		L 1.4(2,3)	A 4,2(4,3)	S 4,0(5,6)
FETCH	L 1,4(2,3)	A 4,2(4,3)	S 4,0(5,6)	N 3,4(5,6)
EXEC			B F,4(2,3) ⌐	—
ADDR.		B F,4(2,3)	A 4,2(4,3)	—
FETCH	B F,4(2,3)	A 4,2(4,3)	S 4,0(5,6)	↳ Next
EXEC			L 1,4(2,3) ⌐	—
ADDR.		L 1,4(2,3)	—	↳ A 4,2(1,3)
FETCH	L 1,4(2,3)	A 4,2(1,3)	—	S 4,0(5,6)
EXEC			STC 1,(7)	—
ADDR.		STC 1,(7)	A 4,2(4,3)	—
FETCH	STC 1,0(7)	A 4,2(4,3)	S 4,0(5,6)	S 4,2(4,3)

Fig. 1.73. Problems that may occur in pipelining. The pipeline must be cleared in the case of GOTO statements or conditional IF statements. Contention appears if one instruction requires the result of the previous instruction or if one instruction changes the following instruction.

example of contention occurs when one instruction computes and stores a result into a specific register. If this register is used in the address calculation of the following instruction the pipeline must first be emptied or delayed. Also if one instruction in a program changes the following instruction the hardware must recognize this and must clear the pipeline. Otherwise the unmodified instruction that is already in the pipeline will be used (Fig. 1.73). From the problems mentioned above one can see that it is impossible to estimate how much speed is gained for a given application because programs have GOTO and IF statements.

1.7.2.4 Multiple functional units. With multiple functional units one can achieve high computing speed within a single processor. Each unit is activated by a separate instruction specifying the corresponding operands. The units overlap in execution, which results in problems of synchronization and bus contentions. The complexity involved in solving synchronization problems can be imagined for cases in which the input parameters to one function may result from others.

1.7.2.5 Many-processor systems. Many-processor systems use many processing elements of the same kind. In order to add two vectors of n elements n additions could be performed in parallel in a single cycle if n parallel processors were available *(vector processing)*. For problems that can be organized in such a manner this technique produces a substantial increase in speed over traditional serial machines. But the performance diminishes considerably for problems that are not vectorizable or if the result of each addition is examined individually. Many physical problems such as initialization and collecting statistical information appear to be non-vectorizable. There is a considerable difference between adding two vectors of 64 elements with 64 processors and summing up 64 numbers with 64 processors. In the first case one needs one cycle, while in the latter at least $\log_2 64 = 6$ cycles are required.

Another approach is to use multiple computers or processing elements, which do not perform the same operation on different data at the same time but work on different tasks of a calculation simultaneously. The multiprocessor systems on the market are designed to run separate problems and do not work collectively on the same program. These processors achieve higher throughput but an individual job requires as much time as on a sequential processor. The main problem in coordinated processing on a single problem lies in multiprocessor control, coordination, and interprocessor communication. If several processors write to the same memory location in a single job the result depends on the timing. The information written from the first processor is lost if a second processor later modifies the same location.

Another problem is contention between processors for memory and bus access. This can easily reduce the performance to 70% for a two-processor system even if the processors work on different programs (Maples 1984). A simulation of the expected performance for a hydrodynamic test code is given in Table 1.5 (Axelrod, Dubois and Eltgroth 1983). An eight-processor system is expected to perform with only 48% efficiency. If one wants to avoid memory contentions with the help of individual memories, a large amount of interprocessor communication can degrade the system's performance.

Decomposition into a serial part and a vector part has a crucial influence on the system, as can be seen in Ware's model of multiprocessors (Maples 1984) in which one assumes that either one processor (for serial code) or all processors (for vectorizable code) are in operation. Table 1.6 shows the way in which the fraction of serial code in terms of time can degrade the performance of the system. A very small amount of 1% serial computation has little effect for a few processors but it can degrade the performance by 50% in a system with 100 processors.

For P processors working in parallel and part of the time in serial, the speedup $S(P)$ and the efficiency E of each processor can be calculated according to the formula:

$$S(P) = \frac{1}{\alpha + (1 - \alpha)/P} = \text{Speedup} \qquad (1.82)$$

Table 1.5. *Simulated performance of a hydrodynamic test code*

No. processors	Speedup	Efficiency
1	1.00	1.00
2	1.77	0.89
4	2.93	0.73
6	3.56	0.59
8	3.84	0.48

Table 1.6. *Ware's model of multiprocessors*

% serial code (α)	Relative speedup (efficiency) Number of processors (P)			
	2	8	16	100
1	0.99	7.5(0.93)	14.0(0.87)	50(0.50)
5	0.95	6.0(0.74)	9.1(0.57)	17(0.17)
10	0.91	4.7(0.59)	6.4(0.40)	9(0.09)
20	0.83	3.3(0.42)	4.0(0.25)	5(0.05)

$$E = S(P)/P = \text{Efficiency} \qquad (1.83)$$

where α is the amount of time in serial calculation. In a typical high-energy physics program the different events can be analysed or generated independently of each other in separate processors. The I/O is sequential because data come from a single file and are written on a single file. Also the collection of statistics and their display in histograms is a sequential task, which must be done by one processor (or vector processor).

1.7.2.6 Programming aspects for parallel processing. When programming a parallel processing system, some new aspects have to be considered that are not encountered in sequential programming. We will restrict our discussion to systems with several processors sharing a common file or memory, neglecting here the optimization of data structures or programs for vector processors.

As a first example let us assume that a host computer reads events from a file and transmits them to different processors for further processing. As soon as one event has finished it is transferred back to the host, together with its result, and written to the output file. This kind of operation has the consequence that *the output events are not in the same order as the input events.*

As another example we consider the generation of Monte Carlo events by a multiprocessor system. All processors are loaded with the same Monte Carlo program and generate events, but as the program in all processors is identical this system will produce the same event sequence on all processors. Therefore *the random number generator must be set individually on each node after the program has been loaded.*

In these two examples independent events are processed by independent processors. The program in each computer is a sequential program. But the host computer must handle requests from all computers and has to synchronize them. We will now consider the situation in which several processors share a single task and process concurrently. As an example let us assume that a list of a thousand numbers must be sorted. In this case ten processors can sort a hundred numbers each. Afterwards the 'pre-sorted' numbers must be merged.

This kind of operation leads to concurrent programming (Ben-Ari 1982). Concurrent programming is motivated by the problem of constructing operating systems. Two or more sequential programs are not independent of each other. They must communicate in order to exchange data or to synchronize for external input or output. In distributed systems synchronization can be done by sending and receiving signals, which is called 'message passing'.

Concurrent programming leads to new programming problems that are unknown to sequential programming.

1. *Mutual exclusion* is the abstraction of many synchronization problems. Activity A_1 of process P_1 and activity A_2 of process P_2 must exclude each other if the execution of A_1 may not overlap the execution of A_2. The most common example of the need for mutual exclusion is resource allocation. For instance, two tapes cannot be mounted simultaneously on the same tape unit. The abstract mutual exclusion problem will be expressed as: *remainder, pre-protocol, critical section, post-protocol.*

2. *Robustness* or *fault tolerance* describes the property of a system that a bug in one process does not propagate to a system 'crash'. It should be possible to degrade the performance if an isolated device fails ('soft failure').

3. *Correctness* of a program is assured when a program prints the correct answer and then stops. In concurrent programming one distinguishes two types of correctnes properties: *safety* and *liveness* properties.

4. *Safety* properties describe the static portion of a program. Mutual exclusion is absolute and does not change during execution. In a

producer–consumer problem the consumer must consume each piece of data produced by a producer.

5. *Liveness* deals with the dynamic behaviour of a system. Liveness means that if something is supposed to happen then eventually it will happen. The most serious breach of liveness is the global form known as 'dead lock'.

6. *Fairness* is the concept that a process wishing to execute must get a fair chance relative to all other processes.

7. *Timing* is ignored in the concurent programming abstraction. One makes no assumptions about the absolute or relative speed. It is a dangerous style of programming to assume that mutual exclusion is not needed because process P_1 should finish its critical section before process P_2 has finished its uncritical part. It might be that process P_2 will run in future on a faster processor. Then this assumption is no longer true and the system might crash.

We will finish this subsection with a short description of *Dekker's algorithm*. This algorithm can be used to describe mutual exclusion and is shown in Fig. 1.74. Two processors have a common memory for exchanging information. Three words in the memory are used: $C1 = 0$ if process P_1 is in its critical section. $C2 = 0$ if process P_2 is in its critical section. If $C1$ and $C2$ are both 0 (both processes want to enter the critical section), $Turn = 1$ indicates to process P_1 that it must check periodically the state of P_2 until P_2 has finished the critical portion of its program. At the end of its critical section P_2 indicates by $C2 = 1$ that it is outside the critical part. One can show that this algorithm is correct. It also fulfils the requirement of robustness. It is assumed in all algorithms that a progam does not abort in its critical path.

If the critical section only contains a single word like a counter or a synchronization flag as in the following example

```
LOAD A
MODIFY A
STORE A
```

one can use special hardware: one uses a processor that allows read–modify–write cycles. During this cycle another processor will be delayed when it wants to access the memory.

1.8 Communication lines, bus systems

The data acquisition of an experiment consists of scalers, ADCs, TDCs, latches, registers, memories, interrupt generators, and so on. These devices

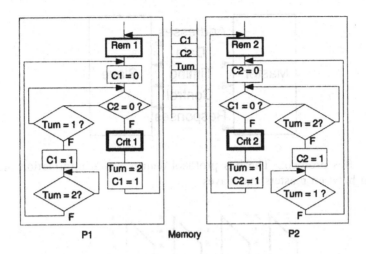

Fig. 1.74. Dekker's algorithm for two processes that should exclusively enter a critical section.

must be connected to a computer for readout. In principle, one can implement all required paths by point-to-point wiring. But this approach is expensive and clumsy compared with linking the components by a few shared data paths or buses. The drawback of this philosophy is that the modules have to be redesigned if a different computer system and bus are used.

This has led to the development of instrumentation buses like CA-MAC, VME, MULTIBUS II, or FASTBUS, which house plug-in modules in standard crates (EUR 4100 1972; EUR 4600 1972; VMEbus 1987; MULTIBUS II 1984; FASTBUS 1983; FASTBUS 1985, PCI 1993). VME and compact PCI are based on the Eurocrate mechanics specification IEC 297-3 (IEC 1994). The instrumentation bus is then interfaced to a computer. If the computer is replaced one only has to change the interface.

Information exchange is initiated by a *bus master*. There may be several masters on the bus, but only one is allowed to be active at a given time. *Arbitration* processes are needed to resolve *contention*. A master can exchange information with a slave by placing an address on the bus. All slaves compare this address with their own individual address and become connected if a correspondence is found. During this connection a master can exchange information with a slave via the bus. At the end of the transfer the connection is broken to release the bus. Transactions between masters and slaves involve the transfer of address and data information over the bus, together with timing and control information for synchronization and specification of data flow. The set of rules needed to exchange this information is called a bus protocol (Fig. 1.75).

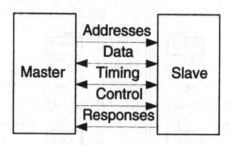

Fig. 1.75. Bus protocol. The bus protocol defines which information must be exchanged between masters and slaves.

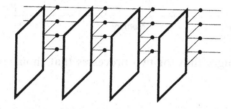

Fig. 1.76. Bus connections of slaves. With the help of tristate gates only one device is active on the bus while the other devices are passive but still connected to the bus lines.

When one designs a bus one is confronted with the following problem (Fig. 1.76). A master *A* wants to read information from a slave *C*. What is the electrical influence of the bus gates and drivers of module *B*, which remains connected to the bus? If one uses a normal AND gate to disable *B*, the output of the gate is a logical 0, which is also connected to the bus like the 0 or the 1 from slave *C*. There are two possibilities for solving this problem.

1. One can use inverse logic (5 V = logic 0, 0 V = logic 1 for a TTL bus). The bus lines are then connected via a *pull-up* resistor of typically 330 Ω to 5 V. The device connected to the bus uses electronic circuits to open the collector output. The device that wants to transfer a logic 1 then pulls the line down to 0 V.

2. Another solution is the use of electronic circuits with *tristate* output. The output is then either logical 0 or logical 1, or it has a high resistance. Modules that do not participate in the transfer are disabled and are then only connected to the bus via a high resistance, which does not disturb the logic signals of the other modules.

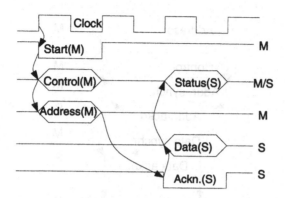

Fig. 1.77. Timing for a synchronous bus. The master asserts address and control information. The slave recognizes its address, finds the requested data, puts data and status on the bus, and marks their presence with an acknowledge signal. There is no handshake between master and slave. The master assumes that the slave is fast enough to get all information.

1.8.1 Synchronous and asynchronous buses

A master puts an address on the bus and exchanges information with a slave after a connection has been established. This is called a bus cycle. A transaction can take one bus cycle or several cycles. Many cycles are used in some systems to transfer first address information followed by one or more data transfers. For each cycle it will be necessary to synchronize the transfer. There are two general ways to carry out bus cycle timing: either *synchronously* (CAMAC in the crates, MULTIBUS II) or *asynchronously* (VME, FASTBUS). Synchronous transfers (Fig. 1.77) require fixed timing between master and slave. A master assumes that a slave can accept or provide information within a certain time. There is no handshake from the slave to acknowledge the cycle. The length of the cycle is determined by the master. It must be adjusted to the slowest device in the system or the master has to check the slave's ready signal at a fixed time. The slave can generate a *wait state* if it is not ready.

Asynchronous transfers (Fig. 1.78) need a timing handshake between master and slave. The master puts valid information on the bus and waits until the slave sends an acknowledgement. There is no fixed cycle time. The timing between master and slave is adjusted according to the needs of the master–slave pair. On the other hand the handshake mechanism involves a time overhead. In addition to the propagation time between master and slave there are additional electronic delays in generating the handshake mechanism. In addition, the master needs some time-out logic in case the slave does not respond to avoid bus hang-ups.

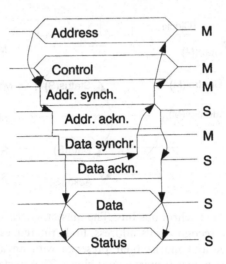

Fig. 1.78. Asynchronous bus timing. The master asserts address and control lines and after some time the address synchronization. The slave acknowledges the address, waits for data validation by the master, asserts data and status information together with the acknowledge signal. The master reads data and removes address synchronization and clears address and control lines. The slave then removes data and status information from the bus. In this way devices with different speeds can operate on the bus. Asynchronous buses are slower than synchronous buses due to the handshake mechanism.

To summarize: a synchronous bus reaches higher speed and should be preferred if all the connected devices have the same speed; asynchronous buses better suit the need when connected devices have different speeds.

1.8.2 Addressing

Address information is put onto the bus to establish a connection to one or more slaves. This is always the first part of the transaction to select the slave that will participate in the bus operation. There are essentially three addressing modes in use:

1. geographical addressing,

2. logical addressing,

3. broadcast addressing.

In *geographical addressing* each slave is addressed by its position in the crate. A CAMAC crate has a crate address, and a module in one of the slots 1–23 is addressed by its slot number. An individual selection line

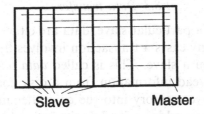

Slave   **Master**

Fig. 1.79. CAMAC addressing. Each slave module is connected by a separate line to the master station.

runs from the crate controller to every slot. Slave modules are enabled when this line is asserted (Fig. 1.79). Replacing a module in a CAMAC crate does not require any switch settings on the board. In FASTBUS every slot receives from its backplane connector a five-bit coded station number. Because of the position-dependent form of addressing it is not possible for two modules to have the same address and respond as slaves. Trouble shooting is easy.

From the programming point of view, geographical addressing is tedious. Suppose that one uses a crate with two ADCs, one TDC, three ADCs, four TDCs and five pattern units. A program that should read first all ADCs, then the pattern units and afterwards the TDCs must know the positions of the devices exactly. In addition it must be changed if the positions are rearranged. In this case *logical addressing* is preferable. In systems with logical addressing each electronic module contains a micro-switch register, which represents the address of that particular module. The bus address lines are compared to the module address. In a case of a match the module sends back a connect. Another method of inserting logical address is implemented in FASTBUS. Geographical addressing is used to assign a logical address into a device register. Taking the example mentioned above the ADCs, pattern units and TDCs will have logical addresses in an increasing order so that a simple loop or DMA can read out the devices.

A drawback of logical addressing is that debugging is more difficult. How can one find the two modules with the same address in a system of 1000 devices? The use of light emitting diodes (LEDs) on each address module opens a computer-independent way by visual inspection.

So far we have discussed how a master can establish a connection with a single slave. In many cases, however, it may be useful for a master to send information to several slaves simultaneously, for example to reset the input. This is called the *broadcast mode* in which multiple slaves can be selected by their broadcast class.

1.8.3 Data transfers

After the selection of a particular slave, data are either written to or read from the slave. In many cases a transaction involves the transfer of many data words to or from a slave. This is called *data block transfer*. Block transfers are used to read information from a disk controller or to copy an event from the slave memory into the computer memory. Instead of sending each transaction address, its data, its next address and next data, etc. it is much faster to send the initial address once, followed by a stream of data. The slave and master may contain internal registers to point to the next address in the memory. At each data transfer in the memory address registers are incremented by one to enable the next memory cell for the next transfer.

1.8.4 Control lines

So far we have described operations such as reading or writing one or several words. We know that the instruction repertoire of a computer for transferring data is much wider. There are instructions like store halfword, store byte, set bit, clear bit. Parts of these operations are also required in bus systems. These operations are distributed by control lines. In VME one can specify different widths of logical addresses, 16, 24 or 32 bits. In FASTBUS the transfer mode can vary from random data, asynchronous block transfer to synchronous block transfer. CAMAC allows many different functions in one cycle such as read, read and clear, read complement, test interrupt, clear interrupt, write, test, enable.

1.8.5 Responses

All the peripheral equipment of a computer contains status registers to inform the computer whether a device is busy, off-line, ready, or has found an error. The status register can be read by a computer in a separate cycle. A similar mechanism is needed for the slaves to inform the master that there are no more data. This information is given to the master during the data cycle and is called response. Some of these responses are as follows.

- Operation OK. If this does not appear there is no slave.

- There are no more data. This response is required if the master does not know at the initialization of a block transfer how many data should be read.

- Illegal operation, such as read information from an output device, e.g. a line printer.

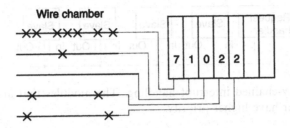

Fig. 1.80. Use of Q response to read variable length information. On the left hand side is an experiment with several wire chambers that contain some information. The crate on the right-hand side should only read as much information from each chamber as possible. This is called sparse data scan.

- Parity error.

- Data not ready. Please wait.

Responses play an important role because of the wide variety of devices that can be attached to a bus. Responses permit the implementation of scanning algorithms for reading data from several modules. Figure 1.80 shows an example where slaves are connected to several drift chambers. The readout is initialized in such a way that each module transfers as many data as there are hits in the drift chamber. In CAMAC a slave transfers a response $Q = 1$ if there are data and $Q = 0$ if there are no more data. If there are no more data ($Q = 0$) the master ignores the information on the data lines and addresses the next slave. In VME for physics applications a chained block transfer (CBLT) allows the fast readout of large blocks. At the start the master knows neither which of the slave modules contain data nor the quantity of data in each of them (sparse data scan).

1.8.6 Interrupts

There are many situations in which a slave has to inform the master or the computer that something has happened. Such situations include the following.

- There is a trigger. Data must be read out.

- There is a bus error or a parity error. Please check.

- The preselected time is over. Clock interrupt.

- The block transfer has finished.

- The operator has pressed a knob on the touch panel.

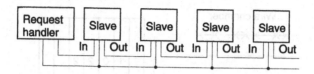

Fig. 1.81. Daisy-chained interrupt handling. The modules that are near to the interrupt handler have higher priority.

There are two ways for a computer to get informed that a device requires special service. The computer can periodically read the status registers of all devices and check whether some action is required. This is called a *polling loop* or *sending a token*. A drawback of this solution is the amount of computer time to check all the bits and pieces. The advantage is that the program can easily use priorities and read significant status registers more often. There are no unknown interrupts due to pick up, no illegal interrupts and no interrupts that appear at high rates to saturate the computer. In order to make efficient use of computer time *program interrupt* facilities are established. An interrupt to a computer causes the program to transfer control to an interrupt service routine. This routine will analyse the interrupt and then execute a subroutine. At the end of the service the interrupted program will continue. In CAMAC, the mechanism to request an attention interrupt is very simple. Each module is connected via an individual Look-At-Me (LAM) line to the controller, which immediately knows which station needs service (see Fig. 1.79).

A different scheme uses a simple request line. Requests from all devices are serviced by a request handler, which has to identify the request. Either it polls all slaves to identify which one has an outstanding request or it sends an acknowledgement to a special bus line that is *daisy-chained* through all the devices. When a device sees an acknowledgement it checks whether it has an outstanding request for attention. If not, the acknowledgement is passed on. If it has a request it will place an identifier on the bus that will be read by the interrupt handler (Fig. 1.81). In a system that has several computers connected to the bus there is the general problem: which computer should be interrupted if a slave gives a request for attention? There are several schemes. Either the request handler decides with the help of the identifier to which computer the interrupt should be addressed. Or the bus, as VME, allows the transfer of interrupts on seven interrupt lines to several computers.

Another approach is used in FASTBUS, which does not use extra interrupt lines. If a slave requires attention it has to become bus master and transmit the attention message to the required destination in a special receiver block. The advantage of this scheme is that it does not require extra lines and the procedure is position independent. But is requires extra electronics on the slave to be able to become bus master.

1.8.7 Multiple masters, bus arbitration

The use of multiple masters causes some problems in ensuring that the interrupts are transferred to the right master. Another problem is that several masters may compete for the bus. There must be a procedure to decide who may use the bus at any time. This is called *arbitration*. A master sends out a bus request and receives a bus grant allowing it to use the bus. In centralized arbitration a single hardware unit is used with request lines to all masters. This arbitrator returns a bus grant to the master. If two masters send a bus request at the same time, one needs rules to decide who goes first. One possibility is priority arbitration, which is widely used. No lower priority grant will occur as long as there is a higher priority request present. To avoid the possibility of a high-priority master locking the bus one can alternatively implement a democratic arbitration scheme. Mastership is allocated in a cyclic way to all competing masters. Which scheme is used depends on the application, but most buses use the priority scheme.

To avoid the extra lines to a central arbitration unit one can use alternative solutions similar to those we have discussed for interrupt handling. If a master wants to use the bus it asserts a bus request. The arbitrator returns a bus grant, which is daisy-chained through all masters and taken by the first master issuing a request (Fig. 1.81). This form of arbitration gives highest priority to the device nearest to the master. When there are many masters this method can be slow because the bus grant must ripple through all the masters. It is also tedious if one wants to add, remove or replace modules. One has to interrupt the daisy-chain and add more cables.

In FASTBUS a scheme was developed to allow for priority without a daisy-chain. Each master contains an internal priority vector. All masters requiring bus access use a common request line. The arbitrator then asserts an arbitration grant. The competing masters then put their vector on common bus lines, which will contain the logic OR of all vectors. The masters then compare this OR on the bus with their internal vector and, if the values differ, remove the low-order bits one after the other. After a certain time there is a stable vector corresponding to the winner. The masters with a vector indicating lower priority have removed all their bits. A simplified version of this system is shown in Fig. 1.82. This system has the advantage that one can remove, add, or replace modules without any recabling. We see that one has to add to the actual data transfer time the time needed by the master to get control on the bus. This time is

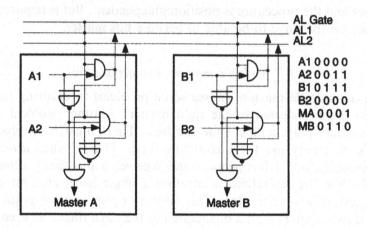

Fig. 1.82. Arbitration scheme for FASTBUS. This scheme works without daisy-chain cables. Each master puts its vector on the bus if the gate is available. The bus then contains the OR of all vectors. The masters then remove their lowest bits if they don't match. The master with the highest vector gets the bus.

called the *latency* of the access. It must be taken into account when a multimaster system is designed. It could happen that a single master can read data from a disk but that the same system may fail if more masters are added. At the end of the transaction a master can release the bus, but as bus arbitration may take some time it might be more efficient to keep the bus unless another master initiates a bus request.

1.8.8 Characteristics of buses used in physics experiments

In this subsection we list some characteristics of typical buses used in physics experiments. CAMAC is still being used in small experiments. Many specific modules (ADCs, TDCs, pattern units) are available on the market. FASTBUS is used by very large experiments. The investments in personnel and hardware are high. As FASTBUS is not widely used, modules must be developed by the experiments and are available only in limited variants. VME is widely used in industry and in physics experiments. Mechanically it is based on the successful Eurocrate mechanics. The specifications are continuously adapted to higher speed and larger boards. The PCI bus is well known from PCs. It is now also adapted to Eurocrate mechanics as compact PCI (CPCI).

1.8.8.1 Characteristics of CAMAC

- The data bus is 24 bits wide. The speed is up to 3 Mbytes per second. A crate has 25 slots. Eighty-two pins are connected to the backplane. A FASTCAMAC version is under study and may reach 7.5 Mbytes per second for block transfer.

- Up to seven crates can be connected by a branch highway.

- Transfers are synchronous within the crate and asynchronous on the highway.

- A crate controller can control the slaves in one crate. It is not possible for a controller to access a slave in another crate.

- The addressing is geographical. Each controller has an address in the range 1–7. Within a crate each slot has an extra line to the crate controller. Within each slot 16 subaddresses can be generated.

- The functions define the operation. Thirty-two functions are available. For example, F0 means 'read data' from slave, F16 means 'write data' to slave.

- Each module replies with a Q response to indicate the status of the module.

- Interrupts (LAM = Look-At-Me) are sent from a slave on an extra line via a transformer (LAM-grader) to the crate controller. The computer can test the LAM with an F8 function command.

1.8.8.2 Characteristics of FASTBUS

- The bus is 32 bits wide. Addresses and data are multiplexed. The speed is up to 100 Mbyte per second. A crate has 26 slots. The backplane has 130 bus lines. The bus is built in ECL technology.

- Normal transfers are asynchronous. For long distances one can use synchronous block transfer to avoid slowing down due to handshakes.

- The basic unit is a segment. One distinguishes crate segments and cable segments. Segments can be connected by segment interconnects.

- A master is capable of requesting and obtaining control of a segment. In addition he can communicate with slaves in other segments.

- Each segment interconnect contains address recognition circuits. This is mainly a list of preloaded routing tables. For a bidirectional operation two lists must exist. Segment interconnects are complicated devices.

- Modules can be addressed geographically (position dependent) or logically (address dependent). The logical address is loaded into the device address register by geographical addressing. Addressing can be performed on the same segment or on another one.

- Broadcasting can be used to synchronize devices or to clear counters.

- For sparse data scan each device uses the T-pin on its connector. The T-pin is set if the device has some data. The T-pin pattern is readout and decoded by the master.

- Devices requesting an interrupt write their address and up to 16 words into the interrupt sensing control register of the processor. The master then has all the information to handle the request.

1.8.8.3 Characteristics of VME

- The bus is 32 bits wide. Addresses and data are not multiplexed. The speed is up to 40 Mbytes per second. VME64 can reach 80 Mbytes per second and uses 32 or 64 multiplexed address and data lines. A crate has 21 slots. The connection to the backplane is via one (single height) or two (double height) 96-pin DIN connector(s).

- VME is defined on a crate basis only. In addition, a VME extension (VSB) bus can be used to connect CPUs with local memories at high bandwidth, and a VMS bus is available for long-distance applications with serial links.

- Data transfers are asynchronous. Eight, 16-, or 32-bit wide transfers are supported.

- In block transfer mode the master sends a single address to a slave followed by a block of data.

- Four request lines with different priorities are provided. On each level the priority is further defined by the position.

- Seven interrupt lines allow the slaves to express their requests for attention.

- VSB: the VSB bus has a 32-bit data path and a 32-bit address path. Up to five boards can be connected to each VSB bus.

- VME is based on Eurocrate mechanics.

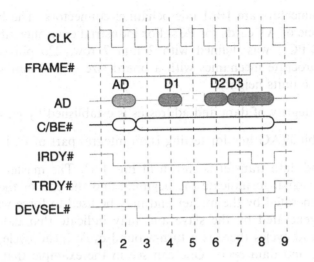

Fig. 1.83. Bus protocol for a PCI read operation. Cycle 2 is the address cycle, cycle 3 is an empty cycle after the address cycle for read, cycles 5 and 7 are wait cycles. AD = address/data bus, IRDY# = data ready from the master, can force a wait cycle, TRDY# = data ready from the slave, DEVSEL# = device select.

1.8.8.4 Characteristics of PCI

- The PCI bus was developed for fast I/O with PCs (PCI 1993). It may have up to 10 electrical loads or five add-in cards and is therefore limited to smaller systems. For add-in boards both the connector and the component connected to PCI via the connector count as a load. The bus is 32- or 64-bit wide. Data and addresses are multiplexed. The peak transfer speed is 132 Mbyte per second for 32-bit and 264 Mbyte per second for 64-bit wide bus.

- Transfer is synchronous, with a maximal clock rate of 33 MHz. The latency time is 60 ns for a write access (two cycles at 33 MHz).

- The arbitration phase can be overlapped with data transfer cycles.

- A PCI slave requires 47 signal pins, a PCI master 49 signal pins. The basic 32-bit connector contains 120 pins and the 64-bit extension has 184 pins.

- PCI is available with two voltages: 5 V and 3.3 V. Keys in the plug define which voltage is used.

- PCI devices must have a 256-byte configuration space that contains device information for automatic configuration support.

- The connectors are IBM microchannel connectors. The board size is the one of AT cards for PCs. For industrial environments *compact PCI* (CPCI) was defined with 2 mm 7 row, 220 pin connectors and Eurocrate mechanics with a board size of 160 mm × 233 mm (6 U = 6 units height).

- Data integrity of data and addresses is established by parity bits.

- The 4-bit JTAG bus for testing is an integral part of PCI.

A typical read data transfer is shown in Fig. 1.83. The master generates a FRAME# signal to indicate the begining and the end of the transfer. IRDY# is generated by the master and may be used to force wait cycles. TRDY# is generated by the slave and may indicate that data are not ready. For read cycles there is a turn-around cycle (dead cycle) between address cycle and data cycle. One can see in the example that the peak transfer rate in PCI is only realized for the transfer of very long data frames. If one reads in a random way from different addresses one needs three clock cycles for each read transfer, which brings down the speed from 132 Mbyte/s to 44 Mbyte/s.

1.8.9 Standardization of data buses

Standardization tends to be a slow process because of the time it takes to reach a consensus from initially widely disparate viewpoints. The problem is aggravated by rapid turnover of the committee membership. Recognizing the time required for this process and the desirability of having a small number of standards a committee was set up to define the 'future bus'.

The process of defining a new standard from scratch in a committee is much slower than the introduction of a data bus by industrial design teams. These teams work in close connection with engineers and customers and in several cases these buses are optimized for a specific microprocessor. The buses available at present are based on the assumption that microprocessors would be their main users. Some of the buses and their characteristics are summarized in Table 1.7

Table 1.7. *Characteristics of data buses*

Item	CAMAC	FASTBUS	VME	PCI
Bandwidth [Mbytes/s]	3	40	160	264
Address width	7/25/4	32	64	32/64
Data width	24	32	8/16/32/64	32/64
A/D multiplexed	no	yes	yes/no	yes
Board size [mm^2]	183 × 305	366 × 403	233 × 160, 366 × 403	311 × 107
Number of connectors	1	1	1/2/3	1
Type of connector	82 pin direct	132 pin	96/120 pin	Microchannel
Arbitration	central	distributed	central	central
Interrupt	LAM pattern	message passing	7 levels	4 levels
Bus protocol	synchr/asynchr	asynchr	asynchr	synchr
Serial bus	extra system	yes	yes	no
Geographical addressing	yes	yes	yes/no	no
Chip set available	no	private	yes	yes

2

Pattern recognition

Pattern recognition is a field of applied mathematics and makes use of results in statistics, cluster analysis, combinatorial optimization, and other specialized branches. The goal of pattern recognition is the classification of objects (Andrews 1972). The range of fields from which objects might be chosen is virtually unlimited: satellite pictures, electrocardiograms, coins, printed or handwritten text, and position measurements along particle trajectories may suffice as examples of physical objects. Non-physical objects can be found in applications such as linguistics, where certain expression patterns may provide clues as to the authorship, or authenticity.

In the data analysis of high-energy physics experiments, pattern recognition problems have to be solved in three different detector types. In track detectors, the signals generated by charged particles have to be grouped into track candidates; this is usually called *track finding* or *track search*. There is a large variety of methods from which to choose, depending on the properties and on the geometry of the detector and of the magnetic field. Sections 2.1 to 2.4 will be devoted to track finding. In calorimeters, the task of pattern recognition is to group the signals to *showers* and to compute certain properties of the shower. This will be discussed in Section 2.5. Finally, in ring-imaging Cherenkov counters (RICHes) images formed by converted photons have to be analysed in order to find *rings*. This is the topic of Section 2.6.

2.1 Foundations of track finding

Track finding usually means finding charged tracks in a track detector. Since most track detectors are able to measure the position of a charged particle only in a fairly small number of points, the task of track finding is equivalent to a partition of these measurements into disjoint sets, allowing for the fact that some measurements are noise or belong to tracks that the experimenter is not interested in, for instance very low momentum tracks. The choice of the algorithm depends heavily on the type and on the quality of the measurement the track detector is able to deliver. Some types of detector give only two-dimensional information, some give space points, others give entire vectors (point plus directions). In addition, the shape of the detector and of the magnetic field determine the mathematical model of the track; this in turn may have a decisive influence on the selection of the most suitable procedure. On the other hand, pattern recognition should be facilitated by the detector design chosen by the experimenter. This interplay is crucial for a successful track reconstruction. In order to avoid unpleasant surprises, the design and the evaluation of the pattern recognition strategy has to take into account imperfections of the detector such as noise, inefficiencies, measurement errors, and outliers.

2.1.1 Track detectors

The technology of track detectors has been dominated for a long time by gaseous detectors. Only recently have solid-state detectors begun to invade the field on a large scale, mostly as high-precision vertex detectors. They will play a crucial role in the inner trackers of the future LHC experiments.

2.1.1.1 Gaseous detectors. Until recently, the great majority of signals for track finding in high-energy physics have been delivered by gaseous detectors, in particular by wire chambers, which have undergone considerable evolution over the last 25 years, as have the pattern recognition methods (Grote *et al.* 1973; Grote and Zanella 1980; Bischof and Frühwirth 1998). In the present book, there is only space to outline the most basic physical mechanisms of wire chambers; the reader interested in more details is referred to the papers by Charpak (1978) and by Marx and Nygren (1978). Recent developments can be found in the proceedings of the Wire Chamber Conference (Bartl *et al.* 1995, Krammer *et al.* 1998).

The basic unit of a wire chamber is a thin metallic wire of typically 10–50 μm diameter. In its simplest form, a detector consists of one such anode wire surrounded by a cylindrical cathode, with a voltage of several thousand volts between them. The chamber is normally filled with a noble gas such as argon at atmospheric pressure; in modern wire chambers other gases are added, for instance isobutane. The isobutane absorbs part of the ultraviolet radiation and in this way keeps the ionization region localized.

When ionizing radiation passes through this cylinder, the light ionization electrons liberated drift towards the anode wire, while the much heavier positive ions stay behind, moving comparatively very slowly towards the cathode. Since the electric field grows inversely with $1/r$, where r is the distance from the wire centre, the electrons gain more and more energy over their mean free path between inelastic collisions. Eventually they initiate an electron avalanche through further ionization in the multiplication region at a distance of the order of the wire diameter. Thus the main charge multiplication occurs in the immediate vicinity of the anode wire. Therefore the path of the travelling electrons is very short, and the corresponding signal cannot be detected by large-scale standard electronics. A larger, but slower, signal is created by the ions travelling across a substantial fraction of the chamber gap. This signal, which has a typical rise time of 10 ns, is detected by the electronics connected to the wire; this wire is then said to have 'fired'.

The precise operational mode of such a detector depends upon the value of the electric field potential. At low voltages, the electrodes collect all the electron–ion pairs, and the pulse height is independent of the applied potential; a detector operated in this mode is called an *ionization chamber*. At higher voltages, gas amplification occurs when additional ions are created from the primary ionization in the strong electric field around the wire. The magnitude of this effect depends upon the electric field strength around the wire, and the pulse height is proportional to the original number of ions created. For this reason, when operated in this mode the detector is known as a *wire proportional counter*. Increasing the applied potential beyond this region causes charge to accumulate around the wire (space charge effect) and the strict correlation between the applied potential and the pulse height is lost, i.e. the charge collected on the anode wire is no longer proportional to the original ionization. At still higher electric fields one reaches the transition region to the *Geiger–Müller* mode of operation. Here the signal is stronger by a factor of about 1000 than in proportional mode. The region of electron–ion pairs is no longer limited to the avalanche, and the rise time of the signal is determined by the speed of the electrons.

Proportional chambers. The mechanism described above remains practically unchanged when many parallel anode wires are arranged in one plane between two cathode planes, at a wire spacing of the order of 1–2 mm. The electric field is then almost homogeneous everywhere except for the region near to the wires where it resembles that in the tube. Such a detector is called a multiwire proportional chamber (MWPC). The chamber therefore works like many independent tubes: an electron cloud created somewhere in the chamber will drift towards a specific wire and lead to a pulse there. A particle passing through this chamber and creating regions of ionization along its path will lead to signals on one or several adjacent wires, depending on its orientation with respect to the chamber plane. By spacing the wires at a suitable distance d, one can in this way measure the position of the impact point on the plane of the wires with a precision of about $d/\sqrt{12}$. The theoretical lower limit of the precision is $d/2\sqrt{12}$ if one assumes that a track close to the wire fires only that wire, and that a track in the intermediate region between two wires fires both wires.

The invention of the MWPC was one of the milestones in the history of particle detectors and experimental high-energy physics (Bouclier *et al.* 1974; Charpak 1978; Sauli 1978). By a fortunate historical accident it arrived just in time to be employed at the first hadron collider, the Intersecting Storage Rings (ISR) at CERN. Detectors with over 50 000 wires (Bouclier *et al.* 1974) have been constructed with chambers operating in the proportional mode. A MWPC has the following advantages:

- it is mechanically simple and not very sensitive to small changes of the gas mixture;

- it has good spatial resolution and good two-track separation;

- even the multi-hit efficiency can exceed 99%;

- event rates of more than 10^6 Hz are possible;

- as the chamber is d.c.-powered, it can be used in a fast trigger requiring processing times below 100 ns;

- the dead time is short and the dead region is localized;

- good signal to background ratio is guaranteed;

- it is hardly affected by strong magnetic fields.

Since a plane of wires measures only one of the two avalanche coordinates in the plane, the following method has been developed to measure the second coordinate as well and thus the position in space: the cathode planes are divided into parallel *strips*, which are a few centimetres wide and run perpendicularly to the wires, or at a suitable angle (Fig. 2.1). On these strips the avalanche induces a positive pulse, which is measured and gives the second coordinate; the third coordinate of the space point is the position of the wire plane.

Another method consists of measuring the amount of charge flowing into amplifiers at both ends of the wire. If the wire has a suitable resistance, the ratio of the charges will depend on the position of the avalanche along the wire. With this *charge division readout*, a precision of 1% (at best) of the wire length can be reached for the second coordinate (see also Section 1.5.10).

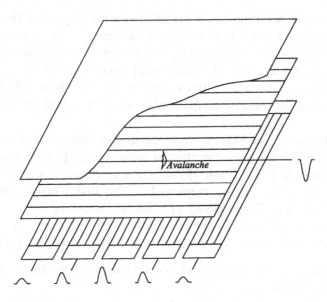

Fig. 2.1. Reading the induced pulse on surrounding electrodes. The avalanche pulse at the anode wire induces pulses in both adjacent cathode planes, one of which, in this example, is subdivided into strips, which are equipped with an analog readout system. The barycentre of the pulse charge distribution gives an accurate avalanche position along the wire, and hence a second coordinate. When there is more than one avalanche in the wire plane there is no way of correlating the two projections.

Drift chambers. To achieve accurate results, large proportional detectors need many wires, each of which needs its own amplifier. In order to reduce the cost of this, to improve the accuracy, and to reduce the amount of matter in the particle's path, detectors have been constructed that exploit the same arrangement of anode wires between cathode planes in a completely different mode. To this end, the chamber is rotated by 90°, such that the particle path runs more or less parallel to the wire plane, but is still orthogonal to the wires (Fig. 2.2). Since the electron cloud drifts initially at constant speed of some 50 μm/ns towards the anode wire, the drift time, i.e. the time lapse between the impact of the particle and the avalanche is a direct measure of the distance from the wire. The impact time is normally determined by separate scintillation counters. With a time resolution of the order of 1 ns, one can then measure positions with an accuracy of 50 μm. However, the apparent resolution can be considerably worse because of correlation effects (Drijard, Ekelöf and Grote 1980). A drift space of many centimetres can be obtained, which leads to a considerable reduction in the number of wires. However, since the chambers measure only the drift distance, there is no information as

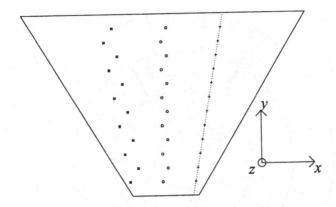

Fig. 2.2. Drift chamber cell; one cell of a cylindrical drift chamber surrounding an interaction region in a colliding beam machine. The staggered sense wires along the symmetry axis of the cell extend in the z direction. This staggering has been very much exaggerated here in order to make it visible; in reality it is typically 200 µm whereas the wire distance is typically a few centimetres. The electric field is oriented in the x direction towards the sense wires, and is homogeneous in y and z. A charged particle passing through the right half of the cell leads to x position measurements along its path as well as the 'ghost' points on the opposite side, which are staggered twice as much as the sense wires thus making their rejection easier.

to which side of the wires the particle passed. This *left–right ambiguity* represents an additional difficulty in track finding. The remedies employed, e.g. staggering the wires slightly, and offsetting drift chamber cells, cannot normally be used during the initial part of the track finding to reject the *image* or *ghost* tracks, but may perform this task in a subsequent step: staggering the sense wires leads to a higher error when a track is fitted to the ghost track, and offsetting drift chamber cells in most cases prevents ghost track sections from adjacent, but offset, chambers being linked into one track. A typical low-multiplicity event in a cylindrical drift chamber arrangement is shown in Fig. 2.3.

Time projection chambers (TPCs). A TPC is in some sense an improved type of drift chamber. The main differences are the following (see Fig. 2.4).

(1) The drift space extends mainly to one side only of the sense wires, thus curing the left–right ambiguity problem. This is achieved with the help of additional field wires.

(2) In many cases, a magnetic field parallel to the electric field leads to an additional confinement of the drifting electron cloud (Marx and Nygren 1978); this, in turn, allows for longer drift lengths of 1 m

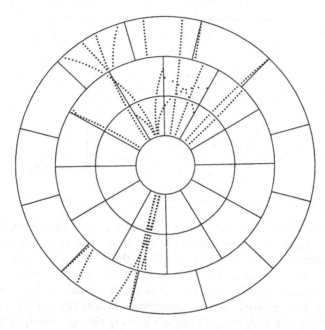

Fig. 2.3. Event in a cylindrical drift chamber detector. The picture shows a cylindrical drift chamber layout similar to the one used in the JADE detector at DESY. The solid lines represent the drift cell boundaries. The sense wires along the cell symmetry axes are not shown, but only the signals and their ghost images. The effect of staggering the outermost layer of cells with respect to the other two is clearly visible: whereas the true particle paths lead to contiguous lines of signal points, their ghost images are disrupted. The picture shows seven tracks of charged particles.

and above. With a time resolution of 1 ns one can measure the drift path length, and therefore the particle position with a theoretical accuracy of 50 μm. The longitudinal diffusion of the electron cloud tends to increase this value to typically a few hundred micrometres.

(3) Opposite each sense wire, a row of small *pads* running along the wire, and oriented orthogonal to it, receives the induced pulse from the avalanche at the sense wire. The barycentre of the charges distributed over the pads can be reconstructed with a precision comparable to that of the drift length measurement or better.

The TPC delivers genuine space points along a particle trajectory since there is no ambiguity as to the association of signals in the two projections, the association being provided naturally by the arrival times. Because of this feature TPCs have already been used, and will be employed in many experiments that intend to measure high-multiplicity events and particle

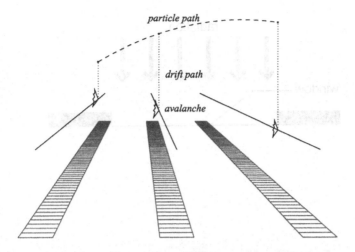

Fig. 2.4. Schematic view of a time projection readout system. The perspective view shows sense wires above rows of pads. The electric field makes the electrons drift vertically towards these sense wires where they give rise to avalanches as in the case of standard drift chambers. The induced pulses can then be read out on the pad row, where each pad is connected to its own amplifier. The scale of the picture is grossly distorted: whereas the pads are typically 5 mm by 8 mm, and sense wires are mounted 4 mm above them, the electron drift path may be 1 m or more.

jets. For instance, two of the LEP experiments (ALEPH and DELPHI) have TPCs as their central tracking devices. TPCs have also been used by heavy-ion experiments. At the LHC, however, both the event rate – nearly a billion times higher than at LEP – and the track density are much too large to allow TPCs to be employed.

Microstrip gas chambers. To overcome the limitations of conventional gas detectors (limited counting rate capability and limited position resolution) a new type of gaseous detector came up in the late 1980s (e.g. Oed 1995): the microstrip gas chamber (MSGC).

An MSGC is a proportional chamber where the anode and cathode strips are on the same chamber wall, ordered in an alternating structure (see Fig. 2.5). They are both thin metallic layers (strips) deposited on an insulating or semi-insulating substrate. The opposite chamber wall consists just of a gas-tight window at a distance of 2–6 mm, usually on ground potential defining a region of charge collection.

The width of the anode strips is usually equal to or less than 10 μm, the cathode strips have widths ranging from 50 to a few hundred micrometres, and the pitch width lies between 50 μm and 1 mm. The production of a wide variety of different structures, even of such with circular electrodes

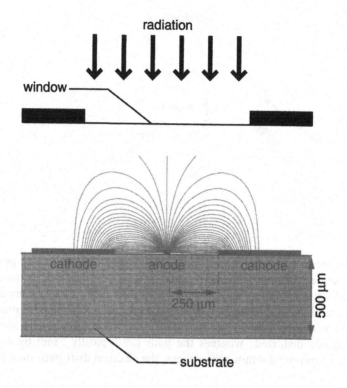

Fig. 2.5. Microstrip gas chamber. Alternating anode and cathode strips are deposited on an insulating substrate. A gas-tight window is used as drift electrode. The region of gas amplification is just near the anode, where a high electric field is produced by the voltage between anode and cathode.

(Micro Dot Chambers), is possible. In an MSGC the ionization electrons induce an avalanche amplification just as in a proportional chamber. The typical rise time of the signal is of the order of 10 ns and the position resolution achieved is of the order of the theoretical limit (pitch width/$(2\sqrt{12})$, see Subsection 3.3.2.1), under optimal conditions. Owing to the small gaps between anode and cathode, the rate capability is larger than 10^6 counts/mm^2 s. The energy resolution is constant near the statistical limits for the entire sensitive area. Moderately high operation voltages and the possibility of industrial mass production at reasonable costs are additional advantages. The ageing and long-term stability behaviour are still subjects of controversial discussion (Schmidt 1998).

2.1.1.2 Solid-state detectors. Although solid-state detectors have been used for nuclear spectroscopy for many decades, substantial progress has been made in the development of solid state detectors for high-energy physics experiments in the past 15 years. These new structures are also used in many other fields like medicine or (with a suitable converter) as detectors for low-energy neutrons. In high-energy physics they have been in use in fixed-target experiments for the detection of short-lived particles for many years. Nowadays they are mainly used as high-precision position detectors in storage ring experiments, in most cases for the detection of secondary vertices from the decay of very short-lived particles. Although rate capability and radiation hardness are extraordinarily high, the behaviour in future experiments with an extreme environment, e.g. at the LHC, must still be proved. A drawback with respect to wire chambers is the limited size of a single detector element, owing to the lack of availability of large, high-purity, single-crystal semiconductors.

The most commonly used semiconductor material is silicon. The large-scale use of these detectors in high-energy physics was made possible by the standard industrial mass production of large silicon wafers with high resistivity.

To understand the physical properties of semiconductors it is convenient to look at the band model. For silicon condensed in crystalline form the energy levels are merged into energy bands, where they are essentially continuous. The valence band and the conduction band are separated by a band gap in which all energy states are forbidden (in Si: $\Delta E = 1.1 \, \text{eV}$). At room temperature a substantial number of electrons are lifted from the valence band into the conduction band (creation of electron–hole pairs). Therefore the direct use of semiconductors (at room temperature) for the detection of minimum ionizing particles passing through the detector is prohibited by the intrinsic noise. To overcome this effect a silicon pn-junction is used, allowing the depletion of the detector bulk by applying a bias voltage (see Fig. 2.6).

The early solid-state detectors were *silicon microstrip detectors*. Many techniques are available nowadays, and the pitch width varies from 25 μm to 200 μm. By the use of intermediate strips and with pulse height interpolation, even a 200 μm pitch allows a position resolution of 30 μm (RMS) for particles passing the detector orthogonal to its surface, and even 10 μm or better for an inclination of 20°.

Recently a new type of detector came into large-scale operation: the *pixel detector*, a diode matrix with pixel sizes of side length of a few hundred micrometres. This concept does not put too difficult demands on the detector itself but rather on the electronics and connection techniques.

For higher particle multiplicities pixels have a substantial advantage compared even with double-sided strip detectors, namely a virtually total suppression of ambiguities. Semiconductors are also in use as solid-state drift chambers, based on a double diode structure.

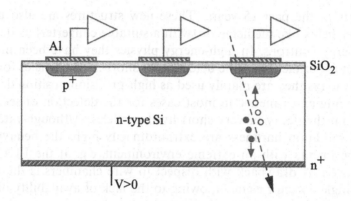

Fig. 2.6. Basic principle of a silicon strip detector. Free electrons and holes, generated in the depleted bulk of the detector by a through-going ionizing particle, drift to the electrodes and thus generate a signal. The detector shown here is an a.c.-coupled single-sided strip detector. The metallic readout strips (usually Al) have no direct connection to the p^+-implants but are electrically coupled to them via internal coupling capacitors, consisting of a thermally grown SiO_2-layer.

In the vertex region of future high-energy physics experiments, silicon detectors will suffer heavily from radiation damage. Therefore the use of diamond as detector material has been proposed (Trischuk 1998). Because of the large separation of valence and conduction bands, the intrinsic noise is sufficiently small even without depletion. However, a substantially higher amount of energy deposit is needed to create an electron–hole pair, leading to a very small signal per particle track. In recent time, substantial progress in the artificial growth of high-quality (detector grade) diamond wafers has been made.

2.1.1.3 Tracking systems. Large experiments at current and future colliders contain several different types of track detectors, so it is appropriate to speak of tracking systems. As an illustration we will briefly describe the tracking systems of the DELPHI detector at the LEP collider and of the projected ATLAS experiment at the future LHC collider.

The DELPHI tracking system (Aarnio *et al.* 1991; Abreu *et al.* 1996) is remarkably complex. Its core is a time projection chamber with a radius of 1.2 m and a length of 3 m. The momentum resolution is improved by other track chambers, both in the central (barrel) region and in the forward region. In the barrel region a charged track first crosses the three silicon layers of the microvertex detector. It then enters the Inner Detector, a cylindrical drift chamber giving up to 24 $R\Phi$-points per track. Surrounding the drift chamber, there are five cylindrical layers of straw tube detectors giving $R\Phi$ information. These measurements are used for triggering and for resolving the left–right ambiguities inherent in the drift chamber. After leaving the trigger layers the track crosses the TPC, which delivers up to 16 space points per track. It then traverses the barrel RICH detector. In order to provide a precise track interpolation in the RICH, there is another track detector surrounding the RICH, the Outer Detector. It consists of five layers of drift tubes giving $R\Phi$ information. The z coordinate is determined by timing the signals at both ends of the anode wires.

In the forward region the microvertex detector has recently been extended down to a polar angle of about 11° by the Very Forward Tracker, a silicon tracker with two pixel and two double-sided ministrip layers. Tracks with polar angles larger than about 20° are measured by the Inner Detector and by the TPC, albeit with a small number of points. Behind the end plate of the TPC there is Forward Chamber A, consisting of six layers of drift tubes. A track leaving this chamber then crosses the forward RICH and enters Forward Chamber B, a planar drift chamber with 12 wire planes.

The complexity of this design is mirrored in the design of the track reconstruction algorithm, which has to take into account the characteristics of each component and has to combine the information delivered by each of them in an optimal fashion (see also Section 3.5).

The design of the ATLAS inner tracker (ATLAS 1994; ATLAS 1997) is heavily influenced by the high repetition rate, the high multiplicities, and the high background radiation at the LHC collider. It combines high-resolution detectors at smaller radii with continuous tracking elements at larger radii. The innermost part of the barrel tracker is made up of silicon pixel detectors, followed by silicon microstrip detectors. At least three pixel layers and four microstrip layers are crossed by each track. In the barrel, the high-precision detector layers are arranged on concentric cylinders, while the end-cap detectors are mounted on disks perpendicular to the beam axis.

The silicon tracker is surrounded by a transition radiation tracker (TRT) made up of straw tubes that provides the possibility of nearly continuous track following (typcially 36 points per track) at lower cost. The barrel TRT straws are parallel to the beam; the end-cap straws are again mounted on disks perpendicular to the beam axis. The combination of the two techniques gives robust pattern recognition and high precision.

For information on the tracking system of the CMS experiment, which is the second general-purpose experiment at LHC, the reader is referred to the literature (CMS 1994; CMS 1998).

2.1.2 Some techniques of track modelling

2.1.2.1 Circles, polynomials, and splines for curve approximation. On many occasions, the particle track is conveniently approximated by a simple curve during the process of track finding. This arises when the particle moves in an inhomogeneous magnetic field, in which case there is no analytical description of the track, or even when the projected track is an ellipse or a circle, namely when the speed of the calculation is important.

The simplest and fastest approximation to a curved track is the parabola:

$$y = ax^2 + bx + c \tag{2.1}$$

which can actually be expressed as a linear function in three of the y values (the y values could be the measured coordinates along a track):

$$y = a_1 y_1 + a_2 y_2 + a_3 y_3 \tag{2.2}$$

with

$$a_i = \left[(x - x_j)(x - x_k) \right] / \left[(x_i - x_j)(x_i - x_k) \right] \tag{2.3}$$

and (i, j, k) being a permutation of $(1, 2, 3)$.

This formula is a special case of Lagrange's formula, Equation (2.4), for polynomials of any degree, the correctness of which follows immediately from the fact that two polynomials of degree n are identical if they share $n + 1$ points:

$$y = \sum_{j=0}^{n} \frac{\pi_j(x)}{\pi_j(x_j)} y_j \tag{2.4}$$

with

$$\pi_j(x) = \prod_{k=0, k \neq j}^{n} (x - x_k) \tag{2.5}$$

Thus, if a hit has to be predicted in a detector plane as function of three points already on the track, Equation (2.2) can be used, where the coefficients a_i, which depend only on the plane positions and are therefore constant, have been stored in a table for all combinations of three planes that may be used.

Equation (2.2) is remarkably simple when compared to the expression for a circle. The centre (x_c, y_c) and the radius r of a circle through three points are given by

$$x_c = (b_1 - b_2 + c_1 a_1 - c_2 a_2)/(c_1 - c_2)$$

$$y_c = b_1 + c_1(a_1 - x_c) \tag{2.6}$$

$$r^2 = (x_1 - x_c)^2 + (y_1 - y_c)^2$$

with

$$a_i = (x_{i+1} + x_i)/2, \quad i = 1, 2$$

$$b_i = (y_{i+1} + y_i)/2, \quad i = 1, 2 \qquad (2.7)$$

$$c_i = (x_{i+1} - x_i)/(y_{i+1} - y_i), \quad i = 1, 2$$

and the points have to be numbered in such a way as to avoid division by zero. Because of this last condition the above formula is inconvenient and 'risky' to use, since it will, of course, become imprecise when the values in the denominator approach zero. This can be avoided by performing a coordinate system shift followed by a rotation in such a way that (x_1, y_1) coincides with the origin, and y_3 becomes zero.

1. Shift:

$$x_i' = x_i - x_1, \quad y_i' = y_i - y_1, \quad i = 2, 3 \qquad (2.8)$$

2. Rotate:

$$d = (x_3'^2 + y_3'^2)^{\frac{1}{2}}$$

$$c_\phi = x_3'/d$$

$$s_\phi = y_3'/d \qquad (2.9)$$

$$\tilde{x}_i = x_i' c_\phi + y_i' s_\phi, \quad i = 2, 3$$

$$\tilde{y}_2 = y_2' c_\phi - x_2' s_\phi$$

3. Find centre, radius:

$$\tilde{x}_c = x_3/2$$

$$\tilde{y}_c = \left[\tilde{y}_2 - \tilde{x}_2(\tilde{x}_3 - \tilde{x}_2)/\tilde{y}_2 \right] /2 \qquad (2.10)$$

(having checked that $\tilde{y}_2 \neq 0$)

$$r^2 = \tilde{x}_c^2 + \tilde{y}_c^2 \qquad (2.11)$$

4. Rotate and shift centre to original system:

$$x_c = \tilde{x}_c c_\phi - \tilde{y}_c s_\phi + x_1$$

$$y_c = \tilde{y}_c c_\phi + \tilde{x}_c s_\phi + y_1 \qquad (2.12)$$

Up to now, only analytical functions have been considered. We recall that an analytical function has infinitely many derivatives in each of its points of the contiguous open interval where it is defined (Dieudonné 1979).

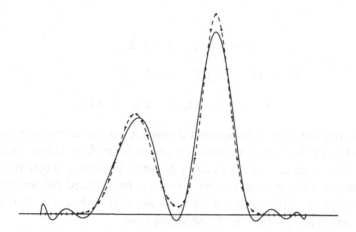

Fig. 2.7. Spline (dashed) and polynomial (solid) approximation. A polynomial of degree 15 is fitted to 51 equidistant points on the curve, which is given by the sum of two Gaussian functions. A cubic spline through the same points as knots is calculated. The polynomial approximation is particularly bad where the curve is flat over an extended area; such a behaviour is, of course, impossible for a polynomial.

A class of very useful non-analytical functions are the *spline functions*. Splines consist of pieces of polynomials linked at certain points called *knots* and are thus piecewise analytical. At the knots, the derivatives of a spline function exist only up to a certain order k, which is specific to the spline in question, such that the derivative of order $k+1$ is a step function. Actually, all spline functions can be expressed as indefinite integrals of step functions.

Since spline curves are not analytical, they are much more readily adapted to follow a path given by a series of points than a polynomial. As an example, both a 15th order polynomial and a cubic spline are fitted to points (marked by small crosses) lying on a curve composed of two Gaussian functions in Fig. 2.7:

$$f(x) = e^{(x+3)^2/3} + 2e^{(x-3)^2/3}$$

The figure speaks for itself. Spline curves are used in many graphical applications, such as the drawing of road maps.

In track finding, two types of spline have mainly been used, simple third-order or *cubic splines*, and B-splines. Third-order splines have been fitted both to the particle track, and to its second derivative; in the latter case their explicit double integration, leading to so-called *quintic splines*, delivers an approximation to the track as given by the equation of motion in an inhomogeneous magnetic field (Wind 1974; see also Subsection 3.3.1).

Spline functions are mathematically defined in the following way. Given n knots

$$x_1 < x_2 < \cdots < x_n$$

and the values

$$x_0 = -\infty, \quad x_{n+1} = +\infty$$

a spline curve $s(x)$ of degree m is defined for all x values through the following properties:

(1) in each interval $[x_i, x_{i+1}]$, $i \in [0, n]$, $s(x)$ is given by some polynomial of degree m or less;

(2) $s(x)$ and its derivatives of orders $0, 1, \ldots, m - 1$ are continuous everywhere.

Splines are conveniently expressed with the help of truncated power functions which are denoted by x_+^m:

$$x_+^m = \begin{cases} x^m, & x > 0 \\ 0, & x \leqslant 0 \end{cases} \tag{2.13}$$

Any spline function s of degree m can be uniquely expressed as

$$s(x) = p(x) + \sum_{j=1}^{n} c_j (x - x_j)_+^m \tag{2.14}$$

where $p \in \prod_m$ (class of polynomials of degree m or less). A *natural* spline is a spline of odd degree $2k - 1$ with $p(x)$ of degree $k - 1$ (rather than $2k - 1$):

$$s(x) = p(x) + \sum_{j=1}^{n} c_j (x - x_j)_+^{2k-1}, \quad p \in \prod_{k-1} \tag{2.15}$$

For a natural spline, the coefficients satisfy the relation

$$\sum_{j=1}^{n} c_j x_j^r = 0, \quad r \in [0, k-1] \tag{2.16}$$

If a function is to be approximated by a natural spline, where function values are given at the knots $x_1, \ldots, x_n$, the number of unknowns $n + k$ matches the number of constraint equations from Equations (2.15) and (2.16). This is probably the reason why such splines are called 'natural'. It can be shown (Greville 1969) that the system of linear equations is non-singular, i.e. that the spline representation is unique.

The popularity of splines becomes obvious if one tries to find the 'smoothest' interpolating function g for n data points given at $a = x_1 <$

$x_2 < \cdots < x_n = b$. 'Smoothest' means here that the kth derivative of g varies as little as possible in the interval $[a, b]$, i.e. that

$$\int_a^b [g^{(k)}(x)]^2 \, dx = \min \tag{2.17}$$

If $C^k[a, b]$ denotes the class of functions for which k derivatives exist on $[a, b]$, and if only functions $g \in C^{k-1}[a, b]$ are considered such that $g(x)^{(k)}$ is piecewise continuous in $[a, b]$, then it can be shown (Greville 1969) that the smoothest interpolating function for a given k is precisely the natural spline of order $2k - 1$, as long as $k < n$. For $k = n$, the solution is, of course, the uniquely defined polynomial of degree $n - 1$, and for $k > n$ there are infinitely many polynomials of degree $k - 1$.

From this it follows that third-order natural splines or cubic splines are the best lowest approximation to a curve looking 'smooth', since the case $k = 1$ leads simply to a connection of the data points by straight lines. For cubic splines, the explicit formulation of $s(x)$ as a sequence of pieces of polynomials still looks relatively simple:

$$s(x) = \begin{cases} a_{00} + a_{01}x, & x \leqslant x_1 \\ a_{i0} + a_{i1}x + a_{i2}x^2 + a_{i3}x^3, & x_i \leqslant x \leqslant x_{i+1}, \ 1 \leqslant i \leqslant n-1 \\ s(x) = a_{n0} + a_{n1}x, & x \geqslant x_n \end{cases}$$
$$\tag{2.18}$$

There are $4n$ unknowns, and $4n$ constraint equations: $1 + 2(n - 1) + 1$ from the function values at the knots, n from the continuity of the first, and n from the continuity of the second derivative at the knots. The matrix of the system of linear equations has a band structure, which has computational advantages.

The formulation of splines with the help of truncated power functions is not satisfactory mainly for two reasons.

(1) Expression (2.15) is biassed in the sense that for increasing x more and more terms enter in the expression. Consequently, the matrix of the system of linear equations resulting from Expression (2.15) has the form of a band matrix plus a triangular matrix, which is inconvenient for large n.

(2) Even worse, it can be shown (Cox 1982) that this system of linear equations is inherently ill-conditioned, which means that it has to be solved using very high precision in the computer or should be avoided altogether.

One would prefer a basis other than truncated power functions that is in some sense 'symmetric' and leads to well-conditioned systems of equations.

The so-called *B-splines* form such a basis. They are defined with the help of divided differences (Abramowitz and Stegun 1970). Given $k+1$ abscissa values $x_0, \ldots, x_k$, and a function $f(x)$ assuming values f_k at x_k, the first divided difference is defined as

$$[x_0, x_1] f = (f_0 - f_1)/(x_0 - x_1) = [x_1, x_0] f$$

and the *j*th divided difference becomes

$$[x_0, \ldots, x_j] f = ([x_0, \ldots, x_{j-1}] f - [x_1, \ldots, x_j] f)/(x_0 - x_j) \qquad (2.19)$$

This can be written in terms of functional values:

$$[x_0, \ldots, x_k] f = \sum_{j=0}^{k} f_j / \pi_j(x_j) \qquad (2.20)$$

where $\pi_j(x)$ is as in Equation (2.5). For a given set of knots x_k with $x_k \leqslant x_{k+1}$ for all k, the normalized B-spline $N_{n,j}(x)$ is defined by

$$N_{n,j}(x) = (x_j - x_{j-n}) [x_{j-n}, \ldots, x_j] (. - x)_+^{n-1} \qquad (2.21)$$

where the expression with the '.' is defined as follows: $[\ldots] f(., x)$ means that the divided difference is to be taken with respect to the variable at the '.', e.g.

$$[x_0 - x_1] (. - x) = (x_0 - x) - (x_1 - x)/(x_0 - x_1)$$

which is equal to 1. This is a special case of the theorem that the *n*th (divided) difference of any polynomial of degree n is a constant.

Applying this theorem, it can be seen from Equation (2.21) that a B-spline extends over a number of intervals that is one higher than its degree. For $N_{n,j}(x)$ this means that it is only different from zero for $x_{j-n} < x < x_j$. This can be seen as follows: for $x > x_j$, the truncated power functions are all equal to zero. For $x < x_{j-n}$, the truncated power functions are all equal to polynomials of order $n - 1$, hence their *n*th divided difference is zero. When a function is interpolated with the help of B-splines, the matrix of the corresponding linear system has consequently band structure.

2.1.2.2 Interpolation and extrapolation. The choice of the track-finding method may depend on the precision with which additional points can be predicted if a part of the track has already been found. A related problem is the linking of track segments that have been found independently.

Interpolation is more precise than extrapolation. For polynomials this can be seen most easily from the Lagrange formula:

$$y(x) = \sum_{j=0}^{n} \pi_j(x) y_j / \pi_j(x_j) \qquad (2.22)$$

with $\pi_j(x)$ as in Equation (2.5); n is the degree of the polynomial y, and y_j is the value at x_j.

If just one of the y values, say y_k, is wrong by an error Δy_k, then the error in y for all x is simply

$$\Delta y = \left[\pi_k(x)/\pi_k(x_k)\right]\Delta y_k = p(x)\Delta y_k$$

with some polynomial $p(x)$ of degree n. All $n + 1$ zeros of $p(x)$ are at x_j, $j = 0,\ldots,n$. Therefore, for $x \in [x_0, x_n]$, $p(x)$ oscillates between its $n - 1$ extrema, which, however, for arbitrary x_j can be made to assume any value, since $\pi_k(x_k)$ can become very small. On the other hand, for equidistant x values, one would hope that the absolute value of $p(x)$ remains below some convenient limit (preferably $= 1$) for $x \in [x_0, x_n]$.

This is not exactly the case, although the maximum absolute values are independent of the x step $h = x_1 - x_0$, and behave fairly well as can be seen from the following list (the degree of p is given in brackets):

$$\max(p(x)) = 1(2), 1.056(3), 1.151(4), 1.257(5),$$

$$1.357(6),\ldots, 6.5(10), \quad x \in [x_0, x_n]$$

This means that for polynomials of low degree, the interpolation error is of the order of the measurement errors.

On the other hand, for x outside $[x_0, x_n]$, $p(x)$ will grow faster than d^n if $d = \min(|x - x_0|, |x - x_n|)$, since there the differences $(x - x_j)$ all have the same sign. This means that the extrapolation error will grow rapidly even for polynomials of low order, and has to be watched carefully.

The interpolation of a circle with $y_1 = y_3$ leads to errors of the same order of magnitude as the measurement error, whereas the extrapolation error is roughly proportional to d^2/s^2, where d is the distance to the centre of the segment, and s the segment length, as long as s is small compared to the radius (Pimiä 1985).

The interpolation and extrapolation errors (introduced by measurement errors) of splines and other functions more complicated than polynomials are normally difficult, if not impossible, to evaluate in a closed form. The approach taken to this problem is in most cases a Monte Carlo simulation of all possible tracks under study, including all errors such as multiple scattering, δ electrons, and other known reasons for measurement errors, and the establishing of road widths with the help of histograms.

2.1.2.3 Parametrization.

Leaving aside quantum-mechanical effects which are very small, the track of a particle in vacuum is completely defined by five parameters, e.g. the five integration constants of Equation (3.65b). The trajectory can then be expressed as a function of one single independent variable, such as the track length s, or the coordinate along an axis of a coordinate system in space, say x. On the other hand, for a plane $x = $ constant, we can consider the values (y, z) on the trajectory for this x value to be functions of the five parameters, describing in this way not just one trajectory, but all trajectories passing through the plane $x = $ constant,

$$y \equiv y(y_0, z_0, y_0', z_0', 1/p)$$

and similarly for z. Finding an expression of this type is called *parametrization*. A different parametrization will normally be necessary for each x value.

As a simple example, consider circles through the origin in which the x axis is a tangent. Then, for a given $x_1 \geqslant 0$, the y values can be expressed as functions of the only free parameter left, the radius r (the formula is valid for circles bending upwards or downwards, if the sign of curvature is included in r):

$$y_1(r) = r[1 - (1 - x_1^2/r^2)^{\frac{1}{2}}], \quad |r| \geqslant x_1$$

In high-energy physics data analysis, parametrization has been applied in three different forms.

(1) To calculate the coordinates of a track in a given detector plane as functions of the five track parameters. This is useful both for Monte Carlo simulation and for track fitting. It is, of course, very much faster than tracking each particle individually through the magnetic field, and then calculating the detector hits.

(2) To parametrize the track hits in some of the 'planes' as functions of track hits in other 'planes'. This is used in track finding to predict coordinates once a part of the track has been found (Lassalle, Carena and Pensotti 1980). This is described in more detail in Subsection 2.4.2.

(3) To express the five track parameters as functions of the hits (= coordinates) in a given detector. These values are normally not good enough to be final, but may represent good starting values for an iterative fit procedure.

In all three cases, the most obvious merit of the method is its speed. Indeed, if the magnetic field is rather well behaved, and, more importantly, if the phase space region in question is small (Aubert and Broll 1974), one can hope to express the desired quantities by polynomials of relatively low order. This high speed makes the method very valuable in real-time applications, see e.g. Subsection 1.5.6. An extra benefit comes from the fact that the magnetic field need no longer be used explicitly once the parametrization has been computed.

The most serious drawback of the method lies in the fact that normally it requires much study and optimization effort. The choice of the phase space region, and the selection of the *training sample* are particularly delicate operations. For example, it is important to choose more points in pattern space near the borders of the trajectory cluster than near the centre. Furthermore it should be clear that any given parametrization is only valid for a specific detector arrangement, and that each change in the detector set-up, and each different subset of planes that must be treated (owing to detector inefficiencies) requires a new parametrization. However, missing points can sometimes be replaced by their predictions (see Subsection 2.4.2.1).

Normally, a linear least-squares fit is used for convenience to parametrize a function of several variables. This can be done with the help of the program MUDIFI (CERN Program Library E5001), which uses for the fit products of monomials, Legendre, or Chebyshev polynomials in each variable. The program keeps only functions that reduce the sum of squares of residuals significantly when added into the fit. In this way, a large number of polynomials can be tried in an automated fashion.

From the remarks on feature extraction (see Subsection 2.2.7) it will become clear that it is advisable to perform such a multidimensional fit preferably after a feature extraction transformation, such as performed by the program LINTRA (CERN Program Library E5002). LINTRA and MUDIFI are designed to work in sequence. In particular, LINTRA will produce the FORTRAN code of a routine performing the Karhunen–Loeve transformation (Brun *et al.* 1979; Brun, Hansroul and Kubler 1980).

Feature extraction followed by parametrization should normally always be considered and even tried out when a detector has to be studied, be it only to gain some further insight into the track sample and the response of the detector.

2.2 Principles of pattern recognition

2.2.1 Pattern space

For numerical treatment objects have to be described by a finite number of parameters that are typically the result of measurements performed on the object, such as grey levels inside small areas of a photograph, frequencies and their amplitudes in a Fourier spectrum, or positions of ionized regions created by a charged particle in a wire chamber. To each of these quantities corresponds a separate dimension in the space formed by all quantities, the *pattern space* {P}. The measurement of an actual object supplies numerical values for the different measured quantities, which are grouped into a vector **x** of length n, the dimension of {P}. Each object gives rise to one point in {P}. The dimension of {P} can be very high, e.g. 10 000 if grey-level measurements are performed on a mesh of the moderate size 100 by 100.

Although in many cases {P} is not a vector space in the mathematical sense, the totality of all measurement values is often called the *measurement vector* for convenience. {P} being a space of some fixed dimension n requires that all n components of the measurement vector **x** exist for all objects in {P}. There are two possible reasons why this might not be the case in a given problem. Firstly, coordinates may be missing because of detector inefficiencies or similar reasons, i.e. they should, in principle, exist, but are not available for a given object. In this case they may be predicted from neighbouring values of the same object, or through some other procedure.

Secondly they may be missing because they are not defined, not even in principle. As an example, consider a track that does not reach the rear part of a detector, either because it leaves the detector through the side, or because it is curved backwards by the magnetic field. In such a case, if one wants to use pattern space transformations such as feature extraction (see Subsection 2.2.7) one is obliged to split the full set of objects into subsets, e.g. by subdividing the phase space in the example above such that several pattern spaces, each with its own dimension, can then be treated separately.

2.2.2 Training sample and covariance matrix

In many different places of this book, a *training sample* is used to tune an algorithm, or to serve as a comparison when geometric acceptance, efficiencies, and other parameters have to be calculated (see e.g. Subsection 1.5.9). A training sample has to be used when there is no mathematical description of the objects studied, and can be used for convenience even if a mathematical description exists. It consists of a set of objects, which in the case of high-energy physics are usually particle trajectories, but can also be single points, point pairs, track segments, or even complete events. A typical training sample consists of several thousand objects. Its careful selection is very often the condition for the success of a method to be applied. It has to be representative for the problem under study. To give a trivial example: if the aim is to find tracks down to a momentum of 1 GeV/c in a detector inside a magnetic field, it is no use employing a training sample with tracks of 5 GeV/c and above. Even a sample of trajectories with uniform population at 1, 2, 3, and so on up to 50 GeV/c will give a strong bias to high momenta. In this case, it is much better to provide a uniform population in intervals of $1/p$, i.e. in equal steps of curvature, since this is a crucial parameter for track finding. A less trivial example is that of event selection: the physics to be studied in a given experiment may just correspond to a small and sparsely populated region of the total phase space, if the training sample is uniformly distributed over the phase space. This could mean that although the total reconstruction efficiency over such a non-representative training sample may be 99.5%, the probability of reconstructing one of the events really looked for could be close to zero.

Closely related to the training sample is the covariance matrix. If $\mathbf{x} = (x_1, \ldots, x_n)$ is a random vector with a known distribution, its covariance matrix $\mathbf{C}$ is defined by

$$c_{ij} = \langle (x_i - \langle x_i \rangle)(x_j - \langle x_j \rangle) \rangle = \sigma_{ij} \qquad (2.23)$$

where '$\langle \ldots \rangle$' denotes the expectation value. The element σ_{ii} is the variance of component i, and σ_{ij} is the covariance of components i and j. The more common case is, however, the one in which the probability distribution is not known explicitly, in which case σ_{ij} is replaced by its estimator s_{ij} which is defined as in (2.23), but now '$\langle \ldots \rangle$' denotes the 'mean over the training sample'.

The covariance matrix is frequently composed of three normally independent, and consequently additive, contributions. The first contribution arises from the correlation between the different measured quantities, i.e. it expresses the fact that the measurements contain redundant information, and the measurement vectors do not fill $\{P\}$, but only a subspace of it (the *constraint surface*). The second contribution results from measurement errors, and the third from the 'random' behaviour of the physical object, e.g. a particle undergoing multiple scattering in matter.

The different contributions to $\mathbf{C}$ can be obtained separately from different training samples, provided such training samples are available, i.e. from Monte Carlo generation: if real measurements are used in a training sample, the effects are, of course, all mixed up. In feature extraction (Subsection 2.2.7) and parametrization (Subsection 2.1.2), a training sample without any errors is used, in order to be able to see the effect of the nonlinearity of the constraint equation. For goodness-of-fit tests, i.e. the calculation of a χ^2, the full covariance matrix has to be used.

In certain circumstances it may be advisable to provide more than one covariance matrix: such is the case when tracks have to be fitted over a large momentum range, when, for low momenta, the multiple scattering and, for high momenta, the measurement errors are very often the dominant contributions to $\mathbf{C}$ (Gluckstern 1963; Mecking 1982). If the utmost precision is required, the covariance matrix has to be computed anew for each track (see Subsection 3.3.2).

2.2.3 Object classification

A *class* of objects is formed by all those objects fulfilling certain criteria that define the class. The goal of pattern recognition is to classify measured objects, i.e. to find a decision function or, in the more general case, a procedure f such that

$$c = f(\mathbf{x})$$

is the number of the class to which the object that is described by $\mathbf{x}$ belongs; c is sometimes considered as a point in the *classification space* {C}.

In high-energy physics pattern recognition we mostly deal with classes that do not share objects. In this case, because to each object there belongs a point in pattern space, the task of pattern recognition consists of finding a set of *hypersurfaces* (surfaces of dimension $n - 1$) that divide {P} into disjoint regions or classes. If, in addition, the different classes form clusters in {P} that are well separated from each other (Fig. 2.8), one may find separating hypersurfaces that are rather smoothly curved, or even linear hyperplanes. The following example from high-energy physics may serve to illustrate the concepts of pattern space and dividing hypersurfaces.

A detector provides measured space points along particle trajectories. If we treat these space points as *objects* and the tracks into which they have to be assembled as *classes* then {P} is simply the three-dimensional Euclidean space. The tracks form *clusters* in {P}. They do not normally share objects (= points) if we exclude the vertex region. Thus, we can define hypersurfaces in the form of tube-like structures around each track (this is often called a *road* in the two-dimensional case), and for straight tracks we can even find linear hyperplanes separating them.

2.2.4 Feature space

The pattern space formed by the raw measurements is often not very convenient to work in. The first reason for this lies in the frequently rather heterogeneous nature of the vector $\mathbf{x}$, which consists of components that are sometimes difficult to compare, like comparing apples to oranges. In addition, these components may have very different significance for the pattern recognition task. Consider for example a vector $\mathbf{x}$ of which the first 10 000 components are the grey levels of a satellite photograph scan on a 100 by 100 mesh, and component 10 001 is the time at which the picture was taken. Clearly this last coordinate is very different from the others and will probably have a much higher importance for the pattern recognition task than any of the others alone.

The second reason is the sometimes very high dimensionality of {P}, which can easily reach values of 10^6 or more. Not only is the manipu-

```
        CHARM-2 Run      528 Event   1175 Code      9
        ----+----1----+----2----+----3----+----4-----
 19- 35 I                                       8        I
 37- 53 I                                       9.       I
 55- 71 I                                       8        I
 73- 89 I                                       *        I
 91-107 I                                    6:          I
109-125 I                                 35             I
127-143 I                              7                 I
145-161 I                                                I
163-179 I                                                I
181-197 I                                                I
199-215 I                                                I
217-233 I                                                I
235-251 I                                                I
253-269 I                                                I
271-287 I                                                I
289-305 I                                                I
307-323 I                                                I
325-341 I                                                I
343-359 I        8*3                                     I
361-377 I        6*.3                                    I
379-395 I        8                                       I
397-413 I        *                                       I
415-431 I        7                                       I
433-449 I        3                                       I
        ----+----1----+----2----+----3----+----4-----
```

Fig. 2.8.   Two clearly separated object classes in {P}. The figure shows an event in the top view of the detector of the CHARM-2 neutrino experiment at CERN. The picture is very much compressed longitudinally in that several planes are grouped into one line (numbers at the left). The first active plane in the detector is plane 21. The picture shows a double event: a beam halo muon has entered the front of the detector (at the top) and stops after about 120 planes after having passed through about 6 m of glass absorber. Further upstream (near the bottom of the picture), a neutrino has interacted in a charged-current event, creating a hadronic shower with an outgoing muon. The total length of this part of the detector is 35 m, the width is about 4 m. The different print symbols here represent groups with different numbers of wires.

lation of vectors of this size tedious and slow even on the most modern computers, but such a high number of dimensions implicitly means that the coordinates must be highly correlated (Fig. 2.9), since the object measured will only have a rather small number of macroscopic or otherwise significant features. High correlation between coordinates is equivalent to saying that the object classes do not fill n-dimensional space regions in {P}, but that they occupy subspaces of much smaller dimensions.

Finally, the third reason can best be illustrated with the help of Figs. 2.10 and 2.11. A simple transformation of coordinates from x, y to r, Φ with

$$r = (x^2 + y^2)^{\frac{1}{2}}, \quad \Phi = \sin^{-1}(y/r) \tag{2.24}$$

converts the dividing hypersurface, a circle in Fig. 2.10, into a linear one in

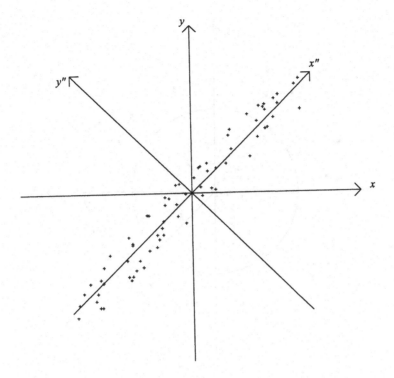

Fig. 2.9. Strongly correlated coordinates in {P}. The figure shows points with strongly correlated x and y coordinates. In a different coordinate system (x'', y'') the correlation almost does not exist, and the absolute values of the y'' coordinate are much smaller than the corresponding y values.

Fig. 2.11. This is an example of a simplification of the problem without a reduction of the dimensionality (= 2). All of this leads to a transformation of the pattern space of dimension n into the feature space {F} of dimension m:

$$z = R(x), \quad z \in \{F\}, \ x \in \{P\}$$

with an arbitrary transformation R. If R is linear, it can be represented by a matrix multiplication (apart from a possible translation):

$$z = Tx$$

with the transformation matrix T of dimension (m, n). The aim is to find a transformation with m as small as possible without significant loss of information about the patterns.

With the help of the three spaces introduced, the task of pattern recognition is then most elegantly formulated as the task of finding the transformations in

$$\{P\} \Rightarrow \{C\} \text{ or } \{P\} \Rightarrow \{F\} \Rightarrow \{C\}.$$

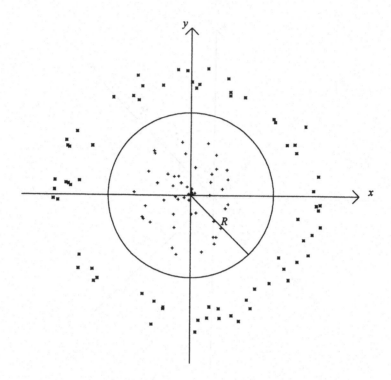

Fig. 2.10. Two object classes divided by a curved hypersurface. The two object classes are represented by points inside a circle (class 1) and outside the circle (class 2) in a Cartesian coordinate system. The hypersurface $x^2 + y^2 = R^2$ is a circle.

2.2.5 Classes, prototypes, and metric

Classes of objects are normally defined by enumerating all of their members, such as in the classification of printed letters with one letter per class, or by specifying a set of *prototypes* in cases where there are too many of them, such as in the case of handwritten letters. The objects themselves can be defined in pattern space or in feature space, either directly as vectors, or as functions thereof. Some or all of the parameters in these functions may depend on the object itself; such is the case if an object, for example a particle track, is required to form a helix in space of which only the direction of the principal axis is known, this being the direction of the magnetic field vector.

In most cases of automatic pattern recognition, some kind of metric is needed to decide whether an object belongs to a certain class or not. A metric is a special kind of distance function d, obeying the following rules:

$$d(\mathbf{x}, \mathbf{y}) \geqslant 0$$
$$d(\mathbf{x}, \mathbf{y}) = d(\mathbf{y}, \mathbf{x})$$

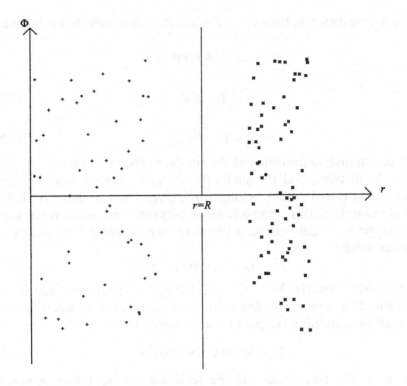

Fig. 2.11. The two object classes of Fig. 2.10 in feature space. The same points as in Fig. 2.10 are now plotted in a coordinate system with the axes r and Φ. The hypersurface $r = R$ is now a hyperplane.

$$d(\mathbf{x}, \mathbf{y}) \leqslant d(\mathbf{x}, \mathbf{z}) + d(\mathbf{z}, \mathbf{y}) \quad \text{(triangle inequality)}$$

$$d(\mathbf{x}, \mathbf{y}) = 0 \Longleftrightarrow \mathbf{y} = \mathbf{x}$$

where $\mathbf{x}, \mathbf{y}, \mathbf{z}$ are vectors in the space considered. Of course, the Euclidian distance

$$d(\mathbf{x}, \mathbf{y}) = \left[(\mathbf{x} - \mathbf{y})^{\mathrm{T}} (\mathbf{x} - \mathbf{y}) \right]^{\frac{1}{2}} \tag{2.25}$$

obeys these rules and is a metric. In order to be useful to the pattern recognition task, the metric must be chosen such that the distance between two objects is in some way proportional to their degree of being different.

As a counterpart one can define a similarity measure between two objects by requesting that a function s exists such that

$$0 \leqslant s(\mathbf{x}, \mathbf{y}) \leqslant 1$$

$$s(\mathbf{x}, \mathbf{x}) = 1$$

$$s(\mathbf{x}, \mathbf{y}) = s(\mathbf{y}, \mathbf{x})$$

$$[s(\mathbf{x}, \mathbf{y}) + s(\mathbf{y}, \mathbf{z})] \, s(\mathbf{x}, \mathbf{z}) \geqslant s(\mathbf{x}, \mathbf{y}) s(\mathbf{y}, \mathbf{z})$$

$$s(\mathbf{x}, \mathbf{y}) = 1 \Longleftrightarrow \mathbf{x} = \mathbf{y}$$

If d is a metric distance, then $s = e^{-d}$ is a metric similarity. If d is bounded, i.e.

$$d_m = \max\left(d(\mathbf{x},\mathbf{y})\right) < \infty$$

then

$$s = 1 - d/d_m \tag{2.26}$$

or

$$s = 1 - d^2/d_m^2 \tag{2.27}$$

would be alternative measures of the similarity (Schorr 1976).

Both the distance and the similarity functions can be used to decide whether a measured object belongs to a given class or not. A slightly unusual example of this is the difference between a measured track $\mathbf{x}$ and its prototype $\mathbf{x}_f$, which has been found through a linear least squares fit to a track model:

$$\chi^2 = (\mathbf{x} - \mathbf{x}_f)^{\mathrm{T}} \mathbf{W} (\mathbf{x} - \mathbf{x}_f) \tag{2.28}$$

with the *weight matrix* $\mathbf{W} = \mathbf{C}^{-1}$, $\mathbf{C}$ being the covariance matrix (see Subsection 2.2.2) known *a priori* either from theoretical considerations, or as defined by a training sample of tracks with errors:

$$\mathbf{C} = \langle (\mathbf{x} - \mathbf{x}_t)(\mathbf{x} - \mathbf{x}_t)^{\mathrm{T}} \rangle \tag{2.29}$$

where $\mathbf{x}_t$ is the true track. If the residuals $\mathbf{x} - \mathbf{x}_f$ follow a normal distribution and if the covariance matrix $\mathbf{C}$ is correct, χ^2 has a well-known distribution, and cuts can be applied at different confidence levels.

2.2.6 *Template matching*

It may happen that for a given classification task, the coordinates of the object can only assume the values 0 and 1, such as in the following example of numerals, where they correspond to the two grey levels *white* and *black*. Each object defining the class then has an associated vector consisting of 0s and 1s, called the *template*. Similarly, a measurement of an object will give such a vector $\mathbf{x}$. If the class is described by a total of m prototype templates $\mathbf{y}_i$, a comparison of $\mathbf{x}$ with all of them has to be performed in order to find the one that it fits best. This procedure is called *template matching* (Young and Calvert 1974). The number of matching pattern bits in template $\mathbf{y}_i$ and object $\mathbf{x}$ is given by

$$n_i = \mathbf{y}_i^{\mathrm{T}} \mathbf{x} \tag{2.30}$$

where i runs over all templates. Grouping the template vectors into a matrix $\mathbf{Y}$, the complete procedure comes down to a single matrix multiplication:

$$\mathbf{n} = \mathbf{Y}^{\mathrm{T}} \mathbf{x} \tag{2.31}$$

```
$0000000000$0000000000$0000000000$0000000000
$0001111000$0000001000$0001111000$0001111000
$0010000100$0000011000$0010000100$0010000100
$0100000010$0000101000$0100000100$0000000100
$0100000010$0001001000$0000000100$0000000100
$0100000010$0010001000$0000000100$0000000100
$0100000010$0000001000$0000001000$0000001000
$0100000010$0000001000$0000010000$0000001000
$0100000010$0000001000$0000100000$0000000100
$0100000010$0000001000$0001000000$0000000100
$0100000010$0000001000$0010000000$0000000100
$0010000100$0000001000$0100000000$0010000100
$0001111000$0000001000$0111111100$0001111000
$0000000000$0000000000$0000000000$0000000000
```

Fig. 2.12. Templates for four numerals. The four numerals 0, 1, 2, and 3 are outlined by print characters '1' in a 10 by 14 mesh with the print background made of characters '0'. The four numerals can be seen best by holding the page at some distance, or under a flat angle with the line of sight.

where the maximum component of **n** defines the best match. Template matching is clearly very well suited to being implemented on vector processors, and has already been applied to track finding in high-energy physics (Georgiopoulos *et al.* 1986). It is also a longstanding method for crude and fast trackfinders in real-time systems (see Subsections 1.5.3 and 1.5.6).

In Fig. 2.12, four numerals are outlined in a 10 by 14 squares mesh. The template matrix **Y** can be directly read off the figure: each column of **Y** is constructed by placing the ten columns of 0s and 1s of one digit below each other, such that the total column length becomes 140.

This method can be generalized in two ways. The first consists of replacing the condition that the variables have to assume the values 0 or 1 by requesting them to lie in some interval, or around a given value with a known error distribution for each variable; in this latter case, the weighted sum of squares of differences between the measured and the prototype values can be used as a matching criterion.

The second generalization can be applied in cases where the number of possible classes is too high to establish a complete dictionary for direct template matching. In this case, one may try to find a solution in two steps, by grouping classes into *superclasses* and performing the template matching on those. Once the superclass has been found, one could then continue by using the smaller directory valid for the classes in this superclass only, or by applying another method such as a combinatorial search. This approach has already been used in high-energy physics track finding (Georgiopoulos *et al.* 1986).

The concept of grouping the template patterns into superclasses can be extended by building a tree of patterns (Dell'Orso and Ristori 1990). The basic idea is to follow a successive approximation strategy and to apply the pattern-matching algorithm to the same data again and again, but each time with increasing spatial resolution. The patterns can thus be arranged in a tree structure, where each node represents a pattern and is linked to all subpatterns generated by increasing the spatial resolution by a factor of two. Consequently, the pattern-matching process can be implemented as a tree search, which is much faster than a purely sequential search. It can easily be implemented on a parallel architecture. It has been shown that in an application to track finding the number of search steps increases very slowly with the number of tracks. On the other hand the algorithm is rather sensitive to noise, because it frequently visits fake branches activated by noise hits at low resolution.

This tree search algorithm still needs a huge amount of memory to store all the patterns, the number of which may easily exceed 10^8 in a fair-sized detector. The number N_p of patterns compatible with straight tracks in a detector with N_1 layers and N_b bins per layer is approximately equal to

$$N_p \approx (N_1 - 1)N_b^2$$

Therefore $N_p \approx 10^8$ in a detector with 8 layers and 4096 bins per layer. In order to avoid storing a large number of patterns it has been proposed to generate the patterns during the search process (Budinich and Esquivel 1990). Of course only valid patterns are generated; in addition, only patterns that are compatible with the current event need to be considered. If, for instance, a certain bin in a certain layer is empty, no pattern with a hit in this bin needs to be generated.

Another obvious way of reducing the size of the dictionary is to use the object description in feature space rather than in pattern space. For example, in order to distinguish and recognize different wave types such as harmonic, rectangular, or triangular, it is sufficient to build a dictionary in feature space (frequency domain) with the first five or so Fourier spectrum components of each wave form rather than to describe each wave in object space (time domain) by a large number of points on it.

2.2.7 Linear feature extraction

If an object such as a particle trajectory is described by an n-dimensional vector $\mathbf{x}$ in pattern space, and the number of free parameters (in this case the number of integration constants of the equation of motion) is f, then the object class occupies a f-dimensional subspace in $\{P\}$, as defined by the $n - f$ constraint equations between the coordinates.

These equations are, however, very often not known explicitly, e.g. if the particle moves in an inhomogeneous magnetic field, and are, in addition, frequently nonlinear, in which case the subspace of the object class is curved.

As an example, consider the points on a circle of radius r_i around the origin to be the objects of class number i. The constraint equation, and accordingly the hypersurface it defines, is not linear:

$$x^2 + y^2 = r_i^2$$

One may, however, subdivide this hypersurface into regions that can be approximated by linear hyperplanes, in this case straight-line segments forming a regular polygon. The actual subdivision will be dictated by the nearest neighbouring classes, and by the precision one has to achieve. This procedure is always possible in regions where the hypersurface is a unique function of the constants of integration. In an actual application, it may be sufficient simply to subdivide the phase space region under study into smaller regions of equal size. Attempts to use curved hypersurfaces have not been successful.

Let us suppose now that a class of objects is defined by a set of prototypes, a *training sample* (Subsection 2.2.2). Then the 'best' linear feature extraction algorithm, which leads to the 'best' linear approximation of the curved subspace, can be defined by requesting that it minimizes the sum of the squares of the distances of all vectors in the training sample to the linear subspace.

This can be written in mathematical terms as follows: let $\mathbf{q}_i$ $(i = 1, \ldots, n)$ be an arbitrary orthonormal basis in the n-dimensional pattern space. Let

$$\mathbf{a}_j = \mathbf{x}_j - \langle \mathbf{x} \rangle, \quad j = 1, \ldots, N \tag{2.32}$$

denote the shifted vectors of the training sample, where '$\langle \cdots \rangle$' stands for the 'mean over the training sample' as before. Then the object vectors $\mathbf{a}_j$ can be written as

$$\mathbf{a}_j = \sum_{i=1}^{n} a_{ji} \mathbf{q}_i$$

$$= \sum_{i=1}^{m} a_{ji} \mathbf{q}_i + \sum_{i=m+1}^{n} a_{ji} \mathbf{q}_i \tag{2.33}$$

The 'best' linear feature extraction now consists obviously of finding the basis $\mathbf{q}_i$ that minimizes the mean squared error if the expansion above is truncated at m, i.e., which minimizes

$$S(m) = \left\langle \left(\mathbf{a}_j - \sum_{i=1}^{m} a_{ji} \mathbf{q}_i \right)^2 \right\rangle$$

$$= \left\langle \left(\sum_{i=m+1}^{n} a_{ji} \mathbf{q}_i \right)^2 \right\rangle \qquad (2.34)$$

Because the basis $\mathbf{q}_i$ is orthonormal, multiplication of Equation (2.33) with $\mathbf{q}_i^{\mathrm{T}}$ gives

$$a_{ji} = \mathbf{q}_i^{\mathrm{T}} \cdot \mathbf{a}_j \qquad (2.35)$$

leading to

$$S(m) = \left\langle \left(\sum_{i=m+1}^{n} a_{ji} \mathbf{q}_i \right)^2 \right\rangle$$

$$= \left\langle \left(\sum_{i=m+1}^{n} (\mathbf{q}_i^{\mathrm{T}} \cdot \mathbf{a}_j) \mathbf{q}_i \right)^2 \right\rangle$$

$$= \left\langle \sum_{i=m+1}^{n} \sum_{k=m+1}^{n} (\mathbf{q}_i^{\mathrm{T}} \cdot \mathbf{a}_j) \mathbf{q}_i^{\mathrm{T}} \cdot \mathbf{q}_k (\mathbf{a}_j^{\mathrm{T}} \cdot \mathbf{q}_k) \right\rangle$$

$$= \left\langle \sum_{i=m+1}^{n} (\mathbf{q}_i^{\mathrm{T}} \cdot \mathbf{a}_j)(\mathbf{a}_j^{\mathrm{T}} \cdot \mathbf{q}_i) \right\rangle \quad \text{since } \mathbf{q} \text{ orthonormal}$$

$$= \sum_{i=m+1}^{n} \mathbf{q}_i^{\mathrm{T}} \left\langle \mathbf{a} \mathbf{a}^{\mathrm{T}} \right\rangle \mathbf{q}_i$$

$$= \sum_{i=m+1}^{n} \mathbf{q}_i^{\mathrm{T}} \mathbf{F} \mathbf{q}_i \qquad (2.36)$$

with the dyadic product matrix

$$\mathbf{F} = \left\langle \mathbf{a} \mathbf{a}^{\mathrm{T}} \right\rangle = \frac{1}{N} \sum_{j=1}^{N} \mathbf{a}_j \mathbf{a}_j^{\mathrm{T}} \qquad (2.37)$$

If $\mathbf{e}_i$ is a normalized eigenvector of $\mathbf{F}$ corresponding to the eigenvalue e_i, then

$$\mathbf{e}_i^{\mathrm{T}} \mathbf{F} \mathbf{e}_i = \mathbf{e}_i^{\mathrm{T}} \left\langle \mathbf{a} \mathbf{a}^{\mathrm{T}} \right\rangle \mathbf{e}_i = e_i \qquad (2.38)$$

holds. From this follows

$$\langle (\mathbf{e}_i^{\mathsf{T}} \cdot \mathbf{a})(\mathbf{a}^{\mathsf{T}} \cdot \mathbf{e}_i) \rangle = e_i$$

or

$$\langle (\mathbf{e}_i^{\mathsf{T}} \cdot \mathbf{a})^2 \rangle = e_i \tag{2.39}$$

Since '$\langle \ldots \rangle$' stands for 'mean over training sample', e_i is equal to a sum of squares and therefore is not negative. If the vectors $\mathbf{a}$, i.e. the original track vectors, span the full pattern space, then the eigenvalues are all positive, since

$$e_i = 0 \quad \text{would mean} \quad \langle (\mathbf{e}_i^{\mathsf{T}} \cdot \mathbf{a})^2 \rangle = 0$$

leading to

$$\mathbf{e}_i = 0 \tag{2.40}$$

because of the condition above, namely the $\mathbf{a}$ spanning $\{P\}$. Equation (2.40) however, contradicts the assumption that $\mathbf{e}_i$ is normalized.

In conclusion, in the case in which the tracks span the full original pattern space $\{P\}$, $\mathbf{F}$ is positive definite, otherwise it is semidefinite with its rank equal to the dimension of the subspace spanned by the track sample. Consequently, since $\mathbf{F}$ is at least semidefinite, minimizing Expression (2.36) means minimizing each term separately. Introducing Lagrange multipliers λ_i then requires that the terms

$$\mathbf{q}_i^{\mathsf{T}} \mathbf{F} \mathbf{q}_i - \lambda_i \mathbf{q}_i^{\mathsf{T}} \cdot \mathbf{q}_i + \lambda_i \tag{2.41}$$

have to be minimized (remember that $\mathbf{q}_i^{\mathsf{T}} \cdot \mathbf{q}_i = 1$).

Setting the derivatives of this expression with respect to the components of $\mathbf{q}_i$ to zero gives the minimum condition for $S(m)$:

$$\mathbf{F} \mathbf{q}_i = \lambda_i \mathbf{q}_i \tag{2.42}$$

i.e. $\mathbf{q}_i$ is an eigenvector of $\mathbf{F}$ with eigenvalue λ_i. The minimum of $S(m)$ is then simply

$$S(m) = \sum_{i=m+1}^{n} \lambda_i \tag{2.43}$$

By taking the eigenvectors of the largest m eigenvalues as the new basis, one has now found the 'best' subspace obtainable with linear feature extraction. The average distance of any track in the sample to this subspace is given by $S(m)$. This method is named after Karhunen–Loeve (Young and Calvert 1974). It is also known as *principal component analysis*.

In the general case a certain number, m (greater than or equal to f, the number of free parameters), of the components of $\mathbf{y}$ will be significant, and the remaining $(n - m)$ components will have small absolute values (Brun, Hansroul and Kubler 1980).

The result is that the eigenvectors of the covariance matrix of the training sample

$$\mathbf{F} = \langle (\mathbf{x} - \langle \mathbf{x} \rangle) \cdot (\mathbf{x} - \langle \mathbf{x} \rangle)^{\mathrm{T}} \rangle \tag{2.44}$$

form the rows of the transformation matrix $\mathbf{T}$, such that the new coordinates become

$$\mathbf{y} = \mathbf{T}(\mathbf{x} - \langle \mathbf{x} \rangle), \quad \mathbf{y} \in \{\mathbf{F}\}, \ \mathbf{x} \in \{\mathbf{P}\}$$

As an example, the eigenvalues and eigenvectors for the transformation (x, y) into (x'', y'') of Fig. 2.9 have been calculated to be:

> eigenvalues:
> 5.4774, 0.0385
>
> eigenvectors:
> 0.720613   0.693337   (1)
> 0.693337   −0.720613   (2)

where the eigenvectors have been written row-wise such that the above matrix is the transformation matrix. The slope of the x'' axis with respect to the (x, y) system comes out as 0.962.

2.2.8 Minimum Spanning Tree (MST)

The MST is used in cluster analysis, and as such is of some interest for track finding, if we consider tracks to be clusters of points. It has been applied successfully both to track finding (Cassel and Kowalski 1981), and to calorimeter analysis in high-energy physics (CHARM-II experiment, CERN). To understand it, a minimum knowledge of the terminology of graph theory is necessary.

A *graph* consists of *nodes* and *edges*. A node can represent any object, and is often graphically represented by a point (see Fig. 2.13). An edge can be symbolized by a line connecting two nodes, and expresses the fact that some well-defined relation between these two nodes exists. As an example, consider the objects to be space points, and the edges connecting the nodes belonging to the same track.

If a positive *weight* (this could be a metric distance or a metric similarity, see Subsection 2.2.5) is assigned to each edge, the graph is called *edge-weighted*. For space points, the distance between them may serve this purpose. An *isolated point* is a node without edge. A *connected graph* is a graph without isolated points. In a *fully connected* or *complete* graph, all nodes are directly connected with all other nodes. A *path* between two nodes is a sequence of edges connecting them. A *loop* or *circuit* is a closed path, i.e. a path connecting all nodes in it to themselves. A *tree* is a graph without circuits. A *spanning tree* is a

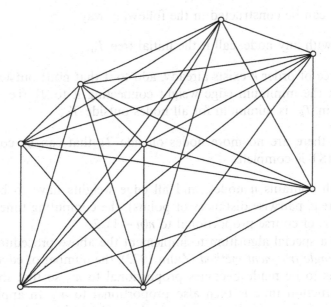

Fig. 2.13. A fully connected graph. The *nodes* of the graph are represented by small circles, the *edges* by connecting lines.

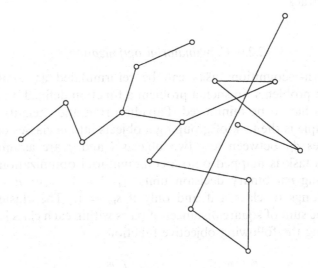

Fig. 2.14. A spanning tree. Each node is connected to at least one other node. From any given node there is only one possible path to any other node. This is equivalent to the fact that no closed paths (loops or circuits) exist.

connected graph without circuits (Fig. 2.14). Finally, a *Minimum Spanning Tree* (MST) is a spanning tree for which the sum of the edge weights has a minimum for a given graph. If all edge weights are different, the MST is unique.

An MST can be constructed in the following way:

- start with any node, call it the initial tree T_0;

- if a tree of order k exists already, add to it that node outside T_k for which the minimum edge weight connecting it to T_k (i.e. to some node in T_k) is minimum for all nodes outside T_k;

- when there are no more nodes outside T_k that can be connected, the MST is complete.

If the graph contains n nodes, and all edge weights have to be calculated once (e.g. pairwise distances of points), the computing time for this calculation is of course proportional to $n(n-1)/2$.

By using a special algorithm to implement the above procedure, called the *Prime single fragment method* (Zahn 1973), the number of edge-weight comparisons to be made becomes proportional to n^2, so that the whole MST construction time is then also proportional to n^2. In applications such as track finding this time can be further reduced by setting an upper limit for the edge weight, i.e. limiting the distance between consecutive points on a track.

2.2.9 Combinatorial optimization

Many pattern-recognition tasks can be reformulated as combinatorial optimization problems. In such a problem a function defined on a discrete set of points has to be minimized (Papadimitriou and Steiglitz 1982). A typical example is the task of grouping n objects into m classes or clusters. The distances d_{ij} between any two objects i and j are assumed to be known. This task is mapped onto a combinatorial optimization problem by considering nm binary decision units s_{ia}, $i = 1,\ldots,n$; $a = 1,\ldots,m$. Object i belongs to class a if and only if $s_{ia} = 1$. The clustering that minimizes the sum of square distances of pairs within each class is obtained by minimizing the following objective function:

$$E = \sum_{a=1}^{m} \sum_{i,j=1}^{n} d_{ij}^2 s_{ia} s_{ja} + \sum_{i=1}^{n} \left(\sum_{a=1}^{m} s_{ia} - 1 \right)^2 \tag{2.45}$$

The first term of E is called the cost term; it requires minimization of the sum of square distances of pairs within each class. There is no contribution from pairs belonging to different classes. The second term is called the constraint term; it is minimal if $\sum_{a=1}^{m} s_{ia} = 1$ for all i, i.e. if every object belongs to exactly one class. E can be rewritten as a symmetric quadratic form:

$$E = n + \sum_{a=1}^{m}\sum_{b=1}^{m}\sum_{i=1}^{n}\sum_{j=1}^{n} T_{ia;jb} s_{ia} s_{jb} \qquad (2.46)$$

the coefficients $T_{ia;jb}$ being given by

$$T_{ia;jb} = \delta_{ab} d_{ij}^2 + \delta_{ij} - 2\delta_{ij}\delta_{ab} \qquad (2.47)$$

The total number of possible states of the nm decision units is equal to 2^{nm}; obviously the best solution cannot be found by an exhaustive search even for a fairly small number of objects. Many algorithms for minimizing a quadratic objective function like the one above are described in the literature; the discussion here will be limited to a few methods that have been or can be used in track finding.

The Hopfield network. The Hopfield network is a *recursive artificial neural network* (ANN) consisting of binary *neurons* or decision units s_i, $i = 1, \ldots, n$ (see also Subsection 1.5.9). It is fully connected, and the connection weights T_{ij} are symmetric. The diagonal of the weight matrix is equal to 0. Updating is usually *asynchronous*, i.e. only a single neuron is allowed to change its state at a given point in time, according to the following prescription (*Hopfield dynamics*):

$$s_i' = \Theta\left(\sum_{j=1}^{n} T_{ij} s_j - t_j\right) \qquad (2.48)$$

The activation function Θ is the Heaviside function:

$$\Theta(x) = \begin{cases} 0, & x < 0 \\ 1, & x \geqslant 0 \end{cases} \qquad (2.49)$$

The quantities t_j are called the threshold values. The negative threshold $-t_j$ can be regarded as the weight of a connection between neuron s_j and a special neuron $s_{n+1} \equiv 1$. It can therefore be assumed without loss of generality that all t_j are equal to 0.

It can be shown without difficulty that with asynchronous updating the Hopfield dynamics always lead to a stable state. To this end one considers the *energy function* of the network:

$$E = -\frac{1}{2}\sum_{i=1}^{n}\sum_{j=1}^{n} T_{ij} s_i s_j \qquad (2.50)$$

The result of updating neuron s_i is a change ΔE of the energy function. It can easily be calculated that

$$\Delta E = -\Delta s_i \cdot h_i, \quad \text{with } \Delta s_i = s_i' - s_i \quad \text{and} \quad h_i = \sum_{j=1}^{n} T_{ij} s_j \qquad (2.51)$$

The summed input h_i is sometimes called the *local field*. If the local field h_i is non-negative, s_i' is equal to 1, and Δs_i is non-negative as well. If h_i is negative, s_i' is equal to 0, and Δs_i is negative or 0. In both cases it is true that

$$\Delta E \leqslant 0 \qquad (2.52)$$

The energy function is non-increasing and is therefore a Lyapunov function. This means that the Hopfield dynamics lead to a stable state, which is a local minimum of the energy function. The order in which the neurons are updated is usually chosen randomly.

If a combinatorial optimization problem has a quadratic objective function with zero diagonal, it can be immediately formulated as a Hopfield network. Asynchronous updating according to the Hopfield dynamics then leads straight to a local minimum of the objective function. The position of this local minimum depends of course on the starting point.

In most cases the local minimum will not be a satisfactory solution of the original optimization problem. Normally one tries to locate the global minimum, or at least a 'good' local minimum not too far from the global one. A simple approach is to rerun the network with different randomly chosen starting points, and to select the local minimum with the smallest energy. A more sophisticated method introduces random perturbations into the dynamics, allowing the network to escape local minima. It is called 'simulated annealing'.

Simulated annealing (SA). The algorithm begins in an arbitrary state of the network and generates elementary transitions by switching the state of a randomly chosen neuron. If the energy decreases the transition is accepted. If the energy increases, the transition is accepted with probability p, where p is given by the Boltzmann distribution:

$$p = \exp(-\Delta E / T) \qquad (2.53)$$

T is a control parameter, which is called the temperature. At high values of T a rise in energy is frequently accepted, so that the energy landscape can be explored. The successive states of the network form a Markov chain, the Boltzmann distribution being its equilibrium distribution. When the system is in equilibrium, the temperature is lowered and updating starts anew. It has been proved that there exist cooling schedules which guarantee that the global minimum is reached with probability 1 in finite time. In practice, simulated annealing is a rather lengthy procedure. One therefore resorts to a faster, deterministic algorithm that operates with the average values of the states at a given temperature. This is called mean-field annealing.

Mean-field annealing (MFA). Mean-field annealing works with the average states v_i of the neurons which now take values in the interval $[0, 1]$. At temperature T, the prescription for mean-field updating is as follows:

$$v_i = \frac{1}{2} \left[1 + \tanh \left(\sum_{j=1}^{n} T_{ij} v_j / T \right) \right] \tag{2.54}$$

There again exists an energy or Lyapunov function that is non-increasing. Mean-field updating at a fixed temperature therefore approaches a stable state, although this stable state need not be attained in finite time. Therefore the activation of all neurons is monitored. If the maximum change of activation in an entire round of updates does not exceed a certain limit, the temperature is lowered, and the next round of updating starts. In contrast to SA there is no guarantee that the global minimum is actually reached. It will be shown, however, that in applications to track finding MFA gives solutions which are very close to the global optimum (see Subsection 2.4.3).

2.3 Basic aspects of track finding

Given a set of position measurements in a detector, the task of track finding is to split this set into subsets (= classes) such that the following conditions apply.

1. Each class contains measurements that could be caused by the same particle.

2. One class (which may possibly be empty) contains all measurements that cannot be associated with particles with sufficient certainty. These may stem from accidental signals, from distorted measurements, from ambiguities in the association with tracks, from weaknesses of the track model, or from tracks that are deliberately excluded from the track search, for instance low-momentum tracks.

This definition is modest enough to represent a realistic goal (Grote 1981). It reduces track-finding to a cluster analysis problem, where a cluster is described as follows (Andrews 1972):

> A cluster is loosely defined as a collection of vectors or points that are close together.

This very general definition stresses the decisive role of the metric used to define clusters. Points that are 'near' when a specific metric is used ('near' could be defined relative to the mean value of all pair distances) may be 'far' from each other when a different metric is used.

The measurements along a particle trajectory have to fit to a track model, which is given by a function in space obeying the equation of motion. The human brain is very good in recognizing clusters, e.g. a sufficiently dense sequence of points along a smooth curve, even in the presence of kinks (sudden changes in the direction), vertices, overlapping tracks, and background, but it is rather poor in fitting these points to a given curve, except for a straight line, whereas a computer excels in this.

Many track-finding methods are based on these complementary capabilities: first, they try to find *track clusters* as efficiently as the eye does, and then they test them rigorously with a fit to a track model, making use of all the *a priori* knowledge of the detector performance and resolution, the particle motion, and both systematic and statistical errors. This has the consequence that most practical methods basically proceed in two steps. First a subset of measurements is selected, forming a *track candidate*. Second, a decision function is used to check whether or not the track candidate is an acceptable track.

A further consequence of the unmatched superiority of the human cluster-finding performance is the requirement that the tracks should be easy to recognize in a suitable graphical presentation of the coordinates. This means that the number of points per track, and the point density along the track, should be high enough to make this possible. It is of the greatest importance for testing the software with real data. If this goal is not met, the quality of the track finding can only be checked with Monte Carlo events, and its efficiency with respect to measured data of real events will always remain in doubt. Fortunately, this fundamental principle is now generally accepted.

The two-step procedure of track candidate definition and track candidate checking is often split further in order to make it faster. In an initialization phase a certain number of measurements are selected and are either rejected or accepted as a possible start of a track. If they are accepted then additional measurements are processed, in several steps, each time considering a possible rejection, until a final check is performed on the fully assembled track. The gain in speed over the basic two-step procedure, which first selects a full track candidate and then applies a decision function, is three-fold.

Firstly, each reduction in the number of points per track candidate brings a considerable reduction in the number of such candidates, since the number of combinations grows with the power of this value: if there are m detector planes and n tracks, then the total number of combinations is of course n^m. Secondly, the track candidate allows the prediction of an interval in which correct points can be found. This interval should be large enough to allow all points that can possibly belong to the track, to be found, even if this may lead to the temporary acceptance of wrong points. Otherwise, if the interval is too narrow to include all points, one will bias the track-finding procedure in a way that is very difficult to evaluate. Thirdly, the application of the final decision function can be very time consuming, for example if an iterative fit to a track model has to be performed. This results in a sequence of decision functions from simple and fast to precise but slow that improves the speed if at each intermediate test a good fraction of the wrong candidates really is rejected.

Of course, great care is needed to avoid the rejection of good track candidates by any of the approximate and simple tests, because these will normally all be biassed in their acceptance of tracks once they start rejecting good tracks.

For this reason as well, one should carefully avoid the rejection of points at this level. If more than one point is found inside the prediction interval, it is advisable to split the current track candidate into as many branches as necessary in order to include one point in each of them, rather than take for example the point nearest to the prediction (except when the correct track model is being used, and when confidence in the track candidate is very high). The multiplication of track candidates can normally easily be dealt with: in most cases, the wrong track cannot be extended further if a track-following method is used. However, it may sometimes lead to several highly incompatible tracks at the level of the final fit, and only then should one choose one of those that have passed the final acceptance criterion (Subsection 2.3.5).

If, for a given method, all possible tracks have been found, a certain number of points will normally remain unassigned to any track. These form the *background*. If several different methods, or the same method with different cuts, are applied a good measure of the efficiency is essential. Such a measure, as will be shown later, is specific to each experiment (Subsection 2.3.6). It will typically be based on a calibration sample (measured events) or a training sample (Monte Carlo events).

2.3.1 Point removal

Once a track has been found, it is very tempting to remove its points from the pool of points before continuing the track search, in order to have a smaller number of combinations to consider. This can, however, only be done safely if the redundancy is high enough, so that a track will still be found if one or two of its points are missing, since two (curved) tracks may, of course, have points in common. Even so, a slight bias against the second track is introduced by removing some of its points. Of course, the removal of all points of a *wrong* track will normally lead to one or several good tracks being lost completely; because of these risks, a suitable compromise between removing and not removing tracks is normally reached in doing both, insofar as the points are removed (marked in some way) from the track candidate initialization, but not from the set of points that can be added to an existing candidate through prediction. This procedure reduces the number of combinations to be tried, and therefore speeds up the whole procedure, but allows the same point to be found on more than one track.

2.3.2 Track quality

In the case of a conflict between two or more tracks in the sense that they are incompatible, and one of them has to be chosen as the (only) correct one, but also in a more general context, it is desirable to have some measure of the quality of a track. Clearly, a simple and well-defined number such as the χ^2 value of the final fit comes to mind, provided that the fit is really unbiassed and the correct weight matrix is used. However, this is a risky, and in most cases arbitrary, choice, because the χ^2 values are, of course, distributed according to well-known laws, and by no means does a small χ^2 indicate with certainty that a track is 'better' than another one with a higher χ^2 value, as long as this value is 'possible' inside the acceptance range chosen. This remains true even if the confidence levels are compared rather than the χ^2 values, which depend on the number of degrees of freedom. If, in addition, the track model is not absolutely correct, short tracks will on average have higher confidence values than longer ones; preferring tracks with higher confidence values would thus lead to the opposite of the desired effect, since a long track, in general, deserves more confidence than a short one.

On the other hand, the number of points and the absence of *holes* (hits predicted but not found in a chamber) are a rather safe measure of quality, provided the final acceptance test has been passed successfully. This leads quite naturally to a hierarchical search for tracks in which the longest tracks are looked for first, either by starting on the outside of the detector (as seen from the interaction region), or, in the case of interpolation, by spanning the initial track candidate over the longest possible interval. Of course, once the longest tracks have been found and 'removed' in the above sense, shorter tracks will normally have to be looked for in chamber subsets.

This consideration, taken together with the earlier one about the dangers of track candidate elimination in an early stage of the track finding, leads to the following situation: in all cases where tracks are found in separate pieces, be it because the detector consists of independent parts, or because the track-finding algorithm leads to a segmentation of the track (Subsection 2.3.4), it is highly advisable to keep even weaker incompatible partial tracks until the final step, in which the complete tracks are assembled and tested with the final check, normally a fit to some track model.

2.3.3 Working in projections or in space

Particle detectors that measure genuine space points along a particle trajectory emerged about 20 years ago (Subsection 2.1.1). The most important example is the time projection chamber. For such detectors, if the coordinates of the space point are really measured with comparable accuracy, the choice is obvious: the track finding should be done in space. This will become clear in the following, since in this case only the advantages remain: the disadvantages of working in space all arise from space points being reconstructed from measurements in different projections.

However, the number of detectors giving only one coordinate per measurement (or two coordinates with very unbalanced precision) is still high, either because different types of detector are used, or because one of the two coordinates is error prone or imprecise or both, which is frequently the case (Pimiä 1985).

In such a case, a detector surface (a plane or a cylinder in most cases) supplies (apart from the surface position) only one coordinate, whereas two are needed to reconstruct the impact point of a particle in space. If several surfaces with parallel wires or strips are placed behind each other, they provide measurements of one projection of the track. For a reconstruction of the track in space at least two different projections are necessary. Extra projections provide further information, which can be invaluable if the detectors are less than 100% efficient, or if the measured coordinates in different detectors are more or less orthogonal, in which case it is necessary to correlate the two projections of a track. In wire chambers, the different projections can be provided either by cathode strip readout, or by charge division (Subsection 2.1.1), or by wire planes with different wire orientations. In solid-state detectors, they are provided by silicon modules with different strip orientations.

To reconstruct the trajectory in space, as is necessary in the great majority of cases, there are basically the following choices.

(1) Combine several local projections into space points, and perform the track finding on those space points.

(2) Find the tracks in the different projections independently, and sub-
    sequently match the tracks in the different projections.

(3) Find the tracks only in some of the projections, and match them
    with the help of the other projection(s). In this case, one needs fewer
    points in the projections used only for matching, since the tracks
    are not reconstructed there.

Space points can be reconstructed from a wire and one or two high-
voltage strip readout coordinates on the corresponding cathode plane(s).
This point will give the avalanche position at the wire hit, regardless of
the particle's impact angle with the wire plane, since the induction spot
is always opposite the avalanche. This means that the particle does not
necessarily cross the strips that give a signal. Space-point reconstruction
is normally preferred to track finding in the wire plane and high-voltage
projections independently. Since the induction spot size is typically 1 cm
or more, signals from nearby tracks frequently overlap in the cathode
plane, and, as a result, the track-finding algorithms may find fewer tracks
there than among the more precise anode-wire signals.

When the different projections are all provided by similar wire planes,
it is possible to proceed either by reconstructing 'space points' and then
associating these into tracks, or by finding the tracks in the projections
independently, and then matching them. In the first case four wire planes
are necessary to define each 'space point', no three of which may have
identical wire directions, since one needs four straight lines to define a
unique line in space (which cuts all four). This approach not only provides
a space point, e.g. the centre of the track segment thus constructed, but
also the trajectory direction at that point, which is certainly an advantage
(Eichinger 1980). Of course, if the impact angle (the angle of the particle
track with the wire-plane normal) can be limited to small values, three
wire planes are sufficient; three wire planes are also necessary for the
resolution of ambiguities.

Accordingly, space-point reconstruction in most cases requires a higher
number of wire planes than the reconstruction of tracks in projections.
This is because in most cases the point density along the track must
not fall below a certain threshhold that is typical for each detector and
experiment. This threshhold is a result of the generally accepted request
that all detectable tracks should be visible to the human eye when plotted
in a suitable way. If, say, 10 points per meter are required, this can
be achieved by 22 planes (10 with vertical, 10 with horizontal, two with
inclined wires for matching) in the case of separate projections, whereas
for space points 30 or even 40 planes would be required.

This argument is reinforced by the fact that the particle detection
efficiency e (the probability that a charged particle crossing one wire plane

leads to a detectable signal) is smaller than 1. This means that for the typical e values of around 99%, three to four times more points are lost in space than in projections. Missing points, however, normally cause a lot of trouble in the track finding.

One aspect in favour of space points is, however, the disturbing occurrence of overlapping tracks in projections, and their virtual absence in space. If for two tracks $y_1(x), z_1(x)$ and $y_2(x), z_2(x)$ we define an overlap in the interval $[x_1, x_2]$ by

$$|y_2 - y_1| < \delta_y \quad \text{and/or} \quad |z_2 - z_1| < \delta_z \quad \text{for } x \in [x_1, x_2]$$

then the 'and' must hold for tracks defined in space, whereas the 'or' applies to tracks in projection. From this it follows immediately that the probability of having overlapping tracks is high for projections, in particular if the tracks are curved, and small in space. For all practical purposes, overlaps in projections have to be considered as the rule, whereas in space they may be treated as exceptions.

Finally, one may observe that the computing time for both alternatives is roughly equivalent: whereas in the second case, space points have first to be constructed, the track finding takes almost twice as long per track in the first case, since each track has to be found separately in the two projections. It is clear from this argument that reconstructing the tracks in more than two projections is a bit wasteful: if three or four equally good projections can be provided, it is preferable to work in space.

In summary one might then say that for simple event topologies, working in projections is sufficient and cheaper, whereas the more expensive approach of working in space is recommended for complicated event topologies in which tracks crowd in certain regions of the detector. Of course, the second approach may simply be almost unfeasible technically with ordinary wire or drift chambers. Such is often the case in cylindrical detectors at collider experiments, where the only way in which wires can be mounted without great effort is parallel to the beam direction. For this type of application, time projection chambers are an almost ideal solution from the track finding point of view, since they provide genuine space points (Subsection 2.1.1).

At the LHC collider the event rate and the track multiplicity will be much too high for a TPC. The inner trackers of the future LHC experiments will therefore be solid-state detectors (Subsection 2.1.1.2). The inner layers are planned as pixel detectors, which are able to deliver space points in the densely populated region close to the beam pipe. The outer layers are silicon microstrip detectors giving precise $R\Phi$-measurements; the z-information will be rather crude, being given only by the size of the detector element. A few layers will, however, be equipped with double-sided detectors delivering true space points. Thus track finding in the silicon trackers will be done in space, in spite of the poor z-information in some of the points. The transition radiation tracker of the ATLAS experiment (Subsection 2.1.1.3) gives no z-information at all; in this device the track finding and the subsequent linking to the silicon tracker have to be done in the projection on the xy-plane. This is possible because of the lower occupancy.

2.3.4 Treating track overlaps

The most common, and most recommended, approach is to exclude regions where two or more tracks overlap from the primary track finding. (In order for these areas to be found, the primary track finding will already have to iterate once, i.e. recognize that an overlap exists, then restart the track finding with this knowledge.) Supposing for the moment that we know how to exclude such regions, the only other cause for finding more than one hit in a track road is noise, i.e. spurious points that do not belong to any recognizable track.

Such noise can lead to two types of error.

(1) The track search can be initialized incorrectly by including a noise point in the initial track segment or track candidate.

(2) A noise point may be taken instead of the correct point during subsequent stages of the track finding. This may happen if either the correct point is missing, or the point prediction algorithm does not work properly, or both. When a noise point is taken instead of the missing correct one, normally very little harm is done if the point prediction is reasonably precise. In the case in which more than one point may be accepted as a correct solution to the prediction algorithm, one has to open a *branch* for each of these points, and follow both tracks, which in the case of a noise point will normally lead either to a (very short) branch, which can be pruned, or to two highly incompatible tracks, differing in just one point, if the track finding manages to continue with the correct track even after having included a wrong point in the track. Such incompatibilities (and even more complicated ones) can then be solved with the methods described in Subsection 2.3.5.

As has been pointed out before, simply taking the point nearest to the prediction is risky, and can only be done with some confidence if the correct track model is used for point prediction, and not an approximation, such as a parabola, or a spline curve.

Let us now consider the problem of recognizing regions where tracks overlap. There are basically two possibilities.

(1) Just two tracks cross at some sufficiently large angle, and the point prediction is so precise that (on one or both of the two tracks) only one extra point from the other track is found during prediction, precisely at the crossing point. In this case, one can apply the same algorithm described above for noise points, and both tracks will safely be found (this will only be the case when the detector provides undistorted signals even for a region where two nearby charged particles are detected by the same detector element, e.g. a wire).

(2) In all other cases, the prediction in neighbouring chambers will lead to the same problem of multiple solutions, i.e. if the *incoming* track is interpolated or extrapolated over several adjacent chambers, there will be more than one point in the prediction interval for most or all of them, whether this now stems from noise or from tracks. In this case, it is easiest to stop the incoming track right there, and start a new one in a clean region.

In this way, some of the tracks will only be found in segments, and these segments have to be joined together in a subsequent step. This can be done using similarity or distance criteria based on these segments, and a method such as the minimum spanning tree (Subsection 2.2.8), by using inverse similarities as edge weights, with typical similarities such as overlapping parts, similar curvature, direction, and inverse distance of the centre points (Pimiä 1985).

Once two or several segments have been joined into a safe track in this way, which normally implies a fit to the correct track model, one can then interpolate this track in the overlap regions and simply pick up the 'nearest to prediction' points. Most importantly, in this way one can clean the overlap area so that one is able to find short, or very curved, tracks after the complete cleaning operation, tracks that do not extend far enough into originally clean areas to be found there.

2.3.5 Compatibility of track candidates

A further graph theory method has been used to resolve a problem in track finding, which arises from the uncertainty with which points are assigned to tracks, and the subsequent result that the same point may be assigned to more than one track. The problem can be formulated in the following general way. For a given graph of nodes and edges, find the subset of a maximum number of nodes that are not linked with each other, and are therefore said to be *compatible*. In track finding, this problem arises when hits or complete track segments belong to two or more track candidates at the same time. If such tracks are represented by nodes, and

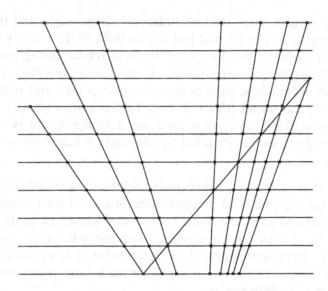

Fig. 2.15. Tracks from a vertex plus a spurious combination. Straight tracks are plotted in an arrangement of parallel detector planes. If the track-finding program does not use the vertex as an extra point on the track, it is likely to find the ghost track that crosses the others. Such a case can be treated with graph logic (see Fig. 2.16).

are called *incompatible* if they share hits or track segments, and this fact is expressed by an edge connecting each pair of such tracks, then we have exactly the situation described above, since it can be argued that the most plausible underlying event structure is the one that allows for a maximum number of compatible tracks (see Fig. 2.15). In this figure, straight tracks coming from a common vertex have been found in a set of ten planes. In addition, the hypothetical track-finding algorithm has found an additional ghost track consisting of a spurious combination of hits from the actual tracks. If two tracks are called incompatible when they have at least one signal in common, and if incompatible tracks are connected, then the graph in Fig. 2.16 is created from the tracks in Fig. 2.15. Obviously, in this case the maximum number of compatible tracks can be obtained by deleting the ghost track represented by the node in the centre. By linking compatible tracks rather than incompatible ones, the problem is then to find all maximum size complete (fully connected) node sets in this complementary graph. Algorithms to find all maximum complete subsets of a graph exist, but are too lengthy to be explained here in detail (Das 1973). A FORTRAN subroutine using the algorithm by Das can be obtained from CERN (CERN Program Library V401).

Looking for a largest set of compatible tracks may not give a unique result; moreover, it might be desirable to take into account the track

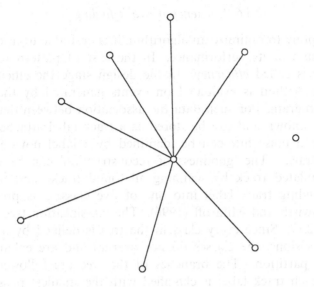

Fig. 2.16. The tracks of Fig. 2.15 in graph representation. Each track is now represented by a node. Tracks are called *incompatible* if they share at least one signal in Fig. 2.15. Incompatible tracks are connected by an edge. The requirement is to remove a minimum number of nodes until all tracks are compatible. Obviously, this leads to the removal of the ghost track (in the centre of this graph), which is incompatible with most others.

quality (Subsection 2.3.2). This approach has been pursued in the track finding in the Forward Chamber A (FCA) of the DELPHI detector (Frühwirth 1993). FCA, having only six planes of drift tubes, has very little redundancy in addition to its inherent left–right ambiguity. Stand-alone pattern recognition therefore produces a number of ghost tracks, most of which are incompatible with true tracks. A quality index is assigned to each track, which combines the χ^2 probability of the track fit with information on the track direction. The 'best' maximal compatible subset is the one that maximizes the sum of all quality indices. The problem of finding the 'best' compatible subset is solved by a Hopfield network (Subsection 2.2.9). A comparison with the exact solution obtained by algorithm V401 of the CERN Program Library shows that the network delivers nearly optimal solutions. For a problem size of 20 tracks the network is faster by a factor of 4 to 20, depending on the frequency of incompatibilities. In addition, the weights of the network depend on a parameter that can be tuned to find either large compatible subsets or subsets with high quality tracks. Results on simulated data indicate that this technique finds more true tracks and suppresses more ghost tracks than the standard sequential approach to selecting compatible tracks.

2.3.6 *Efficiency of track finding*

When attempting to optimize an algorithm, it is desirable to have a quantitative measure of its performance. In the case of pattern recognition this measure is called *efficiency*. In the design stage the efficiency of a track-finding method is evaluated on events generated by the detector simulation program. For such data the association between detector hits and tracks is known and can be stored as a track label attached to each hit. Additional noise hits can be identified by a label not corresponding to any track. The 'goodness of reconstruction' can be computed for each simulated track by scanning all found tracks and by putting the corresponding track label into one of five classes, as proposed by Regler, Frühwirth and Mitaroff (1996). The classification tree is visualized in Fig. 2.17. Since every class in the tree is defined by a sequence of yes/no decisions, the classes do not overlap and are exhaustive, i.e. they form a partition. The branches of the tree are followed in such a way that each track label is classified with the smallest possible class number.

A found track is called perfect if it contains only hits with a given label. If a perfect track contains all hits with that label, it is called complete, and the label is put into class 1. If some of the hits are missing, the track is called incomplete, and the label is put in class 2. Simulated tracks with labels in classes 1 and 2 are reconstructed without bias from wrong hits, although for labels in class 2 the reconstruction is suboptimal due to missing hits. Such tracks can be considered to be found correctly.

If no perfect tracks are found for a given label, there still may be imperfect tracks containing hits with that label. If the label is in the majority in an imperfect track, it is put in class 3; if not, it is put in class 4. If a label does not occur in any found track, it is put in class 5. Simulated tracks with labels in class 3 are reconstructed with a bias because of wrong hits ('outliers'). If there are only few outliers or only a single one, the track may be corrected in the subsequent track fit, if the fit procedure is equipped with the proper outlier rejection mechanism.

There are now several possibilities for defining an efficiency.

(1) For a number of events with a fixed number n of reconstructable tracks each, the average efficiency $\langle e \rangle_n$ is $\langle f \rangle / n$, where f is the number of tracks found correctly in each of these events. This number normally decreases slowly with increasing n. Sometimes an overall efficiency is quoted, which is based on an 'event mix' with varying multiplicities, being either $\langle \langle f \rangle / n \rangle$ or $\langle f \rangle / \langle n \rangle$. The first of these expressions gives higher values than the second one.

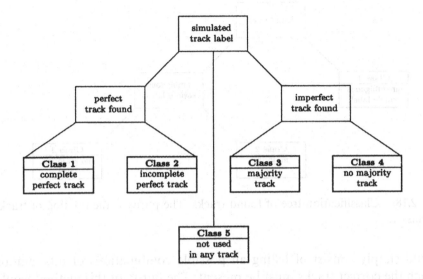

Fig. 2.17. Classification tree of simulated tracks. Simulated tracks in classes 1 and 2 are reconstructed without bias. Tracks in classes 3–5 are reconstructed either with a bias or not at all. The efficiency is the fraction of simulated tracks in classes 1 and 2.

(2) The fraction of events with *all* tracks found correctly can be used. This represents a very severe measure, and will decrease rapidly with $\langle n \rangle$.

(3) The fraction of events in which at least a minimum fraction, say 90%, of the tracks has been found correctly.

(4) Any of the above metrics, but in which certain tracks that are notoriously difficult to find are excluded, such as tracks below a certain momentum, tracks with too few hits, or tracks in particularly tricky detector regions.

To be of real use, however, this definition has to be complemented by a measure of purity of the sample of found tracks. The classification of found tracks is much simpler (Fig. 2.18). Class 1 comprises unambiguous tracks in which all hits carry the same track label. Ambiguous tracks are put in classes 2 and 3, according to whether or not there is a majority label. Tracks in class 2 can possibly be corrected by the track fit; tracks in class 3 are most likely ghosts. The purity of the algorithm can then be defined as the fraction of tracks in class 1.

A satisfactory algorithm should have both high efficiency and high purity; otherwise, the best method, with a guaranteed efficiency of 100%,

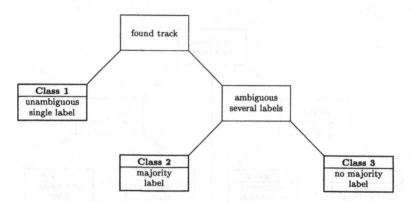

Fig. 2.18. Classification tree of found tracks. The purity is the fraction of tracks in class 1.

would simply consist of listing all possible combinations of hits, among which the correct tracks must be present. The purity of this method would of course be very low indeed. Efficiency is, however, more important than purity. A near-perfect track with a few wrong hits can very likely be corrected in the subsequent track fit, and combinatorial background can be suppressed. A track not found during the track search is lost forever.

The evaluation of efficiency and purity for real data is far more difficult. The correct association between hits and tracks must be found in an independent way, often through scanning of an event sample by eye. This should preferably not be performed by the experimentalist, or even a physicist in the same collaboration, but rather by an independent person, since 'a physicist's eye scan is always biassed by wishful thinking'. Alternatively, if the number of tracks in a measured event can be concluded from some independent source, e.g. from scintillator pulse heights, this may as well serve to define the efficiency and purity of track finding. Except for data with a constant multiplicity per event, global figures such as the overall efficiency defined above should be avoided, and instead the more meaningful dependence on n should be quoted.

Finally it should be pointed out that the computing time used by a given method is also important, and sometimes forces compromises to be made between a good, but awfully slow method, and a fast, but less efficient one. In most cases, however, the efficiency is the more important consideration, and one will normally try to speed up a good method by extending the algorithm that gives the desired results rather than use a less efficient method, or, as Weinberg (1972) has put it, 'Any program that works is better than any program that doesn't'.

2.4 Methods of track finding

2.4.1 A classification

The different track-finding methods can be classified as *global* or *local*. A method is called *global* if all objects (hits or points) enter into the algorithm in the same way. This algorithm then produces a table of tracks, or at least a table in which the tracks can be found more easily than among the original data. The algorithm can be considered as a general transformation of the entire set of the event coordinates or space points. The computing time of a global method should, in principle, be proportional to n, the number of points in the event.

A *local* method, on the contrary, is one that selects one track candidate at a time, typically by starting with a few points only (track candidate initialization), and then makes predictions as to further points belonging to this track candidate, e.g. by interpolation or extrapolation of the current track model based on the track candidate found so far. If additional points are found, they are added to the candidate, otherwise the candidate is dropped after a certain number of attempts, depending on the degree of detector inefficiency the algorithm wants to allow for. Since local methods invariably have to make fruitless attempts in order to find track candidates, and thus use the same point in different combinations, the computing time increases more rapidly than linearly with the number of points.

Purely global methods are independent of the order in which points enter the algorithm, local methods are not, since the treatment of each point depends on the initialization, and the 'track finding history' inside an event in general. However, a good track-finding algorithm gives the same tracks even if the order of the raw data from each chamber is randomized (and therefore the measurements enter the track-finding algorithm in a different order).

2.4.2 Local methods

2.4.2.1 Track following. The method of track following is typically applied to tracks of the *perceptual* type, where the track can be more or less easily recognized by the human eye from the displayed coordinates. This requires the measurement of highly redundant data, and is nowadays aimed at by practically all experiments.

An initial track segment is first selected, consisting of a few points (up to three or four), and this segment is normally chosen as far away from the interaction region as possible, since there the tracks are, at least on the average, more separated than anywhere nearer to it; in addition, in this way the longest tracks will be found first.

In the next step, a point is predicted by extrapolation into the next chamber in the direction of the vertex. This extrapolation may be of 'zero' order simply by choosing the nearest neighbour, first order (straight line), second order (parabola), and possibly higher order, or by other track models such as circles or helices (if working in space). In all cases the aim is to have a fast point prediction algorithm, representing the track locally by the simplest model possible. Speed is of highest concern in real-time applications (see e.g. Subsection 1.5.4). For chambers that are sufficiently close, the parabola extrapolation will be sufficient in many cases with magnetic fields, since it preserves the sign of curvature as does a real particle track. In addition it is very fast, since a parabola through three points is given by a linear expression in the three dependent coordinates (Equation (2.2)), where the coefficients depend only on the plane positions, and can therefore be calculated once and for all for the prediction of hits in any of the chambers.

A rather sophisticated example is the track parametrization that has been applied successfully in the Omega detector (Lassalle *et al.* 1980). From a track sample generated at the target and tracked through the detector, the Karhunen–Loeve feature extraction gives the significant coordinates as well as the insignificant ones, which act as constraint equations. The transformation is applied to sets of successive planes. Since each constraint equation contains the track coordinates in these planes, the coordinate in one particular plane is fully defined when the coordinates in all other planes are known. A track coordinate in a given plane can then be calculated from the part of the track already constructed in the other planes, normally those preceding it. Because sometimes tracks have to be followed in a direction away from the target, and because chamber inefficiencies have to be dealt with at the track initalization stage, an average of three sets of planes have to be parametrized per plane, and the coefficients kept, in total several hundred sets (the detector consists of about 100 planes). This is an example of parametrizing track coordinates as a function of other coordinates. Although the equations are linear, the tracks may be curved as has been pointed out before. The prediction errors are smaller than the intrinsic detector resolution, which is given by the wire spacing.

The tracks are followed in space, although each detector plane measures only one coordinate. During the track-following phase, i.e. after the initialization, missing hits are simply replaced by their prediction, thus avoiding having to parametrize an impractically high number of different sets of planes. Of course this can only be done for two to three successive missing signals at most. It can be argued that in reality this track following does not use extrapolation but interpolation, because owing to the choice of the track sample for the parametrization, the target region is implicitly used as a constraint for the track coordinates. To put it differently, the fact that the tracks have to go through the target region reduces the phase space that has to be parametrized and thus increases the success of the method.

Another example of track following can be found in Mess, Metcalf and Orr (1980), where muon tracks are followed through a calorimeter, this task being made more difficult by considerable multiple scattering and the presence of hadron showers. The target calorimeter in this experiment consists of 78 slabs of marble each 8 cm thick with a plane of proportional tubes at the back of each slab. The wires are oriented vertically and horizontally alternately; there are no inclined wires. Behind the target calorimeter there is an end-calorimeter consisting of 15 circular plates of magnetized iron, each 15 cm thick interspersed with proportional tube planes.

The muon track finding is complicated by the presence of a shower and by multiple scattering in the marble and iron. It proceeds as follows: firstly, those shower signals are removed for which this can be done rather safely without removing (too many) points of the muon track. This removal can only be partial because the difference between the shower signals and that of a minimum ionizing particle are not as high as expected due to space charge effects. Secondly, the track is initialized combinatorially in the magnet system where conditions are cleanest, and then followed back into the target calorimeter, where a concurrent search for other tracks is made. The tracks are fitted to straight lines in the field-free regions, resulting in some of them being found only as pieces because the particle has undergone intense bremsstrahlung somewhere on the track. Finally these pieces are joined together. The track-finding efficiency is over 99%.

In summary, the track-following method is in one way or another concerned only with the local track model, since it always looks only at the next few points, using the most recently found ones to extrapolate the track. This allows a simple, and hence fast, track model. On the other hand, once the distances become too big, the approximate model will not be precise enough; and because of the measurement errors, even an absolutely correct tracking based on a few recently found points is problematic, since most detectors deliver sufficiently precise track parameters only when the full track is used in a fit.

The track-following method uses the combinatorial initialization of candidates. Once a few points have been added to a candidate, it is very likely to be a good track, thus keeping the overhead of following wrong candidates rather small. Accordingly, the computing time is normally proportional to a number between n and n^2 (n is the number of points). In the case of the TRIDENT track finding program in the Omega detector at CERN (Lassalle *et al.* 1980) the computing time is approximately proportional to $n^{1.5}$.

2.4.2.2 Kalman filter. The Kalman filter can be viewed as a statistically optimal refinement of track following. Originally it was conceived as a method of estimating the unobservable states of a discrete linear dynamic system. The state of a charged track at any given surface in the detector can be described by a collection of five track parameters, two for the position, two for the direction, and one for the curvature or momentum. This is called the 'state vector' in the terminology of dynamic systems.

The Kalman filter consists of a succession of alternating *prediction* and *filter* steps. In the prediction step the current state vector is extrapolated to the next detector surface, taking into account multiple scattering and energy loss, if this is required. In the filter step the extrapolated state vector is updated by taking a weighted mean with the measurement. It is now widely used for track fitting, i.e. the estimation of the position, of the direction, and of the momentum of charged tracks (see Subsection 3.2.5).

It has long been recognized that, owing to its recursive character, the Kalman filter can also be employed as a track finder (Billoir 1989; Billoir and Qian 1990a, b). This means that after each prediction step it has to be decided which measurement should be used in the subsequent filter step. Conventionally, the measurement which is closest to the prediction is selected for inclusion in the filter, the distance being expressed by a suitable χ^2-statistic (Billoir 1989). The measurement is marked as used and is no longer available. This guarantees that the final set of track candidates is compatible in the sense that two track candidates cannot have a common measurement (Subsection 2.3.5).

This decision rule is purely local in the sense that no account is taken of the fact that track candidates generated later on might be even closer to the measurement than the current one. If the track density is not very high, this poses no particular problem. In a high-multiplicity environment like at LHC, however, it may be necessary to consider other decision rules which are global in the sense that there is a symmetric competition of all track candidates for the entire set of measurements (see Subsection 2.4.3).

In order to get the filter going a starting track segment or 'seed' has to be computed. As in the case of simple track following, it has obvious advantages to generate the seeds in the less densely populated regions of the track detector. For instance, in a collider experiment the seeds are generated in the outermost layers of the tracking system, and track finding proceeds toward the interaction region. In the case of the ATLAS inner tracker (see Subsection 2.1.1.3), the track candidates found in the transition radiation tracker can be used as seeds for a Kalman filter in the silicon tracker.

2.4.2.3 Track roads. In this case, there is no extrapolation as in the track-following method, but the much more precise interpolation between points is used to predict extra points on the track. By using initial points at both ends of the track (and one point in the centre of curved tracks), a simple model of the track is used to predict the positions of further points on the track, by defining a *road* around the track model. This track model may be (almost) precise, such as a circle in the case of a projection onto a plane that is orthogonal to the field vector of a homogeneous magnetic field, or a straight line in a field-free region; or it may be approximate, in which case the width of the road has to be established by Monte Carlo tracks. In principle, the better the model the narrower the road can be, but rarely can one use the theoretical road width of, say, three standard deviations of the detector resolution. This may be because of systematic errors in the position, to signal clusters, to signals being hidden by background signals, etc.

The method of track roads is slower than the track-following method; but sometimes it is the only workable method available, particularly in the case of widely spaced detector planes (Fröhlich *et al.* 1976), where the redundancy in the coordinate measurements can be very low and where tracks with as few as three space points may have to be accepted. Most modern detectors provide a density of measurements that is high enough to permit the use of faster methods. However, even when there is sufficient redundancy the road method can sometimes be superior in performance and speed when compared with the track-following method, for example in drift chambers with left–right ambiguity. Berkelman, in a description of the CLEO drift chamber (Berkelman 1981) argues rather convincingly (although without proving it) that in this case knowledge of the precise curve, a circle, can only be used profitably if the initial track candidate spans the full track range. The CLEO chamber consists of nine concentric cylindrical drift layers with wires parallel to the cylinder axis. It has been used in an electron–positron collision experiment at the Cornell Electron Storage Ring CESR (see Table 1.1). In the track search, a pair of coordinates (a signal plus its mirror image) is chosen combinatorically in an outer and an inner cylinder, and the origin serves as third point. A first approximation of the curvature is calculated from the raw data. If it is acceptable, a drift time correction is applied, and four roads are established (for the four possible combinations of a signal pair in an outer cylinder and a signal pair in an inner cylinder) inside which points are searched for and added to the track candidate. After this, the candidate with the highest number of signals is accepted as the correct one. When there is more than one suitable candidate, the error of a least squares fit is used to decide between them. In a further step the origin is dropped as a constraint, leading to eight possible candidates from the three initial point pairs. Berkelman believes that his method is less vulnerable to overlap confusion and therefore much faster in execution than the track-following model of the type applied at TASSO (Cassel and Kowalski 1981), which is described in the discussion of the application of the MST in Subsection 2.2.8.

Because (in a magnetic field) a road has to be initialized by three points, and as combinations in different planes have to be chosen either because of detector inefficiencies, or because not all tracks may reach the last plane, and as most initial combinations of three points will be wrong, the computing time of this method is typically proportional to a factor between n^2 and n^3 (n is the number of points). In the case of the MARC track-finding program for the Split Field Magnet (SFM) detector at CERN, the time was proportional to $n^{2.3}$ (Fröhlich *et al.* 1976).

2.4.2.4 Track elements. Here, a track candidate is constructed in two steps: firstly, short track elements are made up of points, normally inside 'natural' subdivisions of the detector, such as drift chamber cells or entire subdetectors. From this track candidate, zero order (nearest neighbour), first order (straight line), or second order (parabola) extrapolations or interpolations are used to define track elements, each of which is then condensed into a *master point* (the weighted average of the cluster) plus a direction. Secondly, these master points are combined using track-following or other track-finding methods.

The great advantage of this method is its speed, compared to using all (up to several hundred) points per track directly. It is as such appropriate for detectors with a very high point density, and was, historically, frequently used for bubble chamber analysis. In addition, the left–right ambiguity of drift chambers can be solved at the track-element level. The reduced number of points, and their wider spacing, are compensated by their higher precision, and the fact that they have a direction associated with them (Eichinger 1980).

This method has been applied in the JADE detector at DESY (Olsson *et al.* 1980). There, the cylindrical drift chamber surrounds the beam tube inside the homogeneous field of a solenoid magnet. The drift chamber sense wires are thus parallel to the beams, and to the magnetic field lines. They are staggered by 200 μm to either side in order to be able to distinguish signals from their mirror images. Three concentric rings of chambers provide up to 48 points for an outgoing track. The detector looks similar to the one sketched in Fig. 2.3.

Charge division on the wires delivers a z coordinate, but it is less precise by two orders of magnitude. Consequently, the track finding is performed in the xy plane of the drift time signals only, which is orthogonal to the wires. Track segments are made out of four or more points in each of the 96 detector cells separately. A parabola is fitted to each segment, and the mirror image tracks are rejected at this stage because of their poor χ^2 values. Track segments are then connected through quadratic extrapolation into full tracks.

Track elements are also used in the track search of the DELPHI experiment (see Subsections 2.1.1.3 and 3.5). In this case the track elements are the result of the local pattern recognition procedures in the various track detectors. They are subsequently combined to track candidates by the global track search. This is probably the most efficient and transparent way to cope with the complexity and diversity of the DELPHI tracking system.

2.4.3 Global methods

A totally different approach is used in the *global methods*, which are also applicable to a wider range of problems in cluster analysis. In these, all points are considered together, and a procedure exists for classifying all tracks simultaneously.

2.4.3.1 The combinatorial method. This method works in the following way.

(1) Split the set of all position measurements into all possible subsets.

(2) Fit a track model to each subset (= track candidate) and call it a track if it fits the model.

Although this method can be applied directly in some simple cases, where there are only very few coordinates in total, for most practical problems it is too time-consuming. Consider five tracks in ten planes, producing 50 measurements, and assume that 1 ms is required to fit the track model to each subset. Even if we ignore possible multiple hits of tracks in the same plane, the processing will take about three hours of computing time, which is normally prohibitive.

2.4.3.2 Global Kalman filter. In the LEP experiments, track finding with the Kalman filter has been implemented with a local decision rule: the measurement that is closest to the prediction is attached to the track candidate and used in the subsequent filter step. This means that no account is taken of the fact that track candidates generated later on might be even closer to the measurement than the current one. In a high-multiplicity environment like that at LHC it may be necessary to consider truly global decision rules such that there is a symmetric competition of all track candidates for the entire set of measurements.

Two ways of implementing such a global competition can be envisaged. In the first one, the prediction steps of all track candidates are computed in parallel. When all predictions have been computed, a global match is sought between all predicted state vectors and all measurements in the current detector surface. Every prediction can match with at most one measurement and vice versa. This is an instance of the general problem of 'matching of broken random samples', which can be solved by a Hopfield network (Frühwirth 1995). Alternatively, a faster suboptimal solution can be found by a sequential algorithm. The restriction on the matching guarantees that the final set of track candidates is compatible in the sense defined above. Unmatched measurements can be used to initiate new track candidates.

In the second method the predictions are also processed in parallel. The decision is, however, deferred to the final stage. In the matching stage therefore we allow all matches between predictions and measurements the χ^2 of which do not exceed a certain cut. Thus a prediction may match with several measurements, and a measurement may match with several predictions. Then the filter step is performed on all valid matches, and the procedure is resumed with an extended set of track candidates, one for each valid match. Clearly the number of track candidates can rise rather quickly if the track density is high. In order to cut down this number somewhat, a test can be done on the total χ^2 of the track candidate. In the final set of track candidates there are usually a lot of incompatibilities. In order to suppress false track candidates an optimal subset of compatible track candidates is computed, for instance by means of a Hopfield network (Frühwirth 1993).

Both global methods seem attractive on the grounds that the prediction steps can easily be parallelized. The performance depends of course on the initial set of 'seeds'. Although care must be taken not to lose tracks by missing a seed, too many seeds result in a proliferation of track candidates and consequently in an unacceptable slowing down of the track search. Using the position of the main vertex in setting up the seeds helps in rejecting false seeds, but biasses the track search against secondary tracks.

It is also important to recognize false track candidates as early as possible. The main criterion is the current χ^2 statistic, but the number of hits assigned to the track candidate should be taken into account as well. As no track detector is 100% efficient, one has to allow that track candidates have a certain number of missing measurements or holes. The actual criterion for dropping a track candidate depends strongly on the particular detector, and no general rule can be given.

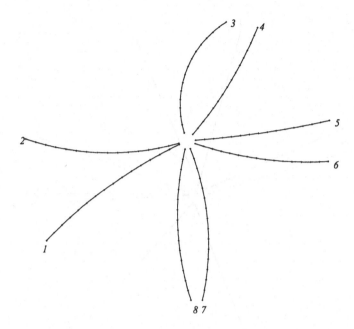

Fig. 2.19. Arcs of circles through the origin. The arcs could be caused by charged particles inside a homogeneous magnetic field, e.g. in the detector of Fig. 2.3.

2.4.3.3 The histogramming method. In this case, one defines a set of n different functions of the point coordinates and enters the function values in a histogram of n dimensions (one dimension for each function), where in practice n is normally 1 or 2, as will be explained later. An n-dimensional histogram has to be visualized as an n-dimensional array of n-dimensional cells; each cell contains a counter which is increased by a given weight when the coordinates of a measurement, consisting of an n-tuple of values, define a point inside that cell. When these n-tuples of function values have been entered, and if the method works correctly, tracks form clusters or 'peaks' in the histogram; these have then 'only' to be found, and the problem is solved.

A simple example may illustrate this method. Suppose that the interactions always take place at the same point, and that there is no magnetic field. In this case, the tracks will form a star around the interaction point, say $(0, 0)$ in a suitable projection. This could, for example, be the case in a colliding beam machine in the projection orthogonal to the beam direction. If we introduce an (x, y) coordinate system, and use as function any of

$$y/x, \sin^{-1}[y/(x^2 + y^2)^{\frac{1}{2}}], \tan^{-1}(y/x) \tag{2.55}$$

then the tracks will appear as peaks at specific function values.

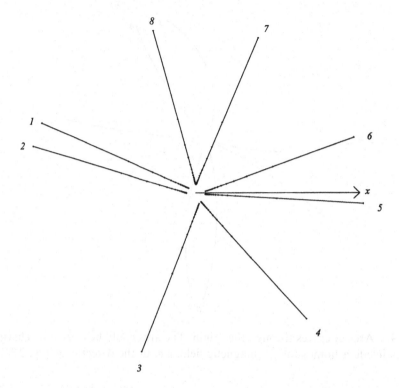

Fig. 2.20.   Arcs of circles after a conformal transformation. The original arcs are given in Fig. 2.19. The nearest distance of any line to the origin is proportional to the curvature of the original circle arc. This offset makes angular clusters of low-momentum tracks in a histogram lower and wider than those of high-momentum tracks (see Fig. 2.21).

Another application of the histogram method is possible if the tracks are circles through the origin (interaction point) as in Fig. 2.19. In this case, the inverse (conformal) transformation

$$u = x/(x^2 + y^2)$$
$$v = -y/(x^2 + y^2) \tag{2.56}$$

will produce straight lines in the (u,v)-plane, their closest distance from the origin being

$$d = 1/(2R) \tag{2.57}$$

with the track radius R (Fig. 2.20).

The histogram method, if applied to this conformal mapping, will tolerate a certain distance of the tracks from the origin and still find the tracks correctly (Fig. 2.21). Overlaps will, of course, have to be treated separately, but are readily recognized because the total point

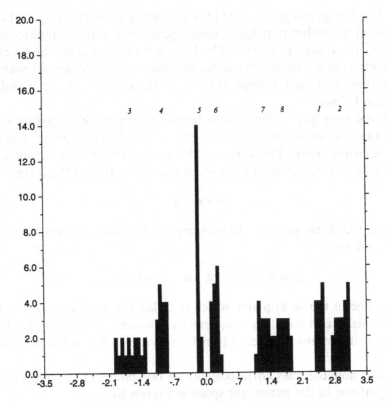

Fig. 2.21. Φ histogram of tracks in Fig. 2.20. The angle with the (arbitrary) x axis in Fig. 2.20 is entered for all points there. The tracks form clusters, which can be found rather easily. The third cluster from the right contains two tracks, easily recognizable from the number of entries in it. A broad cluster (if caused by one track) arises from a large distance of the straight line in Fig. 2.20 from the origin and thus from a track with high curvature according to Equation (2.57).

count becomes too high for one track. The method will at least allow well-separated tracks to be found quickly. This method has been applied in experiment R807 at CERN (Dahl-Jensen 1979) and is based on bubble chamber track-finding algorithms.

One should stress that tracks that do not pass near the point used as the origin for the conformal mapping are not found in this method. This is a strength (if the missed track belongs to the background) and a weakness of the method (if the track belongs to the event and comes from a particle decay). A more general conformal transformation that maps circles in the plane on planes in space is described in Strandlie *et al.* (2000).

In histograms of several dimensions, recognizing the track clusters turns out to be more difficult than finding the tracks directly via a track model. This limits this method to one or two projections.

2.4.3.4 The Hough transform. The histogramming method is a special case of a general algorithm popular in image processing and computer vision. It is called the *Hough transform*. The Hough transform is actually a class of methods comprising deterministic as well as probabilistic algorithms. An overview of recent Hough transform algorithms can be found in Kälviäinen (1994).

The basic principle of the Hough transform can be best explained by considering a set of points (x_i, y_i), $i = 1, \ldots, N$ in the (x, y)-plane, which is called the *image space*. The aim is to detect collinear points, for instance straight tracks. Lines in the image space can be written in the form

$$y = kx + d$$

where k is the slope and d is the intercept. If the line passes through the point (x_i, y_i), then

$$y_i = kx_i + d \quad \text{or} \quad d = -x_i k + y_i$$

This is a line in the (k, d)-plane, which is called the *parameter space*. The slope of this line is equal to $-x_i$, and its intercept is equal to y_i. Thus every point in the image space is transformed into a line in the parameter space.

Consider now the lines corresponding to two points (x_i, y_i) and (x_j, y_j). Their equations in the parameter space are given by

$$d = -x_i k + y_i$$
$$d = -x_j k + y_j$$

It is easy to see that the point of intersection of the two lines in parameter space is equal to

$$k = \frac{y_j - y_i}{x_j - x_i}, \qquad d = \frac{y_i x_j - x_i y_j}{x_j - x_i}$$

These are the parameters of the line in image space that passes through both (x_i, y_i) and (x_j, y_j). Generally, if n collinear points in image space are transformed to lines in the parameter space, these lines intersect each other in a single point (k, d), the parameters of the line in image space which contains the n points.

In practice, the parameter space is divided into an array of discrete cells. When a point in the image space is transformed into a line in parameter space, all cells crossed by the line are incremented. The cell contents are accumulated for all points; therefore the parameter space is also called *accumulator space*. If n points are approximately collinear, the line parameters in the image space correspond to a local maximum or peak in the accumulator space, produced by the approximate intersection of n lines.

If a magnetic field is present, the tracks in a detector are not straight lines any longer. In the simplest case of the field being homogeneous they are circles if projected on a plane perpendicular to the field. If the size of the interaction region can be neglected the circles pass through the origin; such a circle can be parametrized by two parameters (R, ϑ), which are the polar coordinates of its centre. A point (x_i, y_i) in image space is transformed into a parametric curve of the following form in parameter space:

$$R = \frac{x_i^2 + y_i^2}{2(x_i \cos \vartheta + y_i \sin \vartheta)}$$

Again, if n points lie on a circle through the origin, their curves intersect in a single point in parameter space corresponding to this circle. Alternatively, the circles in the image space can be transformed to straight lines by the conformal transformation described above, and the Hough transform has to find these straight lines.

It should be noted that in this version of the Hough transform many cells in the accumulator space have to be incremented for each point in the image space. This is avoided in the combinatorial version of the Hough transform. This algorithm uses pairs of points in image space and computes the parameters of a straight line or a circle through the origin. The corresponding cell in accumulator space is then incremented. In principle all pairs of points have to be processed. In track-finding applications, however, it is frequently possible to drastically reduce the number of pairs, for instance by considering only points in adjacent layers of the detector, and by applying cuts on the direction of the line segments connecting two points. If the number of pairs is still too large, one may resort to sampling only a subset of pairs. This is an example of a probabilistic Hough transform. The combinatorial Hough transform is the standard way of initializing the elastic arm method of track finding (Subsection 2.4.3).

A method that is an extension of the Hough transform is the ExSel++ algorithm discussed in Bischof and Frühwirth (1998). The algorithm was originally developed in the area of computer vision, to fit different types of curves to noisy edge data. It is based on two principles. First, a data-driven exploration produces many hypotheses, or parametric models. Secondly, a selection procedure based on the Minimum Description Length Principle (MDL) selects those hypotheses, which are needed to explain the data. The selection is followed by a robust fitting procedure, which increases the accuracy of the selected hypotheses. Different hypotheses can converge to almost the same hypothesis during the fitting, because the fitting may change the classification of the data. Hence, a final selection is required. The results of the algorithm are a number of models, and a set of outlier or noise points that cannot be explained by any model according to the MDL-principle.

The algorithm is generic insofar as it does not depend on a specific hypotheses generation and fitting procedure. For example, Hough transform techniques can be used for hypotheses generation, and Kalman filtering for fitting the selected hypotheses. Moreover, the exploration can be guided by knowledge about the detector and the tracks to be found. For instance, if one is interested only in tracks with certain properties, e.g. those leaving the detector, those passing through the origin of the detector, or those with high transverse momentum, the hypotheses generation step can be modified such that only tracks with these properties will be generated. However, many types of parametric models can be generated concurrently in the exploration stage, if required.

2.4.3.5 Template matching. This method (see Subsection 2.2.6) requires a dictionary of all possible classes (tracks) and can therefore be applied only in cases where their number can be kept within reasonable limits, somewhere below 10^5, in which case a binary search or a hash algorithm can yield very fast matches (see Subsections 1.5.3 and 1.5.5). The method is typically applied to the case of genuine low redundancy (only a few detector elements) or artificial low redundancy (detector elements are grouped into 'cells' giving one signal each, for the purpose of low-precision track finding).

Used in the track finding for the Mark-III detector at SPEAR (Becker *et al.* 1984), with a total of 12 832 templates, this method is reported to run three times faster than 'conventional approaches'. Each different combination of cells that can be fired by one track creates a separate template. The subdivision of the detector volume into *cells*, which leads to such a low number of possible combinations, means, of course, that there will be regions of confusion where tracks get very near, or cross. These ambiguous regions are then resolved in a 'classical' way, which in the case of the Mark-III is a combinatorial, non-iterative circle fit.

This method is particularly well suited for detectors with cylindrical drift chamber arrangements, since the division of such chambers into drift cells provides a natural basis for the algorithm. In the case of the Mark-III, it works basically as follows. During raw data conversion a cell image matrix of the detector is filled. Combinations of cells that have 'fired' are then compared with the dictionary, once all data of the event have been unpacked. The method is thus global except for the combinatorial search for tracks in regions where they overlap.

For the experiment E711 at Fermilab, a template-matching algorithm was implemented on a fast vector processor. This reduced the computer time per event by a factor of 10 compared to the same algorithm run in scalar mode on the same computer. However, the scalar code had to be rewritten completely for vectorization (Georgiopoulos *et al.* 1986).

2.4.3.6 Minimum Spanning Tree. A variation of the MST method (see Subsection 2.2.8) has been applied to track finding in the TASSO detector at DESY, where it has been shown to be more efficient than the road method (Cassel and Kowalski 1981). The basic element used in this algorithm is no longer a single point, but a pair of points in two adjacent drift-chamber layers. This pair has an associated pair distance and a direction. Pairs are linked into graphs when they share points and when they have similar directions. This second condition is particularly efficient in rejecting image points in drift chambers (left–right ambiguity). In this way, track segments of a certain minimum length are constructed, which allows their track parameters (circle radius and centre) to be calculated. These track segments are then grouped into full tracks by using *similarity* based on the segment parameters.

For a fast search for high-momentum tracks, a modified MST technique is used, where the curvature of a segment defines the edge weight. This also serves the purpose of rejecting arcs containing mirror points, since these will have much higher curvatures than the arcs made up of correct points only. The method is on the whole rather specific to drift chambers with left–right ambiguity, where it works very well.

2.4.3.7 Hopfield networks. Hopfield networks (see Subsection 2.2.9) can solve general combinatorial optimization problems, provided that the objective function is quadratic with zero diagonal. Indeed, they have been sucessfully applied to such diverse problems as graph bisection, graph partitioning, the travelling salesman problem, scheduling and many others. Application to track finding has been proposed independently by Peterson (1989) and Denby (1988). In some points it resembles the MST method of track finding.

The binary neurons v_i of the network represent short track segments connecting hits in the track detector. The weights have to be chosen such that in the final stable state of the network exactly those neurons are active ($v_i = 1$) whose corresponding track segments do actually belong to a track.

It is obvious that generating neurons for all pairs of hits leads to a network with $O(n^2)$ neurons and $O(n^4)$ weights, where n is the number of hits. Even if n is as small as 100, the network becomes prohibitively large and slow. It is therefore mandatory to reduce the number of neurons by applying geometric cuts and by eliminating long segments spanning several layers of the track detector (Stimpfl-Abele and Garrido 1991).

The definition of the weights determines which neurons reinforce or inhibit each other. In order to keep the number of weights manageable, only neurons sharing a hit have a connection weight that is different from 0. In this case there are three different possible configurations (Fig. 2.22). If the two neurons cannot belong to the same track, the connection weight is negative, e.g. equal to -1. If the two neurons may belong to the same track, the connection weight is positive. In the original proposal by Peterson, small angles between adjacent neurons were favoured by larger weights. In the presence of a magnetic field this is clearly a bias against tracks with a small radius of curvature. If the field is homogeneous, the projected track is a circle, and the correct track model can be incorporated at least locally into the connection weights. If a circle is fitted to the three hits of the two adjacent neurons v_i and v_j, a probability p_{ij} that this circle passes through the interaction vertex can be determined. If the probability is above a cut value, the connection weight T_{ij} is then set to

$$T_{ij} = \frac{p_{ij}}{l_i + l_j} \tag{2.58}$$

where l_i and l_j are the lengths of neurons v_i and v_j, respectively. The length appears in the denominator in order to favour short neurons over long neurons. If the detector layers are not equidistant it may be advisable to measure the distance by the number of layers spanned by the neuron. It should be kept in mind that this definition of the weights is strongly biassed against secondary tracks with a large impact parameter.

Different algorithms for minimizing the energy function of the network have been studied (Diehl *et al.* 1997). The conclusion is that mean-field annealing (see Subsection 2.2.9) is by far the best compromise between speed and performance. A comparison with the exact solution obtained by a 'branch-and-bound' algorithm shows that it comes very close to the global optimum.

The first test of track finding with the Hopfield network on real data was done in the TPC of the ALEPH detector (Stimpfl-Abele and Garrido 1991). The performance was comparable to that of the conventional approach via three-dimensional track following. Association of wrong co-ordinates, however, was more frequent with the network. This is probably caused by the fact that no track model was used in the weights. The two methods were also comparable in speed. More than 60% of the time needed by the Hopfield network was spent in setting up the network, even in the extremely simple geometry of the ALEPH TPC. In the much more complicated geometries of the inner trackers of the LHC experiments this percentage will be even higher. Attempts have been made to speed up the

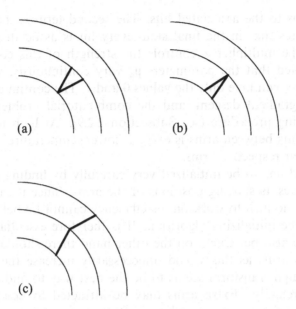

Fig. 2.22. Two neurons sharing a hit. (a) The shared hit is an inner point of both neurons. The connection weight is negative. (b) The shared hit is an outer point of both neurons. The connection weight is negative. (c) The neurons may belong to the same track. The connection weight is positive.

network initialization by using a predefined library of links containing all possible connections between detector elements in the Inner Tracker of the CMS experiment (Scherzer 1997).

2.4.3.8 Deformable templates, elastic arms. This method is also inspired by neural networks, but works with the hits rather than with track segments. The hits are associated dynamically to templates or arms; these are parametrized track models that are fitted to the hits associated to them (Ohlsson, Petersson and Yuille 1992).

Assume that there are N detector signals (x_i, y_i, z_i), $i = 1, \ldots, N$ and M arms with track parameters $\mathbf{p}_a$, $a = 1, \ldots, M$. The squared distance between hit i and arm a is denoted by M_{ia}. In order to specify the association between hits and arms there are NM binary neurons or decision units s_{ia}; $s_{ia} = 1$ means that hit i is associated to arm a. Track finding is equivalent to the minimization of the following objective function:

$$E(s_{ia}, \mathbf{p}_a) = \sum_{i=1}^{N} \sum_{a=1}^{M} s_{ia} M_{ia} + \lambda \sum_{i=1}^{N} \left(\sum_{a=1}^{M} s_{ia} - 1 \right)^2 \qquad (2.59)$$

The first term is the cost term; its minimization performs a least-squares fit of the arms to the associated hits. The second term is the constraint term; it ensures that in the final state every hit is associated to exactly one track. The multiplier λ controls the strength of this constraint. It should be noted that the parameters $\mathbf{p}_a$ vary continuously, whereas the decision units s_i can take only the values 0 and 1. The continuous problem is solved by gradient descent, and the combinatorial problem is solved by an annealing procedure (see Subsection 2.2.9). At high temperature, swapping of hits between arms is easy; at lower temperatures the hits get 'frozen' to their respective arms.

The method has to be initialized very carefully by finding preliminary track candidates as starting positions of the arms. Since it cannot find a track if it has no arm to work on, its efficiency cannot be better than the efficiency of the initializing algorithm. It is therefore essential that there is at least one arm per track; on the other hand there should not be too many surplus arms as this would unnecessarily increase the computing time. The Hough transform seems to be the best way to find initial arms quickly and reliably. Extra arms may be attracted to noise signal, to secondary tracks, or to the same hits as other arms.

The method was used to find simulated tracks in the transition radiation tracker of the ATLAS experiment (Lindström 1995; see also Subsection 2.1.1.3). This tracker is made up of drift tubes giving ambiguous information. The algorithm was extended to cope with this additional combinatorial problem. Initialization was done by a local Hough transform (see Subsection 2.4.3.4). The results are encouraging, although the efficiency of the Hough transform was rather low, about 85%. The cooling schedule required about 200 iterations. The time spent in the annealing phase is about four times as long as is required for the Hough transform, if a fast approximate track parametrization is used. It remains to be seen whether this approach is competitive with other, more conventional, methods of track finding.

2.5 Finding of particle showers

2.5.1 Some definitions

This chapter has the title *Pattern recognition*, a somewhat misleading name in the context of calorimetry: signals in calorimeters caused by elementary particle reactions can very generally be called *patterns*, but *recognition* suggests that the object to be recognized is of a known shape, predictable within narrow limits. We will describe in this section mostly *to what extent* showers in calorimeters are predictable and understood. The objective of *calorimeter analysis* is to deduce from calorimeter signals the phenomenon that is at the origin of a shower. Our objective in this section is to sketch the problems, and to show how detailed analysis proceeds in a number of typical cases, taking care to relate the local solutions to the more general concepts of pattern recognition or analysis.

The development of calorimeters of different types is of a comparatively recent date; they do, however, have a major role in nearly all experiments, particularly at higher particle energies. Their functioning is understood qualitatively and empirically, much of it also at the low-energy level of nuclear interactions.

In this section we will outline the principles of calorimeters and their most characteristic parameters; we will use for the analysis part a small number of characteristic examples. For in-depth reading, we recommend Fabjan (1991), Wigmans (1991a), Cushman (1992), Gratta, Newman and Zhu (1994), and references given there.

2.5.1.1 Total absorption. Calorimeters are devices that entirely absorb, and hence destroy, incident particles, by making them interact. In the process of optimally controlled absorption, cascades of interactions occur, which are called showers; this explains the occasionally used names *total absorption calorimeter* or *shower counter*. Calorimeters measure energy by annihilating it, most of it being converted into heat (*calor* is Latin for heat). Part of the energy is released in the form of a recordable signal such as scintillated light or ionization. The physical phenomena that can be observed with the help of a calorimeter are diverse in nature, but destructive measurement is common to all of them. In order to be usable, a calorimeter must allow the recording of a signal containing adequate detail of the absorption process. From this signal, the original phenomenon can then be inferred, e.g. an incident particle's energy or the energy dissipation in a part of geometrical space.

As shower development is largely independent of the charge of the incident particle, calorimeters also are unique instruments for measuring the energy of neutral particles. They are further unequalled in detecting and measuring particle jets, in which mixtures of neutral and charged particles are present at small spatial separation. Charged particle tracking devices are not useful (or only marginally useful) for either neutral particles or jets. Figure 2.23 shows the development of particle showers of several types, in the detail that only simulation can provide.

The destructive measurement of particle energies in partly or fully active material is of comparatively recent origin. The spread of the method is closely connected with the advent of the availability of high-energy collisions. Calorimeters, first seriously proposed for cosmic-ray physics by Murzin (1967), became popular devices in accelerator experiments as accelerator energies increased, for simple reasons of resolution. Tracking devices reach their precision limits at high energies. They allow the calculation of charged particle momenta using magnetic fields that cause

(a)

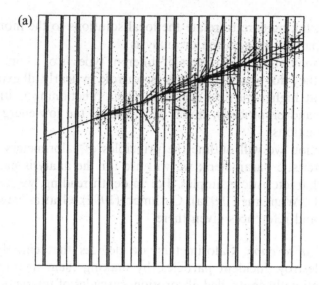

(b)

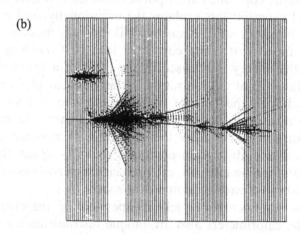

Fig. 2.23.   Electromagnetic and hadronic shower simulation (a) A 5 GeV photon
enters, at a 20° angle, a shower counter as used in SLAC experiment E-137. Eight
modules each consist of wire chambers with walls, a scintillator plane, air gaps,
and an (aluminium) converter. All particles are shown, with neutrals as dotted
lines. (b) Two simulated showers with 100 GeV energy are shown annihilating in
a lead scintillator (50/6 mm) sandwich calorimeter. Air gaps separate the four
modules, each of which is 20 radiation lengths deep. The upper shower is a
single electron, the lower shower a charged π-meson. Note the multitude of local
showers for the hadronic cascade, and the substantial widening by the air gaps.
Photons are shown as dotted lines.

a particle track to bend, as amply described elsewhere in this book. Owing
to the lesser bending, they give inferior relative resolution at high energies
($\Delta P/P \propto P$ for equal field and track length). Calorimeters measure the
energy with smaller relative error at higher energy ($\Delta E/E \propto 1/\sqrt{E}$).

The absorption of an incident particle in a calorimeter proceeds through a multitude of interactions, strong or electromagnetic, each resulting in several secondary particles. These, in turn, will interact again, and the resulting *cascade* or *shower* will contain, as it develops in time, an increasing number of particles of decreasing individual energy.

Some secondary particles, of course, are subject neither to the strong nor to the electromagnetic force. They escape calorimeters undetected (neutrinos). Muons produce an ionization signal not proportional to their energy. Some energy is further lost by nucleus excitation or breakup, and produces no visible signal. It is part of understanding calorimeters to estimate the fraction of energy that thus escapes detection.

2.5.1.2 Calorimeter properties. Depending on the physics context in which a calorimeter is to be used, *calorimeter properties*, i.e. the technical solutions found, will differ within wide limits. The recorded shower aspects will vary accordingly. In other words, calorimeters and data can have very different characteristics, caused both by the variety of possible technical solutions, and by the different physics demands.

Calorimeters are described in terms of *characteristic parameters* such as energy resolution, position resolution, and shower separation, optimization for certain types of particles, dynamical range, spatial coverage. All of these are determined by reaching a compromise between the physics aims of an experiment, the resources needed for construction and analysis, and the mechanical limits resulting from the use of different techniques.

Systematic 4π calorimetry, i.e. full spatial coverage for all types of single particles, charged or neutral, and jets, is today a common goal in large detectors. Often, great care is taken to minimize a calorimeter's uninstrumented areas. Such a *hermetic* detector allows us to deduce the presence of escaping high-energy particles that are not absorbed, like neutrinos.

A further relevant parameter is the *readout and analysis level* at which a calorimeter is expected to act. If used in a fast-trigger processor (Subsection 1.5.7), a signal may be needed so fast that it is limited by the physical processes in the calorimeter, and not by the associated electronics. Fast signals (atomic excitation, ionization) are used to generate a detectable effect, via particle charges. Calorimetry is also the only practicable way to obtain a signal from the neutral particles among the secondary particles produced in a high-energy collision. If used instead as an optimal resolution signal for an off-line analysis program, possibly in combination with other parts of the detector, very different design criteria will be used.

Calorimeter construction allows many more choices than a tracking device. The observed variety of techniques is proof that this type of detector is in permanent search for optimization. Calorimeters are largely empirical instruments, and optimization is absolutely critical. They consequently offer more room for original and creative solutions than other types of detector.

2.5.1.3 Shower properties. Shower properties can fluctuate dramatically for the same energy, depending on the sequence of interactions giving rise to the showers. These fluctuations constitute a characteristic difference between tracking devices and calorimeters. They also are very different for incident electromagnetic particles on the one hand, and for hadronic particles on the other.

Tracking chambers are usually built to interfere as little as possible with the passing track. Once a particle's parameters are given, its track can be predicted within narrow limits. Calorimeters, on the other hand, are devices custom-built for optimal performance on different incident particles, all based on stochastic processes. Only *average* shower properties can be predicted with some confidence and described in terms of *shower parameters*. This stochastic aspect, and the parameters describing showers, will be discussed in some detail below. An important common characteristic of all showers is the statistical multiplication of particles in a shower with increasing incident energy. This multiplication is at the origin of the improved relative resolution at high energy, much like the $1/N$ law, which describes the relative variance of a Poisson-distributed number N.

2.5.1.4 Calorimeter applications. Calorimeters are subdivided so that signals are obtained after integration over a limited physical volume. They therefore allow the energy deposition in a 'cell' of defined position with respect to the interaction point to be deduced. Thus an obvious and simple application of calorimeters is as devices to measure the spatial *energy flow* in an interaction. If the cell structure is fine enough, and showers sufficiently separated, a calorimeter will allow algorithms to associate several cells into 'clusters', and to deduce physical phenomena based on the cluster's properties.

If additional information is available (e.g. from a tracking device), clusters may also be associated to a single track and thus provide a *track energy measurement*. In conjunction with existing understanding of the showering process for different particles, a calorimeter may further be used as a *particle identifier*. Again, it must then be assumed that some independent information establishes the existence of a single particle.

At high energies, the single-particle aspect of calorimeters is often considered less important than its capability to *localize jets* and measure their energy. Jets are particle compounds produced by the hadronization of constituents (quarks or gluons), which manifest themselves in interactions at very short distances.

In order to understand average shower properties and their fluctuations, it is essential to look in some detail at what is known about the properties of the absorption mechanisms. This will allow us to narrow down as best we can the 'patterns' that we want to recognize. This discussion will also help us to understand why such a wide variety of methods of building and calibrating calorimeters exist, and why the data produced by these devices are analysed and used in so many different ways.

Among the relevant showering processes, two major categories can be distinguished: *electromagnetic showers*, which are caused by incident electrons and photons, and *hadronic showers*, which are induced by hadrons. Hadronic showers generally also contain electromagnetic shower components, but, in addition, a large variety of specifically hadronic phenomena determine their parameters. Calorimeters are, of course, not limited to strong or electromagnetic primary interactions; they are particularly valuable instruments in weak-interaction physics, e.g. in neutrino physics, where they serve the double purpose of being the target for the primary particle, and the identifier for secondary leptons or hadrons.

If equipped with suitable electronics (viz. adequate dynamic range to be sensitive to comparatively low signals), calorimeters may also add valuable information to the passage of non-interacting minimum-ionizing particles, like muons.

2.5.2 Physical processes in calorimeters

2.5.2.1 Electromagnetic showers. Electromagnetic showers are those triggered by an incident e^+, e^- or γ. As a π^0, for all practical purposes, decays instantaneously into two photons or a photon and an e^+e^- pair, electromagnetic shower counters, of course, are also relevant in hadronic interactions, in which on average nearly one third of the pionic energy (which dominates other particle contents) will be found in the form of π^0s.

The reactions that predominate in electromagnetic showering all produce further electromagnetic particles. This has two important consequences. Firstly, all the energy in an electromagnetic shower will finally be deposited as ionization loss by electrons. The longitudinal energy distribution of an electromagnetic shower is therefore often given in terms of the *Number of Equivalent Particles* (NEP) i.e. compared with the number of (minimum ionizing) tracks present at any cross section. On the other hand, electromagnetic showers can be quite completely described by quantum electrodynamics (QED), and can be calculated using perturbation theory

(Feynman diagrams). Bremsstrahlung and electron-pair production are the dominant processes for high-energy electrons and photons. Their cross sections become nearly energy-independent above 1 GeV. The more relevant electromagnetic interactions in photon- or electron-induced showers are as follows.

Bremsstrahlung (photon emission in acceleration),
Electron–positron pair creation ($\gamma \rightarrow e^+e^-$),
Compton scattering (photon–electron scattering),
Coulomb scattering (electron–nucleon scattering),
Bhabha scattering (electron–positron scattering),
Photoeffect (electron emission from γ-irradiated nuclei),
Møller scattering (electron–electron scattering),
Annihilation ($e^+e^- \rightarrow \mu^+\mu^-$, or $\rightarrow \gamma\gamma$).

All reactions can be suitably expressed in terms of a scaling variable (i.e. a variable that makes the description independent of the material used), the *radiation length*, usually denoted by X_0. It is defined as the mean path (or attenuation) length of an electron due to radiation, and is given by the approximate formula to better than 20% for $Z > 13$:

$$X_0 \ [\mathrm{g\,cm}^{-2}] \approx 180A/Z^2 \tag{2.60}$$

where A, Z are the atomic weight and the atomic number of the material respectively. If X_0 is to be expressed as a length, it is divided by the density ρ: X_0 [cm] $\approx 180A/Z^2\rho$. These formulae are approximate, but reasonably accurate for high Z. Because of the scaling property of showers with X_0, the dimensions of electromagnetic calorimeters are usually expressed in units of X_0.

Electromagnetic showers start after a small fraction of a radiation length, when track multiplication sets in. This happens somewhat faster for an incident electron than for a photon. Multiplication proceeds until the individual track energies become small, so that ionization and Compton scattering start to dominate and to absorb the low-energy part of the shower. An energy profile (dE/dx) maximum is thus reached at some depth of the shower. As the shower develops further, dE/dx will decrease, eventually exponentially. In other words, towards the end, a shower has a defined attenuation length. For approximate shower parameters see Subsection 2.5.4. Typical electromagnetic showers, from Monte Carlo simulations, are shown in Fig. 2.23.

Numerous authors have discussed the total and differential cross sections of the QED processes relevant for electromagnetic showers. Rossi (1965) has given an extensive analytical treatment of electromagnetic showers (*cascade showers*), albeit with some assumptions. His 'approximation B' resulted in shower parameters that are still a reference result today. In other early publications, similar attempts were pursued, and later more and more were combined with Monte Carlo results (e.g. Hayakawa 1969; Messel and Crawford 1970). Not surprisingly, it is precisely when showers have to be simulated in detail and compared with measurements that a compilation of all available data and theories is most necessary. The slowly converging process of developing complete QED Monte Carlo programs has resulted in a single widely used Monte Carlo code for electromagnetic shower simulation, EGS (Nelson, Hirayama and Rogers 1985), which we will discuss in some more detail in Subsection 2.5.5.

Often, global and average shower properties are sufficient or are more relevant than the detailed understanding of the underlying processes. This is particularly true in simulation, where program complication and computer time restrictions may dictate such simplicity. Global properties are discussed in Subsection 2.5.4.

2.5.2.2 Hadronic showers. Hadronic showers, much like electromagnetic showers, build up through a multiplication process, and decay when individual particles have reached low energies and are eventually stopped ('ranged out'). Unlike electromagnetic interactions, however, hadronic interactions, particularly at lower energies (remember that all showers end up with particles of low energy!), cannot be described with rigour, and a much wider variety of interaction channels is available, depending on the primary and secondary particles' mass and energy.

Hadronic showers result from a mixture of reactions, dominated by multiple particle production, including the entire hadronic spectrum as possible secondaries; π^0s may be produced abundantly in hadronic interactions, electromagnetic showering therefore is a subprocess in hadronic calorimetry. The first hadronic interaction will happen *on average* (with exponential distribution) after one nuclear interaction length, i.e. about 17 cm in iron, 18.5 cm in lead, or about 70 cm in a scintillator. This very first reaction causes, by its varying π^0 content, major fluctuations in the nature of the secondaries. There will be a varying electromagnetic shower component, which will propagate in subsequent interactions and eventually in all relevant shower parameters. Fluctuations in the hadronic interactions result in lower intrinsic performance of hadronic calorimeters compared with electromagnetic calorimeters. A detailed discussion is found in Amaldi (1981) or Wigmans (1991a, b). Examples of hadronic showers are shown in Fig. 2.23.

In contrast to electromagnetic showers, which convert nearly all energy into ionization (track length), hadronic showers produce a variable fraction of *invisible energy*. Some of the available energy is used for the excitation or breakup of the absorber nuclei, and is not fully observable. Nucleus breakup absorbs the binding energy, low-energy neutrons will not always result in an observable signal. In addition, neutrinos or muons may be produced in some reactions. These particles escape the detector without depositing their energy or with only an insignificant signal.

Owing to the invisible energy, a hadronic particle typically generates a signal smaller than that of an electromagnetic particle with the same energy. Without special precautions (we discuss this problem of *compensation* in Subsection 2.5.3.6), the effect is a signal difference of some 20%, dependent not only on the calorimeter construction, but also on the energy and nature of the particle. It is therefore necessary to use a different conversion factor for energy readings if the incident particle is hadronic.

The fluctuations between electromagnetic and hadronic components in a hadronic shower are large. Hence even a correct average compensation, made by adjusting the conversion factor, will result in a worse energy resolution for hadronic showers when compared with electromagnetic showers. This is why so much time has been spent on designing calorimeters that compensate with their hardware, and give the same response for any shower component.

2.5.3 Calorimeter parameters

2.5.3.1 Homogeneous or sampling calorimeters. A general distinction by construction is between homogeneous and sampling calorimeters. In *homogeneous* devices, the role of absorption and signal creation is carried by the same material, typically a pure or doped heavy crystal (like NaI, CsI or BGO, short for $Bi_4Ge_3O_{12}$), or a composite material (like lead glass). Homogeneous calorimeters have effectively longer interaction and radiation lengths, and require more space than sampling calorimeters; the growing and machining of large homogeneous crystals is also not unproblematic. They are, therefore, mostly used for moderate-energy electromagnetic calorimetry, with high-precision requirements.

In (heterogeneous) *sampling* calorimeters, the absorber material is inactive, and interspersed with layers of active (signal-producing) material, typically liquid or solid scintillator. Inactive materials typically are lead (for its short X_0), iron, uranium, or combinations thereof. The ratio of energy loss in active and inactive material typically is of the order 1:10.

Many sampling calorimeters are of the planar sandwich type, with plates arranged orthogonal to the dominant particle direction; however, different ideas have also been implemented, like calorimeter cells made of absorber material interspersed with scintillating fibres ('scintillating fibre calorimeter'), or in undulated form, with liquid scintillator, called accordion calorimeters. Hertzog *et al.* (1990) report a practical implementation of the scintillating fibre type, for electro-magnetic calorimetry;

a large-scale example for the accordion type is the ATLAS liquid-argon calorimeter, presently under construction (see Gingrich *et al.* 1995). Energy resolution has been shown to be largely independent of the direction of the incident particle in such calorimeters, as long as the relative thicknesses of active material and absorber do not vary too much (Fabjan 1991).

2.5.3.2 Granularity. Granularity is the not-so-precisely defined parameter that describes globally the size of the elements in a calorimeter for which an integrated signal will be produced. In practice, calorimeters of very different shapes and sizes have been built, depending on the aims of the experiment and the resources available. Frequently, multiple readout with controlled attenuation allows interpolation within a physical cell, thus improving spatial resolution if a single shower can be assumed.

Before entering a more detailed discussion, some *gross distinctions* must be made. Aspects of granularity that need to be considered are as follows.

- The spatial segmentation normal to the direction of incidence of particles or jets (this determines angular resolution).

- The spatial segmentation along the direction of incidence (this determines the capability to distinguish shower shape, i.e. mostly to differentiate between electromagnetic and hadronic showers); it may also allow leakage corrections, leading to improved energy resolution.

- The ratio of active (signal-producing) material to passive absorber in sampling devices (this directly determines the energy resolution).

- The distribution of the active material along the shower (this must be optimized in function of the mix of energies and nature of incident particles).

In a given experiment, the sizes of cells, their orientation, and their relative order are optimized locally. Frequently, large fractions of the available geometrical space are covered by combined electromagnetic and hadronic calorimeters, optimizing for electromagnetic showers closer to the interaction (only relevant parameter: radiation length X_0) and for hadronic showers at larger distance (two basically relevant parameters: interaction length λ, and X_0).

2.5.3.3 Energy resolution. The energy resolution of calorimeters is usually expressed in terms of $\Delta E/E, \Delta E$ being the overall statistical error (root of variance) including all error sources. At high energies, statistical contributions dominate; the relative resolution therefore improves with

Table 2.1. *Energy resolution for electromagnetic and hadronic calorimeters*

Experiment	Type	Resolution $\Delta E/E$	Reference
	Electromagnetic		
Crystal Ball	NaI (TI)	$0.026/\sqrt[4]{E}$	Bloom and Peck 1983
OPAL	Lead glass	$0.05/\sqrt{E}$	Akrawy *et al.* 1990
Crystal Barrel	CsI (TI)	$0.025/\sqrt[4]{E}$	Landua 1996b
ZEUS	sampling: U + plastic scint.	$0.18/\sqrt{E}$	Behrens *et al.* 1990
D0	sampling: U + liquid Ar	$0.15/\sqrt{E}$	Abachi *et al.* 1994
ATLAS	sampling: Pb + liquid Ar	$0.10/\sqrt{E}$	Gingrich *et al.* 1995
	Hadronic		
CHARM	sampling: Marble + scint.	$0.53/\sqrt{E}$	Diddens *et al.* 1980
ZEUS	sampling: U + plastic scint.	$0.35/\sqrt{E}$	Behrens *et al.* 1990
D0	sampling: U/Fe + liquid Ar	$0.50/\sqrt{E}$	Abachi *et al.* 1994

Note: All energies are to be taken in GeV; the constant terms, to be added quadratically, typically are a fraction of a percent for electromagnetic, and 0.02–0.03 for hadronic calorimeters.

increasing E, owing to the increasing number of particles produced in the showering process. The main tendency for $\Delta E/E$ therefore is to follow the form $k/E^{\frac{1}{2}}$, where k depends on the calorimeter and readout characteristics. This law has been observed to hold up to rather high E, subject to more and more subtle calibration procedures, which eliminate systematic contributions. At low energy, electromagnetic calorimeters can really become better than this law suggests (see Table 2.1). Hadronic calorimetry, however, is not a practical method below, say, 1 GeV; the reactions at lower energies become very dependent on the particle type and on fluctuations, making tracking devices far superior.

Since the energy regime is mostly much higher than the particle masses, the definition of E is not usually given much attention. It should be noted that for most incident particles the best approximation is E_{kin} ($\equiv E_{tot} - M$, where M is the mass of the particle); this is what simulation programs use. If annihilation processes are likely, e.g. for incident $\bar{p}$, the situation is more complicated, and none of the mechanisms is fully understood or experimentally measured.

A general formula to give an approximate resolution is $\Delta E/E = \sqrt{(A^2 + R^2)}$, where A is related to the photon statistics, i.e. the number of photons arriving at the photomultiplier (this contribution is inevitable in most calorimeters), and R deals with the sampling fluctuations in the active material. The fluctuations in sampling calorimeters are of prime importance; a single sampling fraction is a very poor approximation for the large variety of physics processes involved (Brückmann, Behrens and Anders 1988). In Table 2.1, some characteristic resolution functions for single particles are compiled, all of them obtained from experimental data.

Incident $\bar{p}$s have a high probability of annihilating when interacting, at low energy they range out and invariably annihilate. The annihilation energy of $\bar{p}$ and the nucleon is then converted into mostly kinetic energy. Despite the nonannihilation cross section, a $\bar{p}$ should therefore be distinguishable with a good probability at low energy from K or p, by calorimetric methods. The method, however, has not seriously been used.

2.5.3.4 Containment. Containment is a very important boundary condition for any calorimeter. If showers are even occasionally partly converted in the absence of active material, the recorded energies will have a distribution with a tail towards low energy values. Depending on the physics intended, this may result in missing triggers or in severe biases, or simply be detrimental to the energy resolution. The effect of leakage on resolution has been estimated by several authors. Blucher *et al.* (1986) give error terms for a CsI crystal calorimeter. They note a side leakage term scaling with, $\sqrt[3]{E}$, and a longitudinal leakage term scaling with $1/\sqrt{E}$ (both terms refer to $\Delta E/E$). The terms have been derived using the Monte Carlo program EGS for incident photons/positrons, in the energy range 100–5000 MeV.

Generally, calorimeters are built to avoid leakage. Cashmore *et al.* (1985) and Grassmann and Moser (1985), on the other hand, give a simple and unoptimized algorithm, based on Monte Carlo calculations, for how rear leakage can be corrected for, assuming that information is available from the last part of the calorimeter (by longitudinal structure in the cells). It is therefore conceivable to construct cheaper and more compact calorimeters, that accept the degradation in resolution remaining after such corrections.

2.5.3.5 Hermeticity. Containment has another aspect for calorimeters that are built to cover as much of the available phase space as possible, a situation encountered typically in colliding beam experiments. A 4π coverage is of relevance if a calorimeter is expected, beyond the measurement of individual showers, to detect energetic neutrinos. Although not producing any detectable signal themselves, neutrinos carry away their energy, and this can show, via the energy conservation law, as a 'missing energy' in the overall balance. Typically, missing energy can be measured only in the transverse plane (longitudinally, too much energy escapes close to the beams), by adding energy vectors. In other words, every measured energy cell is decomposed into two separate transverse components (using the position of the cell centre with respect to the collision point), and the overall sum of these energy vectors is formed. The resolution is limited primarily by the 'hermeticity', the completeness of coverage under all angles. Corrections for muons (measured in their specific detectors) are required, and often calibrations are differently optimized for calculating missing energy from those used to get optimal resolution in individual particle- or jet-induced showers.

2.5.3.6 Compensation. Attempts were made early (Fabjan *et al.* 1977) to compensate for the energy escaping detection by choosing absorber and active materials with an electron/hadron (usually called e/π) signal factor as close to 1 as possible.

This compensation is of utmost importance for good energy resolution for hadrons, as the electromagnetic π^0 content of hadronic showers fluctuates, and an e/π ratio systematically above 1 (as is the case for a non-compensating calorimeter) translates that fluctuation into worse resolution. Even more important, of course, is compensation for the energy resolution of jets, as there the π^0 content may be considerable even before absorption in the calorimeter begins.

The basic idea proposed originally was to use uranium as the absorber material; this would contribute an additional, i.e. compensating, signal due to nuclear fission caused by nuclear excitation. Indeed, calorimeters were built that had an e/π factor very close to 1 over a wide energy range, and showed experimentally much improved resolution for hadronic showers.

Later, it was understood that fission is only one of the compensating mechanisms in sampling calorimeters. The photon absorption in the (high-Z) absorber material plays a significant role, and so does the conversion of low-energy neutrons into signal, e.g. by detection of de-excitation photons; the hydrogen content in the active medium is relevant here. The optimization of compensation, and hence of energy resolution, in sampling calorimeters, includes the choice of materials, the careful adjustment of passive and active layer thicknesses, and the possibility of shielding the active layers with low-Z material, to absorb soft photons. Wigmans (1991b) gives a detailed discussion.

2.5.3.7 Backscattering. A small fraction of energy in any hadronic calorimeter is backscattered from the initial calorimeter surface. This so-called *albedo* effect (from the astronomical 'whiteness' of planets by backscattering light) has its origin in the nuclear breakup products, which show approximately isotropic distribution. The fraction of backscattered energy is strongly dependent on the incident energy. For hadronic showers which start early in the calorimeter, the number of backscattered tracks is large, but their integrated energy is not an important contribution. It is of the order of 5% at 0.8 GeV incident energy, 3% at 1 GeV, and 1% at 10 GeV. At higher energies, the amount of backscattered energy is reported to become constant at around 150–200 MeV (Dorenbosch *et al.* 1987; Ellsworth *et al.* 1982).

2.5.3.8 Calibration and monitoring. The understanding of the conversion factor between the observed signals and the energy of incident particles is extremely relevant for calorimeter performance. A calorimeter responds to showers by producing light in scintillating material, or ionization in gaseous detectors. The conversion of either into electric signals goes through very different stages, whose detailed *calibration at any point in time* constitutes an important aspect of calorimetry. A high-resolution calorimeter requires substantial care to be taken in understanding every detail of the process of converting and transporting signals through the various stages. Light goes through scintillators, wavelength shifters, light guides, etc., until converted into a signal in a photomultiplier or diode. Electron charges drift to a wire and are collected directly or induce a signal. Electric currents are amplified and stored until triggers have decided that digitization should occur. And the conversion of analog signals to digital numbers itself relies on components whose performance must be understood and monitored. As calibration typically introduces new components, their understanding will also become relevant.

Calorimeters are entirely *empirical* instruments. Their absolute calibration is done by measuring known phenomena, like particles in a test beam. However, usually only a small subset of cells of a calorimeter can be installed for test-beam calibration, and its frequent repetition, which is necessary to follow (monitor) slow changes in detector and/or readout, is outright impossible. Certain reactions to which calorimeters are exposed can be used *in situ* to solve both absolute calibration and the continuous monitoring, but to have them available, easily recognized and abundant is the exception, the most prominent example being Bhabha scattering

$(e^+e^- \rightarrow e^+e^-)$ in electron–positron colliders. Normally, absolute calibrations have to be complemented by frequent intercalibration of cells, e.g. by signals injected from a source or by using the ubiquitous muons contained in cosmic rays. Having significant signals in the desired energy range is difficult, and substantial effort has to be spent to show all aspects of a calorimeter are understood (like linearity of response, impact point and direction dependence). Permanent repetition of some procedures is necessary, e.g. to monitor the degradation of detectors and readout components by radiation.

Different means are used for calibration at different time intervals; the descriptions of the major detectors usually give ample room to these questions. For some more detailed reading we give a few specialized references here: Behrens *et al.* (1992); Antiero *et al.* (1994); Buontempo *et al.* (1995); Zhu *et al.* (1995).

2.5.4 Shower parameters

In the following, we will present what is known about the *average* parameters of showers. It should be understood that they describe *individual* showers rather poorly; in particular, they are of limited usefulness for simulating hadronic showers. Few efforts have been made to describe the important shower *fluctuations* in more than qualitative terms. For the lateral shower shape, even average formulae are not really available. If showers are recognized and used in some specific physics context, the relevant parameters are, of course, those that describe the physical track, the jet, and the energy density in part of the geometrical space. Shower parameters as given in this section may still be useful for interpolating between measured points, or for some aspects of crude simulation.

2.5.4.1 Longitudinal shower shape

Electromagnetic showers Many authors have based an average electromagnetic shower description on analytical calculations published by Rossi (1965). His so-called approximation B (constant energy loss ε per radiation length X_0, valid at all energies of the high-energy approximation for radiation and pair production) results in formulae for the shower maximum (the depth at which dE/dx in the shower reaches a maximum), the shower energy median (the depth at which half the shower's energy has been dissipated), and the shower attenuation (the exponential damping slope in the shower tail). For incident electrons of energy E (incident photons result in a slightly stretched shower), the parameters are given as

$$
\begin{array}{ll}
\text{Shower maximum at:} & X_0\left[\ln(E/\varepsilon) - 1\right] \\
\text{Shower energy median at:} & X_0\left[\ln(E/\varepsilon) + 0.4\right]
\end{array}
\qquad (2.61)
$$

In these formulae, ε is the *critical energy*, which is defined as the energy loss by collisions per X_0. The approximate crossover point, at which an electron loses energy equally by bremsstrahlung and by ionization is at $E = \varepsilon$. An approximate value for ε (good for high-Z materials) is given by $\varepsilon = 550/Z$ [MeV], numerical values are e.g. 7.2 MeV for lead, 20.5 for iron, 29.8 for liquid argon, 87 for plastic scintillator. Iwata (1980) has discussed the attenuation of the shower, and proposed, based on measurements, a shower attenuation length of $3X_0$. See also Barnett et al. (1996).

The constants in Equation (2.61) have slightly different values in the literature. In addition, some authors adjust the constants with respect to the entrance of the absorber, others with respect to the first hard process initiating the shower.

A more complete shower description has been suggested by Longo and Sestili (1975), and has been found to be rather universally applicable. It parametrizes the longitudinal energy density by the ansatz

$$
dE/dx = k_{norm} t^a e^{-bt} \qquad (2.62)
$$

in which t is the shower depth expressed in radiation lengths ($t = x/X_0$), and a, b are parameters. They are meant to be independent of the calorimeter material (which enters, however, the depth variable t) and b has been shown to vary little with the incident particle's energy E. The parameter a is related to the shower peak by $(dE/dx)_{max} = a/b$, and hence varies logarithmically with E, if b is constant and the Rossi approximation is accepted. Normalization to unity is obtained by dividing the density function by its integral from zero to infinity, given by the quotient $b^{a+1}/\Gamma(a+1)$, where Γ is Euler's gamma function defined by

$$
\Gamma(a+1) = \int_0^\infty t^a e^{-t}\, dt
$$

Normalizing to the incident energy is then straightforward, i.e.

$$
k_{norm} = E b^{a+1}/\Gamma(a+1) \qquad (2.63)
$$

The often parametrized normalization to a *total track length* (Rossi 1965) is not as useful. It builds on the notion of 'visible' tracks, i.e. on a parameter describing a minimal energy above which a track becomes visible by ionization. Depending on the active medium in a given calorimeter, this cut-off parameter may be very different. As calorimeters are meant to measure energy, we prefer a direct conversion into MeV. To express dE/dx

in number of equivalent particles (NEP), however, can be useful when calibration procedures include the use of minimum ionizing particles, as provided by cosmic ray muons. It is also in terms of NEP that the statistical properties of calorimeters (the law $\Delta E/E \propto 1/\sqrt{E}$) is intuitively most easily understood.

Longo and Sestili (1975) give numerical values for a and b, obtained by a Monte Carlo program in lead glass for $0.1 \leqslant E \leqslant 5 \, \text{GeV}$. Their values are reasonably parametrized by formula (2.62) with

$$a = 1.985 + 0.430 \ln E, \qquad b = 0.467 - 0.021 \ln E \ (E \text{ in GeV})$$

Measurements in lead and iron calorimeters sandwiched with scintillator, using electrons in the range $5\text{--}92 \, \text{GeV}/c$, have resulted in

$$a = 2.284 + 0.7136 \ln E, \qquad b = 0.5607 + 0.0093 \ln E \ (E \text{ in GeV})$$

This result (Bock, Hansl-Kozanecka and Shah 1981) gives faster tail attenuation than that of Longo and Sestili (1975), but the two agree in a/b (the position of maximum) at the common point $E = 5 \, \text{GeV}$. Compared with Rossi's approximation, both results correspond to a shorter attenuation length, and the shower maximum for $E = 5 \, \text{GeV}$ would translate into a critical energy ε of 4–5 MeV. Multiple similar parametrizations have been done for specific calorimeters; more generally accepted values do not seem available, despite the wealth of data recorded and published in recent years.

Hadronic showers. With a much wider range of physics phenomena involved, and with much larger fluctuations resulting, the description of hadronic showers either in approximate form, or in detail, is not nearly as advanced as for electromagnetic showers. Even the basic cross sections are not fully known down to the small energies relevant for shower development. Further, there is no simple scaling variable, both the radiation length X_0 and the nuclear absorption length λ (the mean free path for inelastic interactions) play a decisive role. Fabjan (1985) gives a few simple formulae for the average shower shape (other, similar expressions are also used)

Shower maximum at: $\qquad \lambda(0.7 + 0.2 \ln E) \ (E \text{ in GeV})$

Shower attenuation length: $\quad \lambda \sqrt[8]{E} \qquad\qquad (E \text{ in GeV}) \qquad\qquad (2.64)$

Full containment lengths $\quad$ maximum + 2.5 attenuation lengths

More attempts at a simple description are collected in Iwata (1980). One should also retain the result of extensive measurements in iron by Holder *et al.* (1978), resulting in

Shower median: $\quad \lambda(0.82 + 0.23 \ln E) \ (E \text{ in GeV})$

A more complete ansatz has led to the parametrization (Bock *et al.* 1981):

$$dE/dx = k_{norm}[wt^a e^{-bt} + (1-w)u^c e^{-du}] \qquad (2.65)$$

where t is the shower depth, measured from the shower origin, in radiation lengths (x/X_0), u is the same depth expressed in interaction lengths (x/λ), k_{norm} is a normalization constant, w and $1-w$ are the relative weights of the two curves, and a, b, c and d are shape parameters. They have been fitted, together with w, using data taken in a variety of materials and with pion beams at momenta from 10 to 400 GeV/c. The quoted results are

$$a = -0.384 + 0.318 \ln E \quad (E \text{ in GeV})$$

$$b = 0.220$$

$$c = a$$

$$d = 0.910 - 0.024 \ln E$$

$$w = 0.463$$

Normalization has to be done separately for the two terms, using two partial gamma functions as in Equation (2.63).

Within the limitations arising from the data used, this description seems applicable to quite different calorimeters. The parameter d is closely related to the hadronic shower attenuation length, and in agreement with numbers given in Iwata (1980) ($\lambda_{att} \approx \lambda$ at $E = 1$ GeV, and increasing with E). As for electromagnetic showers, test results published during the past few years are abundant; and a substantial refinement of this approximation seems possible with a limited effort.

2.5.4.2 Lateral shower shape

Electromagnetic showers Rossi's approximation B explicitly treats showers as linear objects without Coulomb scattering. Hence it does not make any predictions for their lateral extension, which is mainly caused by the multiple Coulomb scattering of electrons. General considerations (Amaldi 1981) lead to a transverse radius of the shower equal to the Molière radius R_m, which is defined as

$$R_m = X_0 E_s/\varepsilon \qquad (2.66)$$

with $E_s = 21$ MeV (from multiple scattering theory), and ε is again the critical energy as in Equation (2.61). The double Molière radius does, indeed, contain the shower rather fully. In practice, however, the assumption of scaling with R_m is a poor one. The lateral extension is determined by competing processes, it depends on depth (showers are cigar-like with a broadening developing slowly), and has a clear energy and material dependence. Further, lateral distributions are very non-Gaussian with at least two components clearly visible, a collimated central (core) and a broad peripheral (tail) component.

An approximation with two exponentials (*core* and *tail*) has been tried (Iwata 1980), where the relative amplitude and at least one attenuation length are functions of depth. Abshire *et al.* (1979) have parametrized test results in lead/scintillator by superimposing two Gaussians. They use electrons at $3\,\mathrm{GeV}/c$ in a detector with a longitudinal segmentation of length $2.6X_0$, $4.1X_0$, and $6.9X_0$. From their data they extract the ratio of the integrated area of the central shower part over the same for the tail, and the widths of these shower parts. For the first (last) segment they obtain the ratio 100 (5). The central part's width is given by $\sigma_{\mathrm{core}} = 0.5X_0$ $(1.0X_0)$, σ_{tail} is of the order of $5X_0$ (in both the first and the last segment). Bugge (1986) has also fitted data with two exponentials, using electron showers at $E = 4\,\mathrm{GeV}$ integrated over $4.7X_0$. If we compare the two exponential curves in terms of their attenuation factor (slope) and their relative contribution (integral), then the core part has an attenuation nine times higher than the tail part, and contains over 80% of the energy.

Also Acosto *et al.* (1992) have used double exponentials and Breit–Wigner distributions to fit experimental data, but all results seem difficult to generalize, being derived for specific calorimeters.

We should note, in particular, that lateral results are strongly dependent on the longitudinal homogeneity of a calorimeter; any gaps will result in a widening of the lateral profile (see Fig. 2.23 for an illustration).

Hadronic showers.   The lateral development of hadronic showers is dominated by the average transverse momentum in hadronic interactions, which is of the order of $300\,\mathrm{MeV}/c$. The variety of interactions and their dependence on material parameters has, however, not permitted a global description of the lateral shower shape as function of depth. Some measurements obtained in test beams with calorimeter set-ups have been published in Holder *et al.* (1978) and Diddens *et al.* (1980). Holder *et al.* (1978) show that a simple Gaussian or exponential will not approximate the measured lateral shower shapes. The hadronic shower width increases linearly with depth, and at least two Gaussian-like curves have to be superimposed to approximate the measured integrated shower profile. If we characterize the curves by their Full Width at Half Maximum ('FWHM', the Cauchy width), then $\mathrm{FWHM} = 0.28\,\lambda$ and $\mathrm{FWHM} \geqslant \lambda$ are characteristic for the two components. Further, the large difference in width between full containment (defined by less than one minimum ionizing particle remaining) and 95% containment is also measured. The depth dependence of the 95% cone can be expressed by its radius

$$R_{95\%} = \lambda(0.29 + 0.26u) \quad \text{at } E = 50\,\text{GeV} \qquad (2.67)$$

and by

$$R_{95\%} = \lambda(0.29 + 0.17u) \quad \text{at } E = 140\,\text{GeV} \qquad (2.68)$$

Here u is the depth in units of λ. The increase with rising energy of the relative contribution from the core part is apparent. Diddens *et al.* (1980) and Jonker *et al.* (1982) show the significant difference in lateral spread between electromagnetic and hadronic showers. In their experiments this difference has been used as prime information for π/e^- separation (see Fig. 2.23, and the case studies in Subsection 2.5.6).

Muraki *et al.* (1985) have reported lateral shower shapes obtained at high energy (300 GeV), when exposing stacks of X-ray film interspersed with iron and lead as absorbers, to a proton beam. They propose a multiparameter formula to fit their data, which underlines the complexity of the parametrization approach. Despite such rather clear evidence, the approximations used in practical applications often do not go beyond the simple Gaussian; see Akesson *et al.* (1985).

2.5.5 Shower simulation

We have mentioned above that shower simulation by Monte Carlo methods, in particular the simulation of hadronic showers, has been a formidable tool in designing calorimeters, and that the completeness of some of the resulting programs is at a very remarkable level. We refer here to programs that embody the complete understanding of the showering process, and have been tuned until all relevant experimental results can be reproduced. Such programs are inherently ideal guzzlers of computer time, and many efforts have been made in the past to approximate locally what the full programs generate in detail. In their crudest form, the approximations do not exceed the formulae given in Subsection 2.5.4. Indeed, many gross detector studies can usefully be done at the level of average showers.

When it comes to full simulation, it becomes necessary to follow all particles to rather small energies; for hadrons, phenomenological approximations for intra-nuclear cascades and intermediate-energy processes have to be made (Ferrari and Sala 1998), and also electromagnetic simulation results are sensitive to multiple low-energy cut-off parameters. The number of particles in a shower is very large, particularly at high energies, so that substantial computing resources will be needed. Simulation being a central tool in optimizing calorimeters, the validity of results and questions of efficiency on computers had led to multiple publications and comparisons (e.g. Fesefeldt 1990). Currently the most mature *full shower simulation* programs, due to their substantial investments in tuning to known physics, are EGS (Nelson *et al.* 1985) for electromagnetic showers, GEISHA (Fesefeldt 1985, Fesefeldt, Hamacher and Schug 1990) and FLUKA (Fassò *et al.* 1994) for hadronic showers.

The tuning of the simulation parameters, in particular cut-offs and integration step sizes, is delicate and cannot easily be generalized, depending as it does on the goal of the simulation; some codes (e.g. FLUKA) use mathematical methods (like importance sampling), which alleviate this problem. In major simulation projects at high energies one has also resorted to relatively high cut-off parameters, using for further shower development randomly selected showers at lower energies, pre-computed in full simulation and stored in 'shower libraries' (Graf 1990). This strategy can save substantial factors in computer time.

These codes are kept in many of the computer centre program libraries of the high-energy physics community, although not always in identical versions. Their use is relatively straightforward, but a high level of expertise is required for parameter tuning and in interpreting results. The use of multiple sensitive cutoff parameters in the infrared (at low energies), requires application-dependent decisions. Also, continuous improvements are being made to these programs, and their evaluation and possibly their introduction by non-expert users is not always trivial. The completeness of all cross sections, particularly at critical low energies, is one area of ongoing work, which the authors of these programs pursue in a painstaking way the merits of which cannot be overestimated.

Another area where specialized experience and expertise are required is detector description, both in structure and content. FLUKA (available from the CERN program library), EGS and GEISHA have been embedded in many local codes that introduce detector details, e.g. in GEANT. Calorimeters come in ever newer shapes and often contain materials whose behaviour is not well understood, i.e. material compounds (molecules or mixtures of materials). Scintillators of various composition or BGO are prominent (and understood) examples. Fesefeldt (1985) discusses ways of calculating the relevant numbers for such compounds.

Several nonauthor comparisons of simulation results with existing data have been made, and some of them are found in the literature, e.g. Fesefeldt (1990) or Ferrari and Sala (1994). Most future improvements are likely to happen *in the framework* of the large existing simulation programs, simply owing to the fact that the major investments in acquiring experience with the applications (and limits) of a system must be maintained, which makes it hard for a radically new approach to succeed.

At very high energies, i.e. for the multi-TeV accelerators that are now only a vision of the future, these codes may not be applicable, because they would be too detailed and prohibitive in computer time. Currently, a number of programs exist for very high energies, written in connection with studies for future accelerators. From various comparisons one can conclude that these programs, used all over the world in accelerator development groups, are in agreement in the relevant results, but they each have their own specific advantages and shortcomings.

When the efficiency of shower simulation is of more concern than the shower detail, general-purpose programs are not what come to mind. *Empirical shower shape parametrizations* have been developed; these are considered satisfactory in many instances where the detailed understanding of the calorimeter is not demanded. Parametrizations as given in Subsection 2.5.4 give an overall shower shape and distribution, but largely fail to provide the fluctuations of showers (Bock *et al.* 1981). They are applicable when a global detector optimization is necessary, the effect of dead spaces has to be studied, or grossly different physics choices are to be compared. In such applications, parametric approximations permit computer time gains by orders of magnitude.

An equally important saving in computer time has been achieved by *pregenerating many individual showers* in all details, and storing the details on a file. Random access to the showers on the file will then result, after some tuning, in satisfactory to excellent simulation, limited mostly by the number of showers pregenerated. The method requires similar starting conditions for many showers. It therefore has been used most successfully, and with major computer time savings, for the shower details after a certain cutoff energy has been reached in individual track energies. Details were published in Longo and Luminari (1985). Also the combination of single-particle simulation (e.g. for high-energy particles) with parametric approximations (at lower energies) has been used. This problem is discussed in some detail in Fesefeldt (1985).

2.5.6 Examples of calorimeter algorithms

The preceding discussions should have established several facts. Calorimeters rely on *stochastic processes* much more than other types of detector. Their precision limits are defined less by the luxury of instrumentation than by the fluctuations of the underlying physical processes. Calorimeters are built and instrumented with a *specific physics problem* in mind. A wide and ingenious variety of solutions have successfully been implemented. Calorimeters are *empirical* devices whose understanding depends largely on test measurements. Calibration procedures with test beams or events of known particle composition play a dominant role. Simulation by Monte Carlo procedures is tuned to reproduce such calibration data correctly, and may then be used to optimize and interpolate.

A consequence of these facts seems to be that a general method for reconstructing particle or jet energies, for identifying particles, and for separating showers of different origin, on the basis of calorimetric data, cannot be given. Our approach, in this subsection, will be to discuss a few calorimeter applications with different characteristics in some detail, suggesting that they are representative for many others. Undoubtedly, calorimetry is a field where ingenious and creative approaches are possible and even necessary, and surprising future optimizations are likely to be found, that are not covered by this short review.

The following examples are taken from calorimeters that have successfully solved the intended physics problem. Our description of the experimental details will necessarily have to be condensed, but most cases are properly documented in the literature for anyone who is interested in more detail. The categories of applications are the following.

Spatial energy flow measurements. This task is the one most directly performed by calorimeters in the absence of any external information. The physical space of interest is subdivided into cells of the desired granularity, with the intercell gaps minimized. Every cell measures an energy value. By suitably weighted sums over energies measured in the cells, one can form total event energy, total transverse energy, asymmetries like left–right or up–down, or escaping (unmeasurable) energy, and use these also in real time, for triggering. Inhomogeneities in the energy flow (i.e. local clustering not explained by the physics model) have also been shown using simple cell energy distributions.

π^0/γ discrimination. Calorimeters attempting to discriminate clearly between the shower left by a single photon and showers originating from two γs from the electromagnetic decay of a π^0, typically need to put the accent on position resolution. They are also in need of vetoing against the presence of a charged particle, i.e. of calorimeter-external information from scintillators or tracking devices. This discrimination is comparatively easy at low energies (< 1–$2\,\text{GeV}$), i.e. for large-angle π^0 decay, but at higher energies, it is necessary to use an optimally discriminating 'feature space' (see Subsections 2.2.4 and 2.4.2), typically found by using training samples of both event types.

Electron/charged hadron identification. This application of calorimeters requires an independent signal from a tracking chamber or equivalent device, which establishes the existence of an isolated charged track. Hadronic showers have large enough fluctuations to mimic, with some finite probability, electromagnetic shower shapes at all energies. As a minimal background of this type, one may visualize the charge exchange reaction $\pi^- + p \to \pi^0 + n$, when this occurs near the beginning of the calorimeter and with small energy transfer to the neutron. The hadronic π^- shower will then indeed be a purely electromagnetic shower caused by the π^0. The tuning of charged e/π discrimination will, therefore, be a delicate one, and will never produce correct decisions for all individual events.

Jet finding. Jets are quark- or gluon-induced compounds of mostly hadronic particles, both neutral and charged, that follow near-by trajectories. Their origins are the partons that emanate from an interaction and 'hadronize' at medium to long distances from the interaction point. Although no firm theory exists as yet for the *hadronization* of partons, phenomenological descriptions exist, and the existence of jets and their most relevant properties have been clearly established and shown to be in agreement with the theory describing the *production* mechanisms of partons. Jets were first observed by e^+e^- experiments at DESY, and described early by global variables (Brandt and Dahmen 1979). Jets originating from hadronic collisions were found at the CERN SPS $\bar{p}p$ collider and at the CERN ISR. For detailed results, refer to Banner *et al.* (1982), Arnison *et al.* (1983b), Akesson *et al.* (1984), and Arnison *et al.* (1986).

Calorimeters are the most prominent devices for detecting jets. Jet algorithms typically define windows in suitable variables and use a maximization procedure to find energy clusters inside such windows. In other words, the problem is that of object classification in the sense of Subsection 2.2.3, with the metric adapted to the measurement variables a calorimeter can provide. Training samples (see Subsection 2.2.2) and physics modelling have to provide the parameters that are used to assign a 'jet likelihood' to clusters so identified, and to guard against random event fluctuations.

We will now discuss examples in more detail.

2.5.6.1 Global flow of energy and missing energy. Energy flow is a natural byproduct of any calorimeter that covers geometrical space with minimal leakage. Individual cells cover a small volume element often delimited by two angles θ and ϕ, and by depth, i.e. distance from the interaction point. For the definition of these variables, see Fig. 2.24. Often, an attempt is made to construct cells bounded such that they cover equal $\Delta\eta$ and $\Delta\phi$, η being the pseudo-rapidity directly related to θ (by $\eta = -\log\tan(\theta/2)$), or at least to group cells bounded by other constraints such that towers allow us to sum cells with approximate bounds in η and ϕ.

All energy-flow variables need to calculate the calibrated energy E_i for each cell i, even when using it as part of the trigger. Usually, each cell energy is multiplied with a factor $|\sin\theta_i|$, to obtain transverse energy E_T, this being the only variable accessible to analysis in colliding hadron beam experiments, owing to the fluctuating longitudinal energy escaping close to the beam directions. In e^+e^- collisions this is not true to the same extent,

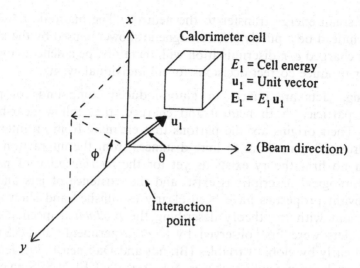

Fig. 2.24. The coordinate system and energy vector definition as used in calorimetry.

and fixed-target experiments do not usually attach the same importance to energy flow or missing transverse energy.

Frequently, energies are thresholded before entering further analysis, viz. low energy or E_T values at the cell level, below a threshold, are set to zero or to an average value, a procedure that avoids errors due to photomultipliers and other sources of noise, and which might accumulate. Thresholding must be studied individually for each detector (including its readout chain) and in function of the physics objective pursued.

Energy flow can then be looked at in function of ϕ, i.e. summing over all θ values and depth; this is how in many experiments the dominance of two jets back-to-back could be made visible in hadronic collisions (e.g. Banner *et al.* 1982). The corresponding analysis for hadronic final states in e^+e^- collisions can look at energy flow in both η and ϕ, as longitudinal energy is well conserved; this has given rise to global jet variables like sphericity or thrust (for more details on jet variables, see Adloff *et al.* (1997), or Bock and Vasilescu (1998)).

Separate sums of two independent transverse energy components, e.g. $E_T \sin \phi$ and $E_T \cos \phi$ are used to detect energy non-conservation in the transverse plane; values clearly exceeding the errors for these sums can be indicative for one or more escaping neutrinos. These transverse energy sums, the missing energy vector, need corrections for observed muons (which deposit only minimum ionizing information in calorimeters), and in most detectors need to be compared against errors that depend on their direction, as the quality of coverage is not isotropic. An early example

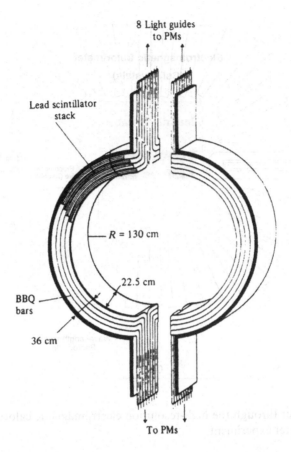

Fig. 2.25. A single electromagnetic calorimeter module ('gondola') as used in the barrel part of the UA1 experiment. The total number of modules is 48.

of this technique is the discovery of the $W^{\pm}$, e.g. in UA1 (Arnison *et al.* 1983a). To demonstrate how diverse calorimeter cells can be used in such analysis, Fig. 2.25 shows two of the 48 crescent-shaped elements of the large-angle electromagnetic calorimeter of that experiment. The calorimeter stack was made of thin sheets of lead and scintillator; each scintillator transmitted light into the wavelength shifter bars, arranged on both sides of the element, summing over four samples in depth, and transmitting eventually to photomultipliers. Missing energy, of course, needed also the adding of electromagnetic calorimeter cells of quite different shape, in the area of larger $|\eta|$, and of the corresponding parts of the hadronic calorimeter.

Total summation of $|E_{\mathrm{T}}|$ over all cells results in a total transverse energy, a quantity that has also been used as an indication for potentially interesting physics.

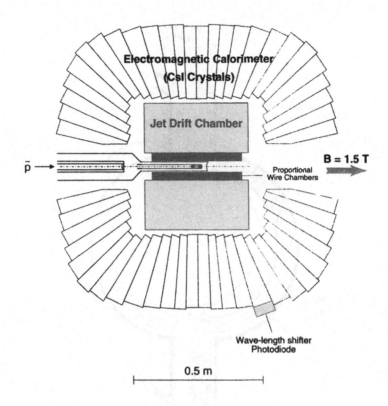

Fig. 2.26. A cut through the high-resolution electromagnetic calorimeter used in the Crystal Barrel experiment.

2.5.6.2 Photon and π^0 reconstruction in the Crystal Barrel experiment.

The Crystal Barrel experiment operated at the Low Energy Antiproton Ring (LEAR) at CERN, from 1989 to the end of LEAR operation in 1996. Most of the data were taken with antiprotons stopped in a liquid-hydrogen target. The main goal of the experiment was the thorough (viz. high statistics) exploration of final states with high multiplicity of neutrals, i.e. π^0, η, and ω, all resulting in signatures containing multiple (from four to eight) γs. Some 10^8 annihilations were collected and analysed; the results published so far have substantially advanced the understanding of scalar and isoscalar mesons, amongst them an isoscalar f_0 (1500), which is most likely explained as the ground-state quarkless glueball predicted by QCD, mixing with other $q\bar{q}$ states lying near in mass. The Crystal Barrel physics has been reviewed by Landua (1996a, b) and Amsler (1998).

The Crystal Barrel detector (Amsler 1998) consists primarily of an assembly of Tl-doped CsI crystals, each of them 30 cm deep (16.1 radiation lengths), as shown in Fig. 2.26. They cover in a regular pattern the area of $12° \leqslant \theta \leqslant 168°$; cells have constant size in $\Delta\theta$, as the rapidity variable does not have the same importance as at high collision energy. Cells cover $6° \times 6°$ in $\Delta\theta \times \Delta\phi$ for $30° \leqslant \theta \leqslant 150°$, and $6° \times 12°$ for the remaining angles; the total number of crystals is 1380.

Photon-induced electromagnetic showers spread over multiple crystals, depending on the incident energy. Up to 20 cells may be touched by high-energy photons. Photons are identified by clusters of contiguous cells with energies exceeding 1 MeV; local maxima are then found by the criterion that all eight nearest neighbours have lower energy (and the maximum cell exceeds a threshold). If only one maximum is in a cluster, this is assumed to represent a single photon, with an energy corresponding to the sum over all cells in the cluster, and a direction given by the centre of gravity, using cell energy as a weight. If, instead, several maxima are present, each of them is assumed to be caused by a photon, and the cluster energy is shared.

2.5.6.3 e/π separation in the Mark III electromagnetic shower counter. This calorimeter at the SPEAR e^+e^- storage ring at SLAC covered the entire solid angle, with the exception of a cone of half-angle 11° around the beam axes. Its central (barrel) part is cylindrical, the end caps are flat, and both follow a very similar design with 24 layers of rectangular proportional tubes alternating with 23 sheets of a lead-antimony alloy ($0.5X_0$ each). All tubes are parallel along the beam for the barrel, and vertical in the end cap. The Mark III calorimeter has been described by Toki *et al.* (1984).

Shower coordinates along the wire are obtained by charge division. The calorimeter is very highly segmented, resulting in excellent position resolution, ±7 mrad/±7 mm perpendicular to the tube, about $\pm1\%$ wire length (±44 mm) along the wire. The electromagnetic energy resolution is comparatively modest ($\Delta E/E \approx 0.17/\sqrt{E}$), owing to the small sampling fraction in the proportional tubes. The energy variable used for photons produced in connection with unstable particles in the later analysis is $\sqrt{E}$, which is preferred to E since it has a more Gaussian-like behaviour. Efficiencies are excellent even at photon energies below 1000 MeV, mostly as a result of the small amount of matter ($0.4X_0$) in front of the first layer of proportional tubes.

Recognition of showers starts in the high-quality projection (e.g. ϕ for the barrel) alone, with clusters defined as at least two cells with hits separated by no more than two cells. Subsequently, the lower-quality projection (e.g. along the beam) is used to check for consistency or for splitting signals. Finally, hits are examined in depth (e.g. radial variable r), and may be split if early hits show a clear two-prong structure (indicative of two close photons from fast π^0 decaying symmetrically). The relevant position information is extracted by fitting a straight line to the detailed hits. The photon entry point is defined to be that of the fitted line at calorimeter entry surface (Richman 1986).

For π/e separation, an algorithm was designed with great care, combining the shower detector information with a time-of-flight (TOF) counter probability for e or π, and with tracking (momentum) information. Eight variables were chosen as the most efficient test statistics to separate π from e, after systematically comparing more than 50 variables. The decision is taken by successive cuts in a *decision tree*, using one of the variables at each node. In short, the test statistics retained are these: total shower energy, longitudinal shower barycentre, shower width, layer-wise energy correlation (using an experimentally determined dispersion matrix), energy/momentum ratio, energy in the early layers (two groups), and TOF probability. The precise tree and the cuts used are energy-dependent; altogether seven different decision trees were defined. Separate tuning was performed for most efficient electron and pion selection. The final performance of both is approximately the same: 20% loss of signal, 4% of background as contamination.

In optimizing the decision tree, a method called *recursive partitioning* (Friedman 1977; Breiman *et al.* 1984) was used. This is based on the notions of sample *purity* and *cost* of misclassification. These two terms describe variables equivalent to *contamination* and *loss*, which we will use in our later discussion. A training sample of events whose classification is known is subjected to a large number of decisions (start with a very large tree, with few events in each decision class). An algorithm then removes more and more decision nodes, attempting to minimize the cost and maintain maximum purity (Coffman 1987).

2.5.6.4 Jet finding as part of the top quark search in the D0 experiment. In March 1995, the two experiments at the Fermilabs Tevatron, CDF and D0, reported evidence for top–antitop quark pair creation in proton–antiproton collisions. Since then, the analysis has been refined, and measured mass and width values constantly improved; for a recent review, see Wimpenny and Winer (1996). The most convincing signature for t ($\bar{t}$) is the decay channel $t \rightarrow W^+ b \rightarrow l^+ \nu b$ (and $\bar{t} \rightarrow W^- \bar{b} \rightarrow l^- \nu \bar{b}$), with l a lepton, but decays of the W into jets have a sixfold branching ratio; jets, therefore, play a major role in finding the top.

Our example here is the algorithm that defines jets, in the D0 experiment. We should note that jets are not well-defined objects even theoretically: next-to-leading order (NLO) perturbative calculations in QCD can result in two partons with substantial separation in $\eta - \phi$. The decision as to whether they should be clustered, viz. considered a single jet, depends on

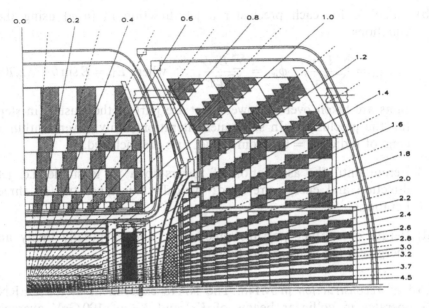

Fig. 2.27. Schematic view of a portion of the D0 calorimeters showing the transverse and longitudinal segmentation pattern. The shading pattern indicates the distinct cells combined for signal readout. The rays indicate the pseudorapidity intervals seen from the centre of the detector.

their distance R defined by

$$R = \sqrt{(\Delta\eta^2 + \Delta\phi^2)} \tag{2.69}$$

If clustered, the vector sum determines the theoretical jet axes.

The experimental jet definition is more involved, as the processes of parton radiation beyond NLO and hadronization contribute to widen the jet, which subsequently generates observable hadron showers. Abbot *et al.* (1997) discuss the problem of comparing the experimentally found and theoretically defined jets in detail.

Figure 2.27 shows a cut through the D0 calorimeter, in particular its segmentation patterns. We deal with a pseudoprojective calorimeter, whose cells are oriented according to mechanical constraints, but sized such that their centres line up along the dominant shower direction of constant (η, ϕ). All cells are uranium/liquid argon, with a dominant cell size of $\Delta\eta \times \Delta\phi = 0.1 \times 0.1$ (cells in the electromagnetic part are halved in both directions, at shower maximum). The fixed-cone algorithm of experimental jet definition relies heavily on the quantity R defined above, and proceeds as follows.

   (a) Sum energies in depth into towers, take the highest-energy tower as
       seed, form a precluster by nearest-neighbour algorithm.

(b) Calculate for each precluster a jet direction in (η, ϕ) using the equations

$$\eta_{\text{jet}} = \frac{\sum E_T \eta}{\sum E_T}, \quad \phi_{\text{jet}} = \frac{\sum E_T \phi}{\sum E_T}, \quad \text{where } E_T = E \sin \theta \quad (2.70)$$

sums are taken over all towers that are part of the cluster (in step (a), the precluster). In subsequent iterations, only towers within a cone of radius $R = 0.7$ from the jet axis are included.

(c) Iterate until the association of towers to a jet is stable, i.e. jet direction and energy are fixed; typically this takes two or three iterations.

(d) Iterate over other possible preclusters, and eventually apply an energy cut (E_T) on the jets found.

2.5.6.5 π^0 reconstruction in NA48. The NA48 experiment at the CERN SPS operates in collinear beams of K_L and K_S of 100 GeV average energy (70 GeV $\leqslant E_K \leqslant$ 170 GeV), produced by a primary proton beam of 450 GeV on beryllium targets. The aim of the experiment is the measurement of the mixing parameter ratio $(\varepsilon'/\varepsilon)$ by observing rates of $K \rightarrow \pi^0\pi^0$ and $K \rightarrow \pi^+\pi^-$, discriminatinig at a 10^{-4} level against three-pion decays, the dominant K_L decay mode.

The role of the electromagnetic calorimeter is, in particular, to discriminate the $2\pi^0$ decay mode from the $3\pi^0$ background with undetected γs. The calorimeter is a liquid krypton ionization chamber, $27X_0$ long, with a readout in a tower structure, giving excellent energy and lateral position resolution (but no longitudinal information). The volume is subdivided, over its full length, by copper/beryllium electrodes, 18 mm wide and 40 µm thick, and with an anode/cathode spacing of 10 mm (the anode being used on both sides). Electrodes are shaped in a zig-zag pattern (with small bending, 50 mrad), so that shower centres and anodes cannot accidentally be aligned over a long distance, which results in distorted signals. Effectively, this results in 13 212 cells of roughly 2 cm × 2 cm over the entire shower length, with a slight widening in track direction by 10 mrad, to point towards the target, 100 m upstream (see Ceccucci (1996) and Barr *et al.* (1996) for details). The structure is 1.27 m long and 2.4 m in outer diameter (see Fig. 2.28).

In shower reconstruction, the first step is to find energy maxima in a 3×3 cell neighbourhood; these seeds are analysed in order of decreasing energies of maximum cells. The algorithm then finds a shower centre-of-gravity by analysing the eight cells around the maximum-energy cell. If a dead cell is in this neighbourhood, a 5×5 region is used to interpolate

Fig. 2.28. A view of the NA48 liquid krypton calorimeter, in front of its cryostat. One can see the support structure and the spacer plates that hold the electric ribbons in place.

the missing cell. The centre-of-gravity position is then used to correct energies, slightly dependent on the distance of the impact point from the anode. As this changes energies, the algorithm is iterative, with rapid convergence. The effective positional resolution is $\pm 4.2/\sqrt{E} \oplus 0.6$ mm, the energy resolution is given by

$$\frac{\sigma_E}{E} = 0.035/\sqrt{E} \oplus 0.04/E \oplus 0.0042 \qquad (2.71)$$

where $\oplus$ stand for quadratic addition and E is taken in GeV.

If two γ-showers overlap in their cells, an energy-sharing algorithm is invoked that finds the two centres and evaluates the two energies, using an energy expected for the cell from the total shower energies and Monte-Carlo-based profiles. As pulses from different events may also overlap in time, a cell sharing in time may also be invoked. Interference by energy depositions from charged particles can also be estimated, as these particles are identified by the tracking devices. The final combining of γs into π^0s selects the pairs closest to the π^0 mass, starting with all possible combinations; it is successful in nearly all cases owing to the excellent mass resolution ($\approx \pm 1$ MeV/c^2).

2.6 Identifying particles in ring-imaging Cherenkov counters

The technique to assign masses to charged particles of equal momentum by using the Cherenkov radiation is a well-established technique in particle physics. It was used for separating protons, kaons and pions in beam lines as well as to identify decay products (e.g. to distinguish between electrons and pions in K_0 decays). Threshold as well as differential Cherenkov counters have been used.

With increasing energies of the particles to be identified, several problems occurred: in order to overcome the Poisson fluctuations, these counters became longer and longer (several tens of metres), and with β approaching 1 the refractive index of gases also had to be increased by high gas pressure. Also aerogel detectors were in use in order to fill the gap of the refractive index between gases ($n > 1.2$) and light gases ($n < 1.002$) at atmospheric pressure and room temperature.

2.6.1 The RICH technique

A new type of Cherenkov counter, the ring-imaging Cherenkov (RICH) counter, is now well understood, and several experiments profit from RICHs in fully analysed event topologies.

The RICH technique was pioneered by Séguinot and Ypsilantis (1977). The idea is a simple one in principle, leading, however, to devices of substantial technical refinement. Somewhat like calorimeters, RICH devices allow quite different implementations.

An overview of current applications of the RICH technique can be found in RICH (1996). It contains conference contributions on results of operational detectors to information on detector research and development. Analysis techniques are discussed in a separate chapter.

Charged particles that enter a medium with a speed greater than that of light in this medium, undergo deceleration due to polarization effects. Polarized molecules of the medium return to their ground state, emitting the characteristic Cherenkov radiation *at a single emission angle* with respect to the incident particle. The relation between the velocity of the particle v, the velocity of light in this medium v_0, the medium's refractive index n, and the Cherenkov emission angle δ, is given by

$$v_0 = c/n$$

$$\cos \delta = v_0/v = c/(vn) = 1/(\beta n) \tag{2.72}$$

where c is the speed of light in vacuum.

If $v > v_0$, i.e. if the velocity β is above the 'Cherenkov threshold', measurement of δ becomes possible and corresponds to measuring β (or v).

If a measurement of momentum p is available for the same track, e.g. from a tracking chamber, the knowledge of both p and β allows us to limit the possible mass assignments to the particle; for a certain range of momenta, this means one can identify the particle unambiguously.

The generated Cherenkov photons all propagate at angle δ with respect to the original particle direction, but with random 'azimuth' around the incident particle; hence for a given point of origin all photons are found on the surface of a cone with a half opening angle δ. RICH devices attempt optimal measurement of δ and therefore of β, by directing the photons into a *photosensitive chamber*. This allows the reconstruction of the cone, if the origin of the photons is also known. Several problems have to be solved or understood.

Photons are emitted with varying wavelength, and the photosensitive chambers as well as the light-transmitting surfaces must be optimized to remain sensitive to the broadest possible wavelength spectrum. This is typically in the ultraviolet region (wavelengths 150–250 nm, or 8.3–5.0 eV of energy). Light output and transmission are major problems. Different wavelengths also introduce 'chromatic aberrations': radiating and transmitting materials have a refractive index that is dependent on the wavelength. A finite wavelength width therefore results in a smearing of the signal, and a resolution limit in measuring β. Further sources of statistical errors result from the multiple Coulomb scattering of the radiating particle in the radiator, from its curvature if a magnetic field is present, from the finite resolution of the photosensitive chamber, and from aberrations of the focussing mirror, if one is used.

The energy loss due to ionization of the charged particle also traversing the chamber is orders of magnitude higher than the energy lost due to Cherenkov radiation. Typically, high quantum efficiencies are needed in order to be able to reconstruct a ring image from the few photons imping-ing on the chamber. The simultaneous existence of a direct track signal in the same photosensitive chamber can be an additional perturbance.

Cherenkov radiation occurs over a certain depth of a radiator, chosen for optimal identification. If this depth is more than a negligible layer, the origin of each photon will not be known, and a δ measurement will not be directly possible. Liquid or solid radiators are typically contained in layers thin enough to define the cone tip with a good approximation, and a cone development zone allows photons to move far enough from the particle axis for a useful measurement. Gaseous radiators, on the other hand, need substantial depth, and large-surface spherical mirrors will be needed to focus all parallel photons onto a single point, thus generating a unique circular image for all particles emitted under the angle δ over a full depth (see Ypsilantis (1981) for details). An illustration is given in Fig. 2.29. If the detection layer and the mirror surface are both concentric

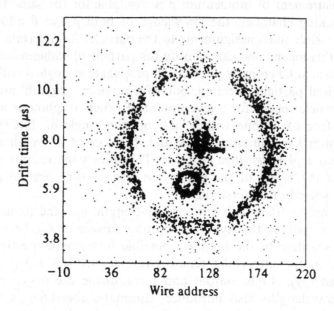

Fig. 2.29. A RICH image, with a centred (unfocussed or 'proximity focussed') ring around the track impact point, and an indirect (mirror focussed) ring slightly off centre. Superposition of multiple tracks with normal incidence.

with normal incidence of the radiating track, the circular image will be centred about the track's impact point in the detection layer.

The magnetic field or the basic design may cause track impact point and the Cherenkov rings to have different centres; an example is shown in Figs. 2.29 and 2.30, both taken from the DELPHI experiment (described in detail in Aarnio *et al.* (1991), and presented also below). A single photosensitive layer provides RICH images from different radiators, one a liquid radiator producing a direct image ('unfocussed' or 'proximity focussed'), the other a gas radiator using spherical mirrors for imaging.

Photosensitive chambers may be proportional or drift chambers, depending on the required signal speed and precision. A small ($< 10^{-3}$) admixture of triethylamine (TEA) or tetra(dimethylamine)ethylene (TMAE) to ordinary chamber gas mixtures will provide the photosensitive effect, but needs careful monitoring of gas cleanliness, concentration, and temperature.

A basic difficulty of photosensitive chambers lies in the fact that weak signals (single electrons) have to be dealt with, and the number of expected Cherenkov photons is small; on the other hand, photons liberated in the avalanche near the anode may cause secondary discharges, deteriorating the positional precision: the gain should thus be limited.

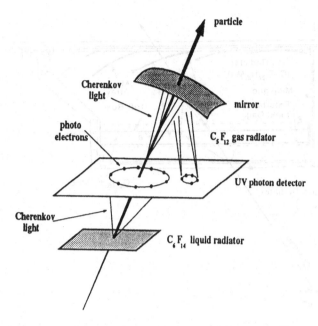

Fig. 2.30. The principle of a RICH detector with two radiators and a single photon detector.

2.6.2 Examples for analysis using RICH detectors

2.6.2.1 Particle identification in DELPHI and SLD. Both the DELPHI detector at CERN's LEP collider and the SLD experiment at the SLAC Linear Collider contain multiple RICH detectors for covering a large part of the available phase space by particle identification over most of the momentum range. They are based on photon detectors that allow reconstruction of Cherenkov rings from both liquid and gaseous fluorocarbon radiators. The DELPHI detector is described in detail in Aarnio *et al.* (1991), the SLD RICH (named CRID) in Abe *et al.* (1995). The cylindrical barrel parts are made in two halves of 175–200 cm length each, made of 24 (DELPHI) or 20 (SLD) sectors; their function is schematically shown in Fig. 2.30. Each sector contains a liquid radiator (1 cm of C_6F_{14} with a quartz window at the outer radius, about 135 cm from the beam axis), drift tubes about 12 cm away in radius containing a TMAE-doped CH_4/C_2H_6 mixture (75/25), with MWPCs at one end (in fact, the chambers act as time projection chambers), and a volume of gas radiator (C_5F_{12}, 40 cm deep in radius, with parabolic focussing mirrors, six or five in each sector). The forward RICH follows very similar principles, viz. construction in sectors, liquid and gas radiators, the latter with parabolic mirrors, a drift box with TMAE-doped gas for photon detection; differences with respect to the barrel RICH are needed

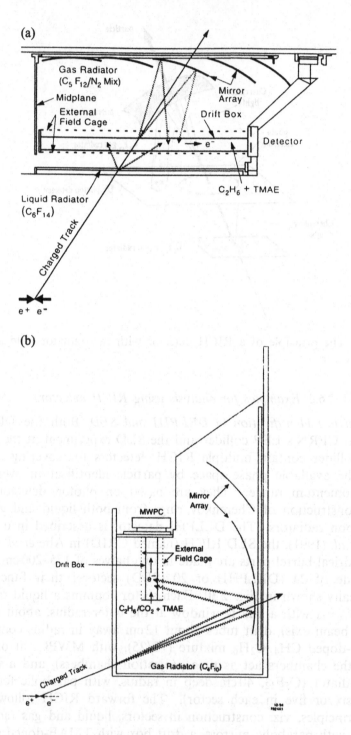

(a)

Gas Radiator
($C_5 F_{12}/N_2$ Mix)

Midplane

External
Field Cage

Mirror
Array

Drift Box

e⁻

Detector

C_2H_6 + TMAE

Liquid Radiator
(C_6F_{14})

Charged Track

e⁺ e⁻

(b)

Mirror
Array

MWPC

External
Field Cage

Drift Box

e⁻

C_2H_6/CO_2 + TMAE

Gas Radiator (C_4F_{10})

Charged Track

e⁺ e⁻

Fig. 2.31. For caption see facing page.

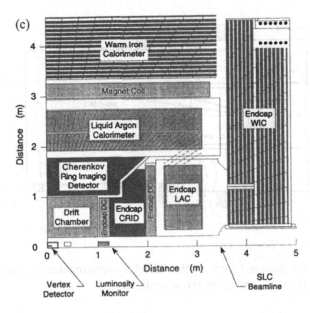

Fig. 2.31 View of the SLD RICH detector ('CRID'). (a) Barrel section. (b) End-cap section. (c) The RICH in the SLD detector. From Pavel (1996).

in geometry because of the different angle of incidence for particles, and the fact that the drift field is at a right angle with the solenoidal field of the detector. The SLD detector is shown in Fig. 2.31, both barrel and endcap. In general, these RICH detectors allow particle identification in the range of 0.7–3 GeV/c (liquid) and 2.5–25 (gas radiator). Electron/pion discrimination works up to some 6 GeV/c.

The analysis of data (Abreu *et al.* (1996) for DELPHI, Abe *et al.* (1996) for SLD) typically starts with a track measured in both momentum and position by the tracking detectors, whose variables are extrapolated to the RICH; ten detected photoelectrons are measured per track on average, in a RICH image; their measured positions are converted individually into Cherenkov angles, and the statistical distribution of angles is compared to that expected for the measured momentum and different hypotheses of particle mass, viz. e, μ, π, K or p. The uncertainties due to the radiator thickness and the magnetic field are discussed in detail by Pavel (1996).

Algorithms are different depending on the physics goal: the accent may be on high rejection (e.g. of $\pi^{\pm}$) or high efficiency. The algorithms, in particular, take a different approach to background rejection (background in the distribution of Cherenkov angles is caused by photoelectrons from nearby tracks, or by non-photon hits): the background may be assumed to be a smooth curve, and the signal (a Cherenkov angle) determined by maximizing a likelihood function. SLD have spent substantial effort on

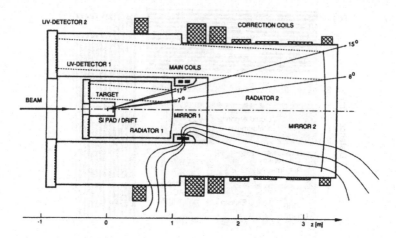

Fig. 2.32. Cross-sectional view of the CERES spectrometer. In the lower part, magnetic field lines are shown in the acceptance.

the definition of likelihood ratios to optimize particle identification, using large samples of Monte Carlo events. In DELPHI, background is also dealt with by grouping hits into clusters, using weights according to quality criteria, which take into account measurement errors and ambiguities with nearby tracks. Eventually, the highest-quality cluster alone is used to measure the average Cherenkov angle. In some cases of physics analysis, only a statistical analysis is wanted: the question to be answered is how many particles of different types are present. In this case, no individual tagging is required, only a continuous estimator of Cherenkov angle is needed.

2.6.2.2 Fast two-photon identification in CERES and HADES. The CERES (NA45) experiment at the CERN SPS is critically dependent on its two RICH detectors: its goal is to identify prompt electron pairs as robust (free of final-state interaction) signatures in proton and heavy-ion collisions, and this has to be done at a fast-trigger level (prompt lepton pairs are at a 10^{-5} signal level compared to trivial e^+e^- pairs from γ conversions or π^0 Dalitz pairs). A sketch of the detector is shown in Fig. 2.32; two RICH chambers with gaseous radiator produce typically 12 photons for each ring, which are detected via an array of pads in a multiwire proportional chamber (preceded by multistep counters: two stages of parallel-plate avalanche counters). There are some 50 000 pads in each RICH, their physical dimension is $2.7 \times 2.7\,\text{mm}^2$ and $7.6 \times 7.6\,\text{mm}^2$, respectively. This systematic arrangement allows the entire detector to be equalled to a (binary) imaging chamber producing 'pixels'.

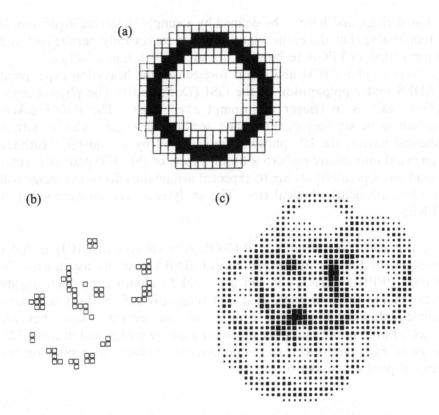

Fig. 2.33. (a) The pixel mask used for the ring correlation of the second-level trigger. The central part of the mask shown in black counts positively, the inner and outer pixels count negatively. (b) Input image of the second-level trigger processor, and (c) the result of a Hough transformation exhibiting maxima at the ring centres.

The analysis (Baur *et al.* 1994) starts with no information about tracks: rings have to be found without guidance; the only simplification is given by the fixed radius, as all interesting electrons are highly relativistic and have the asymptotic maximum Cherenkov radius. The task is solved by pattern matching, in a systematic scan over the entire RICH 'image'. The mask used has a shape as shown in Fig. 2.33. Perhaps the most interesting feature is the fact that the mask contains two patterns: all 'on' pixels in the ring are counted, whereas pixels near the ring are counted negatively; this avoids, to some extent, bremsstrahlung and conversion effects, and rings originating from pions. The systematic movement of the mask over the possible ring centre positions with simultaneous counting (+ and −) of hits results in a new image, whose pixels have gray values. Maxima in this image (that can be called a Hough transform of the original) correspond to found rings, and have to be defined by a simple clustering algorithm. It is remarkable that the entire algorithm was successfully parallelized and implemented, in FPGA technology, for an analysis time of 40 μs.

A very similar RICH analysis is foreseen in the heavy-ion experiment HADES under preparation at the GSI (Darmstadt). The physics motivation again is to trigger on prompt e^+e^- pairs. The RICH covers azimuth ϕ in six sectors, and each sector has a gas radiator with a spherical mirror; the UV photons are captured by a solid photocathode segmented into square pads ($6 \times 6\,mm^2$), each of the 5800 pads per sector is read out separately. Owing to expected astigmatic effects, the image will be a (θ-dependent) elliptical ring. The analysis is very similar to that of CERES.

2.6.2.3 Particle identification in BABAR.

A novel type of RICH is under development for the future experiment BABAR at the asymmetric B-factory (PEPII at SLAC). It will use a solid radiator (very high quality silica) in the form of rods, that will internally reflect the Cherenkov-radiated photons and guide them to an end section, where they are projected onto photomultiplier tubes via a quartz wedge and mirrors. The image of each photoelectron is thus directly available as a pixel (in the array of photomultiplier tubes).

3

Track and vertex fitting

3.1 The task of track fitting

Track fitting and the treatment of multiple scattering have had a long tradition in high-resolution *cosmic ray* and *bubble chamber experiments* (e.g. Rossi and Greisen 1941; Eyges 1948; Moorhead 1960; Gluckstern 1963; Laurikainen 1971a, b; Laurikainen, Moorhead and Matt 1972). The importance and the feasibility of track and vertex fitting in the more complex environment of *experiments with electronic detectors* have been recognized over the past three decades. This was when detectors with good and well-defined resolution came into operation (Charpak *et al.* 1968; Charpak 1978; Regler and Frühwirth 1990), and at the CERN-ISR track and vertex fitting with rigorous attention to multiple scattering were successfully applied in the early 1970s (e.g. Metcalf, Regler and Broll 1973; Nagy *et al.* 1978). Cross sections could be measured over more than seven decades with precise knowledge of the measurement errors on momentum transfer, knowledge of the losses due to χ^2 cuts, and a good estimation of the background. Important applications in set-ups with fixed targets were experiments at the CERN-SPS, e.g. those in the OMEGA spectrometer (Lassalle *et al.* 1980), or one where protons had been scattered on polarized protons (Fidecaro *et al.* 1980). Since then track fitting has been applied to several experiments at e^+e^- and $p\bar{p}$ storage rings where the selection of rare reactions in the presence of high multiplicities requires ultimate track separation and kinematic judgement, which can only achieved by exact track and vertex fitting. On the other hand, the increased computing time required for complex events has sometimes discouraged the application of such rigorous reconstruction methods.

In the introduction to this book the experimental scenario of today's front-line high-energy physics experiments was described.

- High multiplicities (20–100) due to the high collision energy obtained in storage rings, and even much higher multiplicities in heavy ion collisions.

- Momenta of particles in the final state ranging from a few hundred MeV up to several hundred GeV.

- Very long spectrometers (up to 100 m) in fixed-target experiments, but with small spatial separation between tracks in the vertex region.

- Complex modular track detectors combining different techniques having different resolutions ranging from a few micrometres in microstrip detectors, allowing experiments to achieve the necessary precision for high momenta, to a few millimetres or even centimetres in calorimeters.

- Resolutions that vary as a function of the impact point of the traversing particle.

- Multiple scattering in the detector frames, supports, and cables, which competes with the detector resolution, at least, for lower energies and for secondary vertices, which lie inside the beam tube.

- Large background from secondary activities of the particles to be measured, requiring optimal use of the information available for the final rejection of wrong associations of coordinates to tracks as well as of tracks to vertices.

- And last, but not least, high event rates leading to a large amount of data, with a steadily increasing demand in triggering and mass storage devices.

From the physics point of view a few aspects that are closely connected with the necessity for optimal track fitting should be mentioned again.

- Invariant masses had to be determined with optimal precision and well-estimated errors, e.g. to determine the number of types of neutrinos from the width of the Z^0 mass ($p\bar{p}$ experiments), or to resolve the combinatorial problem when identifying short-lived particles from their decay products.

- Secondary vertices must be fully reconstructed; this requires optimal track-bundling capability and optimal geometrical resolution in order to evaluate lifetimes of the order of 10^{-13} s (e.g. D mesons, B mesons, τ leptons).

- Kinks must be located when a charged particle decays into a final state that again has only one charged particle, as in the case of escaping neutrinos (e.g. from the decay of charged pions or kaons).

- Muon identification requires adequate treatment of the enormous amount of multiple scattering inside the muon filter, which is often made of ferromagnetic material (magnet yokes) ranging from one to several metres in thickness.

The association of coordinates and sometimes directions into tracks and the bundling of tracks into vertices, with complete use of the information available from the track detectors to obtain the ultimate track resolution, also require the availability of a precise track model for the path of a charged particle in a magnetic field, and the knowledge of the detector resolution of all the modules involved. Furthermore, the amount of material in the beam tube and the detector traversed by the particle must be known to a good approximation not only for correct and efficient treatment of multiple scattering (Metcalf *et al.* 1973; Regler 1977) but also to account for the energy loss. Furthermore, appropriate matrix algorithms must be chosen, which perform the tasks mentioned above within a reasonable computing time.

Although a detailed track fitting is not performed on all channels, even the selection of rare final states embedded in millions of similar topologies may require the detailed inspection of a high number of events if no clear trigger signal is available.

In this chapter first the use of the least squares method is discussed and justified (Section 3.2). Section 3.3 discusses the basic ingredients of track fitting by the least squares method (track model, weight matrix, and minimization). Section 3.4 describes the bundling of individual tracks into vertices, and Section 3.5 gives two examples of the track reconstruction strategy in large experiments.

For the purpose of this chapter it is assumed that the problem of associating the often many hundreds of coordinates – as measured by different kinds of detector – to track candidates (pattern recognition) has already been solved, with the exception of a few remaining ambiguities, although in special cases the tasks track fitting and pattern recognition are interwoven. The same statement holds for vertex bundling. The task of reconstructing an interesting event out of a mass of measurements is, in general, split into several steps, depending on the complexity and modularity of the set-up (see also the previous chapter).

The final goal of this chapter is to help to obtain the ultimate geometrical resolution using the fastest and smallest program possible, and to confirm the hypothesis that the measurements that have been grouped together represent a particle's track up to the final confidence tests; the same also holds for track bundling into a common vertex. It should act as a guide for the physicist when he or she wants to adopt one or other method, with regard to speed, flexibility and trustworthiness (Eichiner and Regler 1981), and it should also help in the choice of an optimal detector arrangement during the experimental design (Regler 1981) (see also Subsection 3.3.2.3).

3.1.1 Some symbols used in this chapter

(a, b), $[a, b]$	Open and closed interval
$\{c\}$	Region
$\langle c \rangle$	Expectation value
$\equiv$	Identical by definition
$\cong$	Approximately equal
$\sim$	Proportional
$\leftarrow$	Is replaced by
$\mathbf{AB}$	Matrix multiplication (but $\mathbf{A} \cdot (\mathbf{B} - \mathbf{C})$)
$(\mathbf{cc^T})_{ij} = c_i c_j$	direct tensor (or dyadic) product

3.2 Estimation of track parameters

Before discussing track fitting in practice, some theoretical estimation theory background will be reviewed.

3.2.1 Basic concepts

The task of track fitting, as described in Section 3.1, requires consideration of:

- the behaviour of the detector, i.e. essentially its geometrical layout and accuracy ('resolution');

- a mathematical 'model', which approximates the particle trajectories sufficiently well.

A schematic representation of the 'ingredients' and the 'recipe' of track fitting is shown in Fig. 3.1.

Knowledge of the detector behaviour allows an estimate of the track parameters with *minimum variance*. Furthermore, the resolution defines the scale in which decisions with respect to the track search must be taken, either to remove wrongly associated coordinates, or to reject a track string hypothesis as a whole.

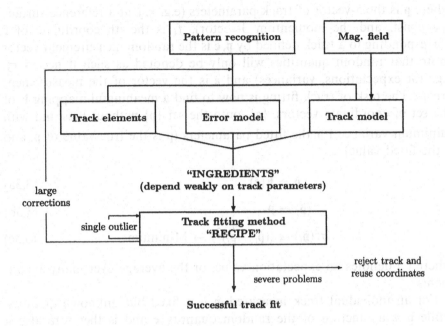

Fig. 3.1. 'Ingredients' and 'recipe' of the track fit.

The detector resolution can be estimated from theoretical considerations, from measurements in a calibration experiment, or from the tracks actually recorded when these have a sufficient number of constraints. In any case, the behaviour of the detector must be checked both during data acquisition and during off-line analysis by appropriate tests (Section 3.2).

In order to obtain the parameters defining a particle trajectory in a magnetic field, the path of the track must be known as a function of these parameters. In general it is appropriate to represent this 'track model' in *measurement space*. This is an *n*-dimensional space in the case of *n* measured coordinates along the track (e.g. if *m* planar detectors at fixed positions measure two coordinates each, then $n = 2m$). Of course, *n* is a widely varying function. A track is then given by a point on a *five-dimensional hyperplane* (a 'constraint surface'), according to the constraints given by the *equation of motion* of a charged particle in a magnetic field.

This 'measurement vector' deviates from the constraint surface due to randomly distributed *experimental errors* (see Fig. 3.2):

$$\mathbf{f} : \mathbf{p} \to f_i(\mathbf{p}), \ i = 1, \dots, n, \quad \text{or} \quad \mathbf{f}(\mathbf{p}) \tag{3.1}$$

i.e. **f** is a deterministic function of **p**, and

$$\mathbf{c} = \mathbf{f}(\overset{t}{\mathbf{p}}) + \boldsymbol{\varepsilon} \tag{3.2}$$

where $\mathbf{p}$ is the 5-vector of track parameters (e.g. x, y at a reference surface $z_r = \text{const}$, and the momentum 3-vector), f_i is the ith coordinate of $\mathbf{f}$ corresponding to a track defined by $\mathbf{p}, \mathbf{c}$ is the random measurement vector (note that random quantities will only be denoted as such if necessary, e.g. for expectations, variances) and $\boldsymbol{\varepsilon}$ is the vector of the measurement errors. The task of track fitting is now to find a meaningful mapping $\mathbf{F}$ of the set of coordinate vectors, $\{\mathbf{c}\}$, onto the set $\{\mathbf{p}\}$ without bias and with minimum variance for the fitted parameters ($\overset{t}{\mathbf{p}}$ is the true value of $\mathbf{p}$, and $\tilde{\mathbf{p}}$ the fitted value):

$$\tilde{\mathbf{p}} = \mathbf{F}(\mathbf{c}) \tag{3.3a}$$

$$\langle \tilde{\mathbf{p}} \rangle = \overset{t}{\mathbf{p}} \tag{3.3b}$$

$$\sigma^2(\tilde{\mathbf{p}}_i) \equiv \langle (\tilde{\mathbf{p}}_i - \overset{t}{\mathbf{p}}_i)^2 \rangle \rightarrow \text{Minimum} \tag{3.3c}$$

where $\langle\ \rangle$ denotes an expectation value, or the average over many experiments.

For an individual track measured, $\overset{t}{\mathbf{p}}$ is a fixed but unknown quantity, while $\tilde{\mathbf{p}}$ is a function of the random quantity $\mathbf{c}$ and is therefore also a random quantity. The variance has to be considered as describing the experimental errors for repeated measurement of the same track, and it may well differ for different $\overset{t}{\mathbf{p}}$. The fact that $\overset{t}{\mathbf{p}}$ itself usually has a random distribution from one track to another does not affect what follows. A simple example will be given at the end of Subsection 3.2.2.

3.2.2 *Global track fitting by the Least Squares Method (LSM)*

It turns out that the LSM best meets the requirements of track fitting, being simple, rather fast, and familiar to experimentalists. Its important statistical properties, together with its numerical simplicity, form the basis of the wide range of its application.

If the track model can be sufficiently well approximated by a *linear model* in the neighbourhood of the measurements, and if the errors vary sufficiently little with the track parameters that they can be considered as being constant in the neighbourhood of an individual track's path, then the LSM *estimation* has *minimum variance* among the class of linear and unbiassed estimates (for more details see Subsection 3.2.3).

In the linear approximation, the LSM is suitable for simple error propagation. Individual track fitting and track element merging can be done before vertex fitting, and in a third step the kinematical constraints can be imposed. If the measurement errors are not too far from 'Gaussian', or asymptotically in the limiting case of a large number of measurements, the square root of the variance can be considered as an interval estimation

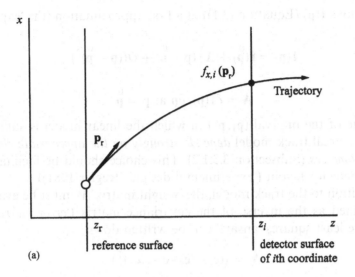

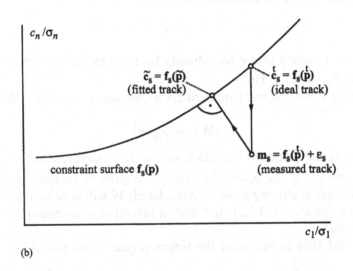

Fig. 3.2. (a) Reference plane, the five-dimensional track parameter $\mathbf{p}_r$ and the 'impact functions' $f_i(\mathbf{p}_r)$, $i = 1, \ldots, n$. (b) The five-dimensional constraint surface; the fit minimizes the weighted Euclidian distance between a real measurement and an ideal track point lying on the constraint surface, $\mathbf{f}(\mathbf{p})$. The scale is the measurement error σ, and subscript S denotes scaled. Uncorrelated measurement errors are assumed for the purpose of graphical representation.

(see Tables 3.1 and 3.2), and a simple test quantity is obtained in parallel (the χ^2).

The track model is, in general, the set of solutions of the equations of motion, whereas the track model in the LSM is the *linear expansion* of the functions $\mathbf{f}(\mathbf{p})$ (Equation (3.1)) at a first approximation (the 'expansion point' $\overset{0}{\mathbf{p}}$):

$$\mathbf{f}(\mathbf{p}) = \mathbf{f}(\overset{0}{\mathbf{p}}) + \mathbf{A} \cdot (\mathbf{p} - \overset{0}{\mathbf{p}}) + O((\mathbf{p} - \overset{0}{\mathbf{p}})^2) \qquad (3.4)$$

with

$$\mathbf{A} = \partial \mathbf{f}(\mathbf{p}) / \partial \mathbf{p} \text{ at } \mathbf{p} = \overset{0}{\mathbf{p}}$$

The range of the interval $(\mathbf{p}_1, \mathbf{p}_2)$ in which the linear model is sufficiently close to the real track model *depends strongly on the appropriate choice of track parameters* (Subsection 3.3.1.2). This choice should be facilitated by a clever detector layout ('experimental design' (Regler 1981)).

In addition to the track model, the 'weight matrix' $\mathbf{W}$ must be evaluated (it is defined as the inverse of the covariance matrix ('*error matrix*') $\mathbf{V}$), before the least squares 'ansatz' can be written down:

$$\mathbf{V} = \langle (\mathbf{c} - \langle \mathbf{c} \rangle)(\mathbf{c} - \langle \mathbf{c} \rangle)^{\mathrm{T}} \rangle \qquad (3.5)$$

with $\mathbf{c} = \mathbf{f}(\overset{t}{\mathbf{p}}) + \boldsymbol{\varepsilon}$ from Equation (3.2), and

$$\mathbf{W} = \mathbf{V}^{-1}$$

where it is assumed that $\mathbf{c}$ has already been corrected for a possible *bias*, (i.e. $\langle \boldsymbol{\varepsilon} \rangle = \mathbf{0}$), or $\langle \mathbf{c} \rangle = \mathbf{f}(\overset{t}{\mathbf{p}})$.

In simple cases the measurement errors are *uncorrelated*, i.e. $\mathbf{W}$ is of the form

$$(\mathbf{W})_{ij} = \delta_{ij} / \sigma_j^2 \qquad (3.6)$$

where σ_j is the standard deviation of the jth measurement ε_j, i.e. $\sigma_j = (\langle \varepsilon_j^2 \rangle)^{\frac{1}{2}}$, and δ_{ij} is the Kronecker delta. In general (mainly in the cases where *multiple scattering* must be considered) $\mathbf{W}$ will also have *off-diagonal* terms (see Subsection 3.3.2), and matrix inversion is necessary to obtain it numerically.

The LSM tries to minimize the function (see Subsection 3.2.3)

$$M = [\mathbf{f}(\overset{0}{\mathbf{p}}) + \mathbf{A} \cdot (\mathbf{p} - \overset{0}{\mathbf{p}}) - \mathbf{m}]^{\mathrm{T}} \mathbf{W} \cdot [\mathbf{f}(\overset{0}{\mathbf{p}}) + \mathbf{A} \cdot (\mathbf{p} - \overset{0}{\mathbf{p}}) - \mathbf{m}] \qquad (3.7)$$

where $\mathbf{m}$ is a '*realization*' of the random quantity $\mathbf{c}$, i.e. a specific measurement. Differentiating M with respect to $\mathbf{p}$ and putting $\partial M / \partial \mathbf{p} = 0$ yields (with rank of $\mathbf{A} \geqslant$ dimension of $\mathbf{p}$, i.e. at least as many independent measurements as independent track parameters):

$$\tilde{\mathbf{p}} = \overset{0}{\mathbf{p}} + (\mathbf{A}^{\mathrm{T}} \mathbf{W} \mathbf{A})^{-1} \mathbf{A}^{\mathrm{T}} \mathbf{W} \cdot (\mathbf{m} - \mathbf{f}(\overset{0}{\mathbf{p}})) \qquad (3.8)$$

As an example, the two track parameters will be evaluated in a Cartesian projection (y, z) for the case in which there is no magnetic field. Detectors are placed along the z axis, measuring coordinate y_i. Then the trajectory (Fig. 3.2a, with y instead of x) is a straight line:

$$y = a + zb \tag{3.9}$$

If the reference plane is chosen to be at $z = z_r = 0$, the linear track model (Equation (3.4)) is

$$f(p) = Ap \tag{3.10a}$$

with

$$p = \begin{pmatrix} a \\ b \end{pmatrix}, \quad A = \begin{pmatrix} 1, & z_1 \\ \vdots & \vdots \\ 1, & z_n \end{pmatrix}$$

or

$$f_i(p_1, p_2) = p_1 + z_i p_2 \tag{3.10b}$$

If multiple scattering is negligible, $(W)_{ij}$ is equal to δ_{ij}/σ_j^2 (Equation 3.6)) and Equation (3.8) has the form

$$\begin{pmatrix} \tilde{a} \\ \tilde{b} \end{pmatrix} = \begin{pmatrix} \sum 1/\sigma_i^2, & \sum z_i/\sigma_i^2 \\ \sum z_i/\sigma_i^2, & \sum z_i^2/\sigma_i^2 \end{pmatrix}^{-1} \begin{pmatrix} \sum m_i/\sigma_i^2 \\ \sum (m_i z_i)/\sigma_i^2 \end{pmatrix}$$

$$= \frac{1}{\det(A^T W A)} \begin{pmatrix} \sum z_i^2/\sigma_i^2, & -\sum z_i/\sigma_i^2 \\ -\sum z_i/\sigma_i^2, & \sum 1/\sigma_i^2 \end{pmatrix} \begin{pmatrix} \sum m_i/\sigma_i^2 \\ \sum (m_i z_i)/\sigma_i^2 \end{pmatrix} \tag{3.11}$$

where

$$\det(A^T W A) = \left(\sum 1/\sigma_i^2 \right) \left(\sum z_i^2/\sigma_i^2 \right) - \left(\sum z_i/\sigma_i^2 \right)^2$$

If the σs do not depend on the impact coordinates and angles, the matrices are constants of the set-up and, for efficient algorithms, must be evaluated only once, which can be an important feature for *fast real-time processing* (Chapter 1).

Note that if W is diagonal most of the matrix operations can be simplified by the following substitution ($a_{ij} \equiv (A)_{ij}$):

$$\begin{aligned} a_{ij} &\leftarrow a_{ij}/\sigma_i \\ m_i &\leftarrow m_i/\sigma_i \\ f_i &\leftarrow f_i/\sigma_i \end{aligned} \tag{3.12}$$

However, some care has to be taken when calculating the 'pull quantities' (Equation (3.24)).

3.2.3 A few remarks on estimation theory

3.2.3.1 Generalities. In order to discuss the properties of the *estimators* appropriate for track fitting, it is assumed that the 'hypothesis' of the track model is correct (Subsection 3.3.1). This implies that:

- the equation of motion can be solved with sufficient precision;

- the magnetic field is measured sufficiently accurately and the information is easily accessible;

- the material traversed is known, thus allowing the evaluation of energy loss and multiple scattering, which also requires in some cases (low-energy particles, electrons) the identification of the particle;

- no wrong measurements have been associated during the process of pattern recognition, a necessary condition for the consistency of the track model with the vector of measurements.

It is further assumed that the detector resolution is understood, in particular that the covariance matrix $\mathbf{V}$ of the measurement vector $\mathbf{c}$ is known (Subsection 3.3.2).

With the definition of the track model in coordinate space (Equation (3.1)) and the randomly displaced measurements (Equation (3.2)), one can obtain a *conditional probability density function d* describing the detector resolution, where the dominant variable is the difference between the measurement vector and the corresponding quantities of the 'undisturbed track' $\mathbf{f}(\mathbf{p})$, namely $\varepsilon = \mathbf{m} - \mathbf{f}(\mathbf{p})$. However, this function quite often also depends explicitly on the measurement vector itself:

$$d(\mathbf{m};\overset{t}{\mathbf{p}}) = d'(\varepsilon,\mathbf{m}) = d'(\mathbf{m} - \mathbf{f}(\overset{t}{\mathbf{p}});\mathbf{m}) \tag{3.13}$$

Before discussing the properties of the estimation of track parameters from a measured coordinate vector $\mathbf{m}$ we shall review a few of the requirements that one may ask of an estimator. The task is to estimate the track parameters themselves ('*point estimation*'), as well as an interval indicating how much the estimated parameters may possibly deviate from the true ones ('*interval estimation*'). Additional tasks will be to check whether the pattern recognition hypothesis associating this set of coordinates into a track segment was correct, and whether the detector behaviour (measurement error) has been understood properly.

In general, one distinguishes between *explicit* and *implicit* estimators. In the first case the estimator is defined by a function of the measurement vector, while in the latter case the quantities to be determined are only implicitly defined.

The most common implicit estimator is given by the '*Maximum Likelihood Method*' (MLM). Its aim is to inspect the joint probability density function of the measurement vector (Equation (3.13)) and to compare the relative chance of obtaining a similar experimental result (measurement vector) for different assumptions for the track parameters $\mathbf{p}$ versus a 'null hypothesis' $\overset{0}{\mathbf{p}}$. To establish this '*likelihood ratio*' R, the quantity $\overset{t}{\mathbf{p}}$ in Equation (3.13) is replaced by a running parameter $\mathbf{p}$:

$$R_{\mathbf{m}}(\mathbf{p}) = d(\mathbf{m};\mathbf{p})/d(\mathbf{m};\overset{0}{\mathbf{p}}) \tag{3.14a}$$

or, since the denominator is constant, just

$$L_{\mathbf{m}}(\mathbf{p}) = d(\mathbf{m};\mathbf{p}) \tag{3.14b}$$

The estimated parameters $\tilde{\mathbf{p}}$ will be obtained by the requirement that $L_{\mathbf{m}}(\mathbf{p})$ (or, equivalently, $\ln(L_{\mathbf{m}})$) is a maximum or

$$\partial \ln L_{\mathbf{m}}(\mathbf{p})/\partial \mathbf{p} = \mathbf{0} \tag{3.15}$$

which asymptotically fulfills the requirements (Equation (3.3)).

If d' depends, at least approximately, only on the difference $\varepsilon = \mathbf{m} - \mathbf{f}(\mathbf{p})$, and if its density function can be approximated by a *Gaussian distribution*, then Equation (3.14) is fully determined by the corresponding covariance matrix $\mathbf{V}$ or by the *weight matrix* $\mathbf{W} = \mathbf{V}^{-1}$ (Equation (3.5)):

$$L_{\mathbf{m}}(\mathbf{p}) = d_{\mathbf{m}}(\varepsilon)$$

$$= (2\pi)^{-n/2}[\det(\mathbf{V})]^{-\frac{1}{2}} \prod_{i,j=1}^{n} \exp[\varepsilon_i (\mathbf{W})_{ij} \varepsilon_j /2] \tag{3.16}$$

where $\varepsilon = \mathbf{m} - \mathbf{f}(\overset{t}{\mathbf{p}})$. In this case the MLM *coincides with the* LSM. In the general case, track fitting with the MLM is often impractical or even impossible, owing to the computer time required.

If $\mathbf{W}$ is not constant but varies only slightly with $\mathbf{p}$, it may be sufficient to evaluate this explicit dependence on $\mathbf{p}$ once for a track at an approximate point $\overset{0}{\mathbf{m}}$, and to keep it constant for the remaining track-fitting procedure.

There are cases such as the Multi-Wire Proportional Chambers (MWPC) where the main variable is the distance from the physical impact to the nearest wire, which is a deterministic step function rather than a probability density (see Subsection 3.3.2.1).

The linearized LSM yields an explicit estimator (Equation (3.8)). For most applications by an appropriate choice of $\mathbf{p}$, $\mathbf{f}(\mathbf{p})$ can be approximated by a linear expansion in the neighbourhood of a 'starting vector' $\overset{0}{\mathbf{p}}$ (Equation (3.4)). (For the nonlinear case an iteration procedure can be adopted, see Subsection 3.3.4.)

Recalling Equation (3.8) (before the realization of a measurement $\tilde{p}$ and c are random vectors)

$$\tilde{p} = \overset{0}{p} + (A^T W A)^{-1} A^T W \cdot (c - f(\overset{0}{p}))$$

we will now discuss some properties of the LSM for a linear model. (Those proofs that are not given here can be found in the literature (e.g. Kendall and Stuart 1967).)

(a) Bias If the measurement vector $c = f(\overset{t}{p}) + \varepsilon$ (Equation (3.2)) is *unbiassed*, i.e. $\langle \varepsilon \rangle = 0$, then $\tilde{p}$ is also unbiassed (3.3a).

Proof: Since $\overset{0}{p}$ can be chosen arbitrarily, it can be set equal to $\overset{t}{p}$ without loss of generality

$$\langle \tilde{p} - \overset{t}{p} \rangle = \langle X \varepsilon \rangle = X \langle \varepsilon \rangle = 0 \tag{3.17}$$

with

$$X = (A^T W A)^{-1} A^T W$$

∎

(b) Variance The covariance matrix of $\tilde{p}$, i.e. the *error matrix* of the *fitted parameters*, is given by

$$C(\tilde{p}) = (A^T W A)^{-1} \tag{3.18}$$

Proof:

$$\langle (\tilde{p} - \overset{t}{p})(\tilde{p} - \overset{t}{p})^T \rangle = \langle ((A^T W A)^{-1} A^T W \varepsilon)(\varepsilon^T W A (A^T W A)^{-1}) \rangle$$
$$= (A^T W A)^{-1} A^T W V W A (A^T W A)^{-1}$$
$$= (A^T W A)^{-1} A^T W A (A^T W A)^{-1} = (A^T W A)^{-1}$$

with

$$V = \langle (c - \langle c \rangle)(c - \langle c \rangle)^T \rangle$$
$$W = V^{-1}$$
$$\langle (X \varepsilon)(\varepsilon^T X^T) \rangle = X \langle \varepsilon \varepsilon^T \rangle X^T = X V X^T,$$
$$\langle \varepsilon \varepsilon^T \rangle = V = W^{-1}$$

∎

(c) The 'Gauss–Markov theorem'   For a linear model, e.g. the linearized track model of Equation (3.4), the LSM is the linear *unbiassed estimator* with *least variance* ('optimal') within this class if:

$\langle \varepsilon \rangle = 0$ (see above);

ε has a non-singular covariance matrix;

the weight matrix **W** used to compute the minimum of M (Equation (3.7)) is the inverse of the covariance matrix, $\mathbf{V}(\varepsilon)^{-1}$.

(d) Consistency   An estimator is called *consistent* if the following relation holds: $\lim_{n\to\infty} \overset{t}{\tilde{p}}_n = \overset{t}{p}$, where 'lim' means convergence in probability, i.e. for any $\eta > 0$ and any $\varepsilon > 0$ (not to be confused with the measurement error) an $N(\varepsilon, \eta)$ can be found such that $\mathrm{prob}\,(|\tilde{p}_n - p| > \varepsilon) < \eta$. However, an infinite number of measurements is only an abstract hypothesis.

(e) The 'Cramér–Rao inequality'   For the variance of an unbiassed estimator $\tilde{p}$, the following inequality holds as a general lower bound for the precision which can be achieved in an experiment:

$$\sigma^2(\tilde{p}_i) \geqslant (\mathbf{I}_c^{-1})_{ii} \tag{3.19}$$

$$(\mathbf{I}_c(\overset{t}{\mathbf{p}}))_{ij} = \left\langle \frac{\partial \ln L_c(\mathbf{p})}{\partial p_i}\bigg|_{\overset{t}{\mathbf{p}}} \times \frac{\partial \ln L_c(\mathbf{p})}{\partial p_j}\bigg|_{\overset{t}{\mathbf{p}}} \right\rangle_c$$

$$= \int \left\{ \left[\frac{\partial L_m(\mathbf{p})}{\partial p_i}\bigg|_{\overset{t}{\mathbf{p}}} \bigg/ L_m(\overset{t}{\mathbf{p}}) \right] \times \left[\frac{\partial L_m(\mathbf{p})}{\partial p_j}\bigg|_{\overset{t}{\mathbf{p}}} \bigg/ L_m(\overset{t}{\mathbf{p}}) \right] \right\}$$

$$\times d(\mathbf{m};\overset{t}{\mathbf{p}})dm \tag{3.20}$$

where $\sigma^2(\tilde{p}_i) \equiv \mathrm{var}\,(\tilde{p}_i) \equiv \langle (\tilde{p}_i - \overset{t}{p}_i)^2 \rangle$ for an unbiassed estimator and $\langle\ \rangle_c$ denotes the expectation with respect to the random variable **c**. L comes from Equation (3.14b) and d from Equation (3.13). $\mathbf{I}_c(\mathbf{p})$ is called the *'information'*. If the measurements are uncorrelated, d factorizes and so does L, and Equation (3.20) is just the sum over the information given by the individual measurements (thus justifying its name) together with Expression (3.19) for the variance. Equations (3.19) and (3.20) show that the information is just the square of the relative change (sensibility) of L due to a variation of **p**, weighted with the probability density of the measurement vector, which accounts for the frequency of occurrence of a measurement. If in Expression (3.19) the equality is true, then $\tilde{p}$ is called a *minimum variance bound estimator* (or the 'efficient estimator') (see

Table 3.1. *Properties of the LSM*

| Model | Errors | |
	Non-Gaussian	Gaussian
Linear	Optimal among unbiassed linear estimators	Equivalent to the MLM with a sufficient statistic and therefore a *minimum variance bound estimator* (*mvbe*) even for a finite sample of measurements
Nonlinear	No general properties for a finite sample of measurements but asymptotically an *mvbe*	Equivalent to the MLM and therefore asymptotically a minimum variance bound estimator

Table 3.1). This is true only for a special class of probability density functions d, e.g. for Equation (3.16). Otherwise, one defines the relative efficiency of an unbiassed estimator $\mathrm{var}\,(\tilde{p})/(\text{minimum variance bound})$. If an estimator has minimum variance but a larger than the minimum variance bound, it is called *optimal*; this also applies when one is restricted to a special class of estimators (see the Gauss–Markov theorem above).

If the measurement errors are Gaussian (e.g. Equation (3.16)), the LSM is equivalent to the MLM. It follows that for the linear model the LSM is then a minimum variance bound (efficient) estimator even for a finite sample of measurements.

Table 3.1 summarizes the properties of the LSM. Asymptotically, in all four cases, the LSM is *normal*, *unbiassed* and *consistent* (e.g. Eadie *et al.* 1971).

(f) 'Robustness' An important property of an estimator is its *robustness*. An estimator is called robust if it is insensitive to measurements that deviate from the expected behaviour. There are two ways to treat such deviating measurements: one may either try to recognize them and then remove them from the data sample; or one may leave them in the sample, taking care that they do not influence the estimate unduly. In both cases robust estimators are needed.

In practice two types of deviating measurements have to be distinguished.

- The measurement is generated by the particle under investigation, but with an error much larger than expected. Such a measurement, for which the probability of occurrence is low, is called an *'outlier'*. It can be caused by several processes, but the most common one is the creation of energetic electrons in gaseous detectors (δ-rays). Sometimes it can be detected by a larger 'cluster size' (Subsection 3.3.2.1).

- The measurement is a noise signal that was associated to the track by the pattern recognition; it can either be genuine detector noise, or picked up from another particle's set of measurements ('ghosts').

In the framework of the LSM the outlier problem could be handled globally by modifying the error matrix, thus ensuring the proper propagation of errors. However, in most cases this would spoil the overall resolution, and the χ^2 would be distorted, so that a χ^2-cut would be nontransparent and could spoil the overall normalization and – if correlated with $\tilde{p}$ – introduce a bias (Subsection 3.2.4).

An algorithm should be made *robust* by taking some reasonable default action when 'standard proceeding' does not seem appropriate. Robust procedures compensate for systematic errors as much as possible, and indicate any situation in which a danger of not being able to operate reliably is detected. As robust estimators are not normally fully efficient, a compromise must be found between robustness and the final resolution of the physical quantities to be estimated.

In particle physics many pragmatic approaches have been invented to solve the outlier problem such as the removal of the largest absolute reduced residual (Equation (3.24)) if it exceeds a few standard deviations for position-sensitive devices ('truncation'), and the removal of a fixed percentage from the signal sample in the case of a pulse height analysis (e.g. for a dE/dx measurement, see Subsection 3.3.1.5). This latter method is called the 'censored mean'.

For a more systematic study on the detection of outliers in track fitting see Frühwirth (1988).

3.2.3.2 The LSM in practice. Many modern particle track detectors collect large samples of track coordinates, and the asymptotic properties of the LSM are therefore of real interest in practice. The convergence properties depend mainly on the approximate linearity of the track model and on the deviations of the measurement error density function from a normal distribution (Equation (3.16)). (Throughout this chapter the terms 'Gaussian' and 'normal' distribution are used as synonyms.) If the distribution has zero mean and unit variance, it is called 'standard normal'.

An important advantage of the LSM is that the covariance matrix of the fitted quantities (Equation (3.18)) is easily obtained for each individual track, thus allowing error propagation and subsequent fits with

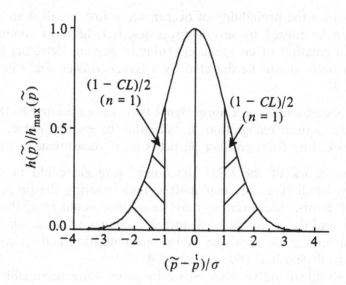

Fig. 3.3. The probability density function for $\tilde{p}$ (i.e. for repeated experiments) in the linear Gaussian case (or in the asymptotic limit). The shaded region is the complement to the confidence level for $\pm 1\sigma$. In practice σ is taken from error propagation (LSM), from the likelihood function, or, in the nonlinear finite case, from Monte Carlo simulation.

additional constraints (e.g. common vertex, relativistic energy–momentum conservation; see also Section 3.4).

Another interesting property in the linear model is the *equivalence* of the LSM to the class of Gaussian errors with respect to the properties of the forms of *quadratic averages* (i.e. of variances, $\langle \chi^2 \rangle$, etc.), which thus allows the definition of reduced test quantities with a variance of 1 or a 'pseudo χ^2 distribution' with an average equal to the corresponding number of degrees of freedom. Using the correct variances, and with the asymptotic property of being normal, in many cases reduced quantities are well approximated by standard normal distributions, and the effective pseudo χ^2 distribution is well approximated by a real χ^2 distribution. These quantities and their use for test purposes will be discussed shortly.

Use of this equivalence has already been made when calculating the error matrix of the fitted parameters by error propagation (Equation (3.18)). For a linear model and Gaussian errors, error propagation again gives Gaussian-distributed quantities (Fig. 3.3), and therefore this matrix is a direct measure of the interval limiting the deviation from the true value with a given probability ('*interval estimation*'), see Table 3.2 with $\sigma_i = (\mathbf{c}(\tilde{\mathbf{p}}))_{ii}$ from Equation (3.18). The estimated interval $(\tilde{p}_i - \sigma_i, \tilde{p}_i + \sigma_i)$ is also called the '*confidence interval*'.

Table 3.2. *Confidence interval for Gaussian-distributed quantities*

n	prob $(\lvert \overset{t}{p}_i - \tilde{p}_i \rvert < n\sigma_i)$ (%)	Confidence level (%)
1	68.3	31.7
2	95.4	4.6
3	99.7	0.3

If no assumption is made about the distribution of the measurement errors, only an upper limit can be given for the probability that $\overset{t}{p}_i$ is not within $\tilde{p}_i \pm \sigma_i$:

$$\text{prob}\,(\lvert \overset{t}{p}_i - \tilde{p}_i \rvert > n\sigma_i) \leqslant 1/n^2 \qquad (3.21)$$

This is called the 'Chebycheff inequality'. In practice this limit is quite unsatisfactory compared with Table 3.2, and more restrictive limits exist; often the final meaning of σ must be gained from Monte Carlo or from the empirical 'χ^2' distribution of the experimental data. For a large number of measurements advantage can be taken of the asymptotic normality.

A classical check for the proper use of the LSM is the inspection of the *reduced residuals* ('stretch functions' or 'pull quantities'). Again it is assumed that the parameter dependence is not too far from being linear: During the fit, the parameters undergo an adjustment, and so does the track model in coordinate space. As the coordinates are the directly measurable quantities, it is advisable to inspect the differences between these measurements **m** and the readjusted or improved ones, i.e. $\mathbf{f}(\tilde{p}) =: \tilde{\mathbf{c}}$. The difference $m_i - \tilde{c}_i$ is called the ith residual. The variance of $\tilde{\mathbf{c}}$ is given by simple error propagation:

$$\mathbf{C}(\tilde{\mathbf{c}}) = \mathbf{A}(\mathbf{A}^{\mathrm{T}}\mathbf{W}\mathbf{A})^{-1}\mathbf{A}^{\mathrm{T}} \qquad (3.22)$$

where **A** is given by Equation (3.4) and **W** by Equation (3.5).

The covariance of the residual vector $\mathbf{r} = \mathbf{c} - \tilde{\mathbf{c}}$, $\mathbf{C}(\mathbf{r})$, is smaller than the measurement error $\mathbf{V}(\mathbf{c})$

$$\mathbf{C}(\mathbf{r}) = \mathbf{V}(\mathbf{c}) - \mathbf{C}(\tilde{\mathbf{c}}) \qquad (3.23)$$

where the minus sign in front of $\mathbf{C}(\tilde{\mathbf{c}})$ reflects the positive correlation between **c** and $\tilde{\mathbf{c}}$ (Equation (3.22)), and approaches the fitted values towards the measurements. (Note that the covariance matrix is always a symmetric, positive definite matrix. Otherwise, the set of coordinates is not sufficient to define a track.)

Proof:

$$r = c - \tilde{c} = f(\overset{t}{p}) + \varepsilon - A \cdot (\tilde{p} - \overset{t}{p}) - f(\overset{t}{p}) = \varepsilon - A\tilde{p} + A\overset{t}{p}$$

$$= \varepsilon - A \cdot [(A^T W A)^{-1} A^T W \varepsilon + \overset{t}{p}] + A\overset{t}{p}$$

$$= [1 - A(A^T W A)^{-1} A^T W]\varepsilon = Y\varepsilon$$

$$C(r) = Y\langle \varepsilon\varepsilon^T\rangle Y^T = V + A(A^T W A)^{-1} A^T W V W A(A^T W A)^{-1} A^T$$

$$\qquad -A(A^T W A)^{-1} A^T W V - V W A(A^T W A)^{-1} A^T$$

$$= V + A(A^T W A)^{-1} A^T - 2A(A^T W A)^{-1} A^T$$

$$= V - A(A^T W A)^{-1} A^T = V - C(\tilde{c})$$

∎

In case of zero constraints the track would just follow the measurements, i.e. $C(\tilde{c}) = V$ and $C(r) = 0$.

The reduced residuals can now be calculated (Equation (3.23)). As mentioned above they should be distributed with mean 0 and variance 1:

$$p_i = \frac{m_i - \tilde{c}_i}{[\sigma^2(c_i) - \sigma^2(\tilde{c}_i)]^{\frac{1}{2}}} \tag{3.24}$$

The reduced quantities are very sensitive to wrong error assumptions or to misalignment and are often used to identify some necessary corrections.

3.2.3.3 The χ^2 distribution.

Finally the χ^2 *distribution* will now be discussed briefly. For measurement errors with a given n-dimensional Gaussian distribution (Equation (3.16)), the χ^2 distribution is the probability density function for the weighted sum of the squares of the measurement error. It is obtained by transformation into spherical coordinates $((\chi^2)^{\frac{1}{2}}, \omega)$ and by integrating over the $(n-1)$-dimensional solid angle:

$$\chi^2 = \sum_{i,j=1}^{n} \varepsilon_i (W)_{ij}\varepsilon_j \tag{3.25a}$$

$$g(\chi^2) = (2\pi)^{-n/2}[\det(V)]^{-\frac{1}{2}} \int e^{-\chi^2/2} \frac{\partial\varepsilon_1, \ldots, \varepsilon_n}{\partial\chi^2 \partial^{n-1}\omega} d^{n-1}\omega \tag{3.25b}$$

where χ^2 is the square of the absolute value of the reduced radius vector (the weighted sum of squared measurement errors) and ω is the angular contribution. After a unitary transformation T diagonalizing $W((TWT^{-1})_{ij} = \delta_{ij}/\sigma_j'^2, \sigma_j' = \sigma(\varepsilon_j'), \varepsilon_j' = (T\varepsilon)_j)$ and with the reduced quantities $\varepsilon_j'' = \varepsilon_j'/\sigma_j'$ one obtains: $\chi^2 = \sum_{j=1}^{n} \varepsilon_j''^2$.

It follows from *spherical symmetry* and from *normalization* (with the 'radial coordinate' $\chi : \chi^{n-1}d\chi = (\chi^2)^{(n-2)/2}d\chi^2/2$ and $|T| = 1$), or from the evaluation of the Jacobian matrix that

$$g_n(\chi^2) = e^{-\chi^2/2}(\chi^2)^{(n-2)/2}/(2^{n/2}\Gamma(n/2)) \qquad (3.25c)$$

and

$$\langle \chi^2 \rangle = n \qquad (3.25d)$$

For non-Gaussian errors, the χ^2 distribution is distorted with respect to Equation (3.25c), but in the linear model the expectation value is conserved. The following relation holds for the probability density function:

$$\chi^2 = \chi_{n_1} + \chi_{n_2} \text{ implies } g_{n_1+n_2}(\chi^2) \qquad (3.26a)$$

Furthermore, in the LSM, the χ^2 distribution is defined by the number of degrees of freedom (proof similar to that of (Equation 3.23)):

number of degrees of freedom = number of (independent)

measurements (plus other constraints) − number of

(independent) adjustable parameters, i.e. $n_f = n_m + n_c − n_p$ (3.26b)

The notion of 'number of degrees of freedom' can be misleading; here it means the number of degrees of freedom for adjusting $\tilde{c}$ other than just choosing **m**, which would be the natural choice if no additional constraints were given by the track model ('zero-constraint fit').

3.2.4 Test for goodness of fit

The test for goodness of fit consists of two separate parts, but with strong interdependence in practice.

(a) Pull quantities In order to obtain a correct estimation of the track parameters, three conditions must be fulfilled (see above).

- The track model must be correct (see also Subsection 3.3.1).

- The covariance matrix of the measurement errors must be correct (Subsection 3.3.2).

- The reconstruction program must work properly (Section 3.5).

A common way to test these requirements is to check the variances of the *pull quantities* (Equation (3.24)). A regular check of these quantities also shows whether the detector behaviour is stable with time or not. Furthermore, the pull quantities also give a good indication of any individual track detector that is beginning to deteriorate, while if only the global χ^2 is known, this recognition may be difficult. A more global but less sensitive check is to observe the mean of χ^2, i.e. $\langle \chi^2 \rangle$, and, in the case of Gaussian errors, even the whole χ^2 distribution itself. In this latter case it is convenient to plot the corresponding value of the cumulative distribution function $G(\chi^2)$, which is uniformly distributed (Fig. 3.4) between 0 and 1

$$G_k(\chi^2) = \int_0^{\chi^2} g_k(x)\,dx \qquad (3.27)$$

When fitting the five parameters of a particle trajectory in a magnetic field, k is usually the number of measured coordinates minus five (e.g. the five initial conditions of a track at a given reference surface), because each independent coordinate contributes one constraint. $G_k(\chi^2)$ allows one to make a single plot independent of the number of degrees of freedom, i.e. independent of the number of measured coordinates of an individual track.

So far it has been implicitly assumed that no wrong coordinates were associated by the pattern recognition, which is, in fact, an additional condition for the correctness of the track model. In order to ensure this in practice, only selected event topologies with little background and low track multiplicities (e.g. elastic scattering) should be chosen to tune the fitting program and to test and calibrate the detector (errors and alignment). However, it must be guaranteed that such a sample is still representative. In the global environment the behaviour of a detector model might be worse than in the 'clean environment' of a test beam. The detector behaviour sometimes also deteriorates when there are high track multiplicities, and so does the behaviour of the pattern recognition program; or the low-energy behaviour (i.e. multiple scattering and energy loss) may not be tested well enough by such a sample; also the event rate might influence the resolution. It is only after all these conditions are fulfilled that one may have confidence in the covariance matrix of the measured quantities, $\mathbf{V} = \mathbf{W}^{-1}$, and – consequently – of the fitted quantities, $\mathbf{C}(\tilde{\mathbf{p}}) = (\mathbf{A}^{\mathrm{T}}\mathbf{W}\mathbf{A})^{-1}$.

(b) χ^2 test The second check concerns a proper statistical test of the correctness of the association of the measured coordinates into a track. The 'null hypothesis', H_0, assumes that all coordinates belong to the track. In track fitting the 'alternative hypothesis' H_1 is usually just the logical complement of H_0. However, in practice one distinguishes two classes of alternative hypotheses: (a) that several coordinates have been associated although there was no track (this is called background, and the track is often called a 'ghost'), and (b) that only one coordinate (or a small fraction of coordinates) has been associated wrongly to a real track.

A suitable quantity to decide if H_0 is tenable (if no alternative hypothesis has explicitly been established) such a test is called a 'significance test'. A suitable test quantity is obtained automatically by the LSM i.e. the χ^2. When the null hypothesis is true, the expectation value of the χ^2 distribution is just the number of degrees of freedom (Equations (3.25c) and (3.26b)); in the linear model and with Gaussian errors the minimum of M (Equation (3.7)) obeys a real χ^2 distribution (Equation (3.25b)) according to Cochran's theorem (Kendall and Stuart 1967, p. 84), and the χ^2 is *uncorrelated to the fitted track parameter* $\tilde{\mathbf{p}}$.

As almost any kind of background has a tendency to have a larger χ^2 (this requires of course an overdetermined track, i.e. more independent measurements than independent parameters (Equation (3.26b))), a common way to test H_0 is to define a decision criterion making use of the χ^2 value of an individual track, e.g. to give up a certain percentage of good tracks with large χ^2, in order to eliminate the background (see also Section 4.5).

A function of random observables is called a 'statistic'. When using a test variable, e.g. χ^2, for the goodness of fit it is called a 'test statistic'. In practice one preassigns a *critical* χ^2 value, dividing the interval $(0, \infty)$ into two regions: an acceptance region $[0, \chi_c^2)$ and a critical (or rejection) region $[\chi_c^2, \infty)$ which implies the selection in the coordinate space of a certain region of acceptance, the set $\{\mathbf{m}\}_{\text{Acc}}$ out of the set of all $\{\mathbf{m}\}$ around the constraint surface (Fig. 3.2(b)), and its complement, the rejection region $\{\mathbf{m}\}_{\text{Rej}}$. The loss α caused by this 'χ^2 cut' (i.e. the rejection of tracks with $\chi^2 \geqslant \chi_c^2$) is given by

$$\alpha = \int_{\chi_c^2}^{\infty} g_k(x)\, \mathrm{d}x = 1 - G_k(\chi^2) = \mathrm{prob}\left(\{\mathbf{m}\}_{\text{Rej}}\right) \qquad (3.28)$$

with $G_k(\chi^2)$ as defined in Equation (3.27).

The probability α of rejecting real tracks, i.e. with H_0 true (prob $(\{\mathbf{m}\}_{\text{Rej}}|$ $H_0 = \text{true}))$ – often quoted as percentage 100α – is called the '*level of significance*' or the '*size*' of the test. In practice α should be kept as small as possible; according to the level of background, it should lie between 10% and 0.1%. If, due to high background contamination, α must be chosen to be larger than this, the danger of a bias must be considered with great care. Tables for $\alpha(\chi_c^2, k)$ are given in most text books on statistics, and practical implementation of Formula (3.28) may be found in scientific computer program libraries (e.g. NAGLIB and the CERN program libraries, which are now transformed into object oriented libraries based on C++ (the LHC++ project)). The quantity $1 - \alpha$ is called the '*confidence level*' or '*confidence coefficient*'. Rejecting the track hypothesis H_0 when it is true is called a 'type I error' or an error of *the first kind* (see also Section 4.2).

The experimental χ^2 distribution g_{exp} is the weighted sum ('mixture') of two probability densities: a real χ^2 distribution and a 'χ^2 distribution as obtained from the background', i.e. of wrongly associated coordinates (g_{bg}) where the χ^2 was evaluated using the LSM under the false assumption $H_0 = $ true (Fig. 3.4):

$$g_{exp}(\chi^2) = g(\chi^2|H_0 = \text{true})\,\text{prob}\,(H_0 = \text{true})$$

$$+g_{bg}(\chi^2|H_0 = \text{false})\,\text{prob}\,(H_0 = \text{false}) \qquad (3.29)$$

where 'bg' denotes background.

Accepting H_0 when it is false is called a 'type II error' or an error of *the second kind*, and the corresponding probability is

$$\beta = \int_0^{\chi_c^2} g_{bg}(x|H_0 = \text{false})\,dx \qquad (3.30)$$

The quantity $1 - \beta$ is called the '*power*' of the test. The total contamination is given by the ratio of background to accepted events

$$\text{Contamination} = \frac{\beta\,\text{prob}\,(H_0 = \text{false})}{(1-\alpha)\,\text{prob}\,(H_0 = \text{true}) + \beta\,\text{prob}\,(H_0 = \text{false})} \times 100\%$$

$$(3.31)$$

H_1 can be more elaborate than just the complement of H_0, e.g. when a coordinate could be assigned to two different tracks or when looking for a primary and a secondary vertex, a second test quantity can be chosen for the LSM fit under the assumption that H_1 is true. This will reduce the risk of a wrong decision, but can easily cause a bias: namely, that the loss is no longer a *representative sample* for the all track topologies of good events, which means that differential cross sections can be biassed by different losses for different momenta.

If the background is such that its suppression would cause too important a loss of tracks, and if no other information is obtainable, the only solution is to improve the overall detector resolution, keeping the χ^2 distribution of good events stable when the correct weights are applied for the LSM, but shifting the pseudo χ^2 obtained from background to larger values. Another way to improve the background separation would be to add additional detectors, but there are fundamental limits to this owing to multiple scattering, energy loss, and secondary processes ('*overinstrumentation*'), and also more practical ones such as the space, money, and the manpower available.

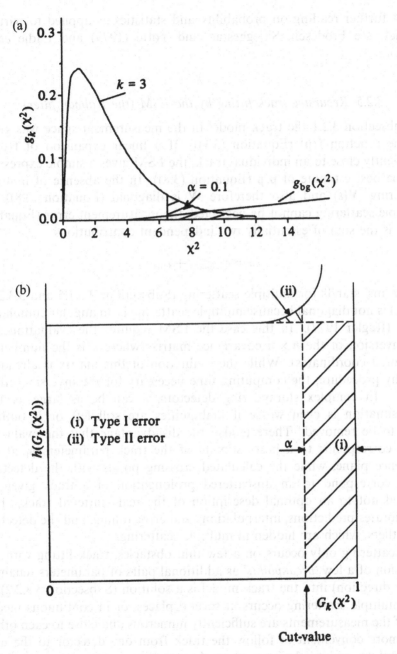

Fig. 3.4. (a) χ^2 distribution for three degrees of freedom. The level of significance α of 10% corresponds to a confidence level of 90%. The experimental χ^2 distribution is the weighted sum of g_k and g_{bg}. (b) The density function as obtained by the inverse of the distribution function $G_k(\chi^2)$, with (the prime denotes derivative) $G'_k(\chi^2) = g_k(\chi^2)$ Equation (3.27). The enhancement on the right side is caused by wrongly associated coordinates, giving an excess of large χ^2s.

For further reading on probability and statistics as applied to particle physics, see Frodesen, Skjeggestad and Tøfte (1979) and Eadie *et al.* (1971).

3.2.5 Recursive track fitting by the LSM (the Kalman filter)

In Subsection 3.2.1 the track model in the measurement space was given by the function $\mathbf{f}(\mathbf{p})$ (Equation (3.1)). If a linear expansion of $\mathbf{f}(\mathbf{p})$ is sufficiently close to an individual track, the LSM gives a simple expression for the best estimate of $\mathbf{p}, \tilde{\mathbf{p}}$ (Equation (3.8)). In the absence of multiple scattering, $\mathbf{V}(\varepsilon)$ and also therefore $\mathbf{W}$ is diagonal (Equation (3.8)). If multiple scattering cannot be neglected, the measurement error (Equation (3.2)) is the sum of essentially two independent contributions

$$\varepsilon = \varepsilon_{\text{detector}} + \varepsilon_{\text{ms}} \tag{3.32}$$

where 'ms' stands for multiple scattering (Subsections 3.3.1.5 and 3.3.2.2). $\mathbf{V}(\varepsilon_{\text{ms}})$ is not diagonal because multiple scattering is an angular cumulative effect (Regler 1977). In this case the LSM requires the evaluation and the inversion of the $n \times n$ covariance matrix (where n is the number of measured coordinates). While the evaluation of this matrix is already a lengthy procedure, the computing time necessary for the inversion grows as n^3. In complex storage ring detectors, n can be as large as 100. The situation is even worse if ambiguities are still left or if outliers have to be removed. There is also the disadvantage that the evaluated track corresponds to a best estimate of the track parameters $\tilde{\mathbf{p}}_r$ at the reference plane, while the calculated crossing points with the detectors, $\mathbf{f}(\tilde{\mathbf{p}}_r)$, correspond to an unscattered prolongation of a track given by $\tilde{\mathbf{p}}_r$, and not to an optimal description of the real scattered track. This deteriorates predictions, interpolations and error tuning, and the detection of outliers, which are hidden in multiple scattering.

If scattering only occurs on a few thin obstacles, track fitting with the inclusion of a few *'breakpoints'* as additional pairs of parameters (changes in the direction) into the track model is a solution (Subsection 3.3.2.2).

If multiple scattering occurs at several places, or in continuous media, and if the measurements are sufficiently numerous and close to each other, it is more convenient to follow the track from one detector to the next one and to update the current track parameters when a measurement is encountered. This method has first been proposed under the name of 'progressive fit' (Billoir 1984; Billoir, Frühwirth and Regler 1985). Subsequently it was shown that it is a special case of the Kalman filter, an algorithm that is well known in time-series analysis and signal processing (Gelb 1975; Brammer and Siffling 1975; Anderson and Moore 1979).

The Kalman filter is a method of estimating the states of dynamic systems. A dynamic system is an evolving stochastic model of some time-varying phenomenon. The notion of 'time' is to be interpreted in a generic sense; it denotes a univariate parameter on which the system state $\mathbf{p}(t)$ is functionally dependent. For an application to track fitting only systems in discrete time need to be considered. The state of the system at time t_k is denoted by $\mathbf{p}_k$.

The application of the Kalman filter to track fitting is straightforward if the track is interpreted as a discrete dynamic system (Frühwirth 1987; Regler and Frühwirth 1990). To this end the detector surfaces crossed by the track are numbered from 1 to n; the state of the track at surface k is given by the 5-vector $\mathbf{p}_k$ of the track parameters at the intersection of the track with surface k.

The trajectory of the particle between two adjacent surfaces is described by a deterministic function on to which is superimposed some random disturbance. This is summarized in the *system equation*:

$$\mathbf{p}_k = \mathbf{f}_k(\mathbf{p}_{k-1}) + \mathbf{P}_k\boldsymbol{\delta}_k, \quad \langle\boldsymbol{\delta}_k\rangle = 0, \quad \mathbf{C}(\boldsymbol{\delta}_k) = \mathbf{Q}_k \tag{3.33}$$

The function $\mathbf{f}_k$ is the track model between surfaces $k-1$ and k (see Subsection 3.3.1). The random variable $\boldsymbol{\delta}_k$ describes the effect of matter on the trajectory, mainly multiple Coulomb scattering (see Subsection 3.3.1.5). In the language of dynamic systems it is called the *process noise*. Its covariance matrix $\mathbf{Q}_k$ can be assumed to be known for all k (see Subsection 3.3.2.2). The matrix $\mathbf{P}_k$ takes into account the fact that the process noise $\boldsymbol{\delta}_k$ often affects only a subset of the state vector. For instance, in thin scatterers only the track direction is influenced by multiple scattering, whereas the offset is negligible.

It is not required that the state of the system can be observed directly, although this is not excluded. In general the observations are functions of the state vector, and they can be corrupted by measurement errors. This is summarized in the *measurement equation*:

$$\mathbf{m}_k = \mathbf{h}_k(\mathbf{p}_k) + \boldsymbol{\varepsilon}_k, \quad \langle\boldsymbol{\varepsilon}_k\rangle = 0, \quad \mathbf{C}(\boldsymbol{\varepsilon}_k) = \mathbf{V}_k = \mathbf{W}_k^{-1} \tag{3.34}$$

The function $\mathbf{h}_k$ maps the track parameters on the measurements provided by the particular detector at surface k. This mapping is often a simple projection on a subset of the state vector. The covariance matrix $\mathbf{V}_k$ of the measurement errors is assumed to be known. In addition it is postulated that all $\boldsymbol{\delta}_k$ and all $\boldsymbol{\varepsilon}_k$ are independent random variables.

The track model $\mathbf{f}_k$ is nonlinear in most cases. As the Kalman filter in its basic form assumes a linear system equation the track model is approximated by a linear function, namely its first-order Taylor expansion with derivative matrix $\mathbf{F}_k$. The function $\mathbf{h}_k$ is frequently linear; otherwise it can be approximated by its first-order Taylor expansion with derivative matrix H_k. It may therefore be assumed that both the system equations and the measurement equations are linear for all k:

$$\mathbf{f}_k(\mathbf{p}_{k-1}) = \mathbf{F}_k \mathbf{p}_{k-1} + \mathbf{c}_k, \quad \mathbf{h}_k(\mathbf{p}_k) = \mathbf{H}_k \mathbf{p}_k + \mathbf{d}_k, \quad k = 1,\ldots,n \qquad (3.35)$$

In the following the constant terms $\mathbf{c}_k$ and $\mathbf{d}_k$ are suppressed for the sake of convenience. The aim is to find a method of estimating the track parameters from the observations. Suppose that an estimate of $\mathbf{p}_k$ is based on the observations $\{\mathbf{m}_1,\ldots,\mathbf{m}_j\}$. Such an estimate is denoted by $\mathbf{p}_{k|j}$. In order not to clutter the notation, the tilde on estimated quantities is dropped from now on. Three cases can be distinguished:

(a) $k > j$: prediction, estimation of a 'future' state;

(b) $k = j$: filtering, estimation of the 'present' state by including the current measurement $\mathbf{m}_k$;

(c) $k < j$: smoothing, estimation of a 'past' state (see below).

In the linearized system, the Kalman filter is the linear least-squares estimator of the state vector. It works recursively. Suppose that an estimate $\mathbf{p}_{k-1|k-1}$ of $\mathbf{p}_{k-1}$ is available, based on observations $\mathbf{m}_1,\ldots,\mathbf{m}_{k-1}$, along with its covariance matrix $\mathbf{C}_{k-1|k-1} = \mathbf{C}(\mathbf{p}_{k-1|k-1})$. The Kalman filter needs two steps to produce an estimate $\mathbf{p}_{k|k}$ of $\mathbf{p}_k$. The first one is a prediction step from $k-1$ to k:

$$\mathbf{p}_{k|k-1} = \mathbf{F}_k \mathbf{p}_{k-1|k-1} \qquad (3.36a)$$

$$\mathbf{C}_{k|k-1} = \mathbf{F}_k \mathbf{C}_{k-1|k-1} \mathbf{F}_k^{\mathrm{T}} + \mathbf{P}_k \mathbf{Q}_k \mathbf{P}_k^{\mathrm{T}} \qquad (3.36b)$$

The predicted state vector $\mathbf{p}_{k|k-1}$ contains the information from all measurements up to $k-1$. Its covariance matrix $\mathbf{C}_{k|k-1}$ is a sum of two terms, the first one being the linear error propagation of the previous covariance matrix, the second one taking into account the process noise. It should be pointed out that the predicted covariance matrix $\mathbf{C}_{k|k-1}$ may fail to be positive definite because of rounding errors, especially if the distance spanned by the prediction step is large and some of the elements of the starting matrix $\mathbf{C}_{k-1|k-1}$ are small because some measurements are very accurate. Experience shows that it is normally mandatory to implement the entire Kalman filter in 64-bit precision. If this is not sufficient it is recommended to use the square-root form of the filter (Anderson and Moore 1979).

The LS estimator of $\mathbf{p}_k$ based on $\mathbf{p}_{k|k-1}$ and $\mathbf{m}_k$ is obtained by minimizing the following objective function

$$L(\mathbf{p}_k) = (\mathbf{m}_k - \mathbf{H}_k \mathbf{p}_k)^{\mathrm{T}} \mathbf{W}_k (\mathbf{m}_k - \mathbf{H}_k \mathbf{p}_k)$$

$$+ (\mathbf{p}_{k|k-1} - \mathbf{p}_k)^{\mathrm{T}} \mathbf{C}_{k|k-1}^{-1} (\mathbf{p}_{k|k-1} - \mathbf{p}_k) \implies \min \qquad (3.37)$$

The estimator $\mathbf{p}_{k|k}$ is a weighted mean of the prediction and the observation, generalized to the multivariate case:

$$\mathbf{p}_{k|k} = \mathbf{C}_{k|k}[\mathbf{C}_{k|k-1}^{-1}\mathbf{p}_{k|k-1} + \mathbf{H}_k^T\mathbf{W}_k\mathbf{m}_k] \tag{3.38a}$$

$$\mathbf{C}_{k|k} = [\mathbf{C}_{k|k-1}^{-1} + \mathbf{H}_k^T\mathbf{W}_k\mathbf{H}_k]^{-1} \tag{3.38b}$$

By using the relation

$$\mathbf{C}_{k|k-1}^{-1} = \mathbf{C}_{k|k}^{-1} - \mathbf{H}_k^T\mathbf{W}_k\mathbf{H}_k \tag{3.39}$$

and after some matrix algebra, the weighted mean and its covariance matrix can alternatively be written in the following form, which is the usual one in the literature on filtering:

$$\mathbf{p}_{k|k} = \mathbf{p}_{k|k-1} + \mathbf{K}_k(\mathbf{m}_k - \mathbf{H}_k\mathbf{p}_{k|k-1}) \tag{3.40a}$$

$$\mathbf{K}_k = \mathbf{C}_{k|k-1}\mathbf{H}_k^T(\mathbf{V}_k + \mathbf{H}_k\mathbf{C}_{k|k-1}\mathbf{H}_k^T)^{-1} \tag{3.40b}$$

$$\mathbf{C}_{k|k} = (\mathbf{I} - \mathbf{K}_k\mathbf{H}_k)\mathbf{C}_{k|k-1} \tag{3.40c}$$

The filtered estimate now appears as the prediction plus a correction term involving the current observation. The influence of this correction term is determined by the gain matrix $\mathbf{K}_k$ of the filter. If the filter were implemented in hardware the gain matrix would be represented by an amplifier. After the completion of the filter step the fitting proceeds by the next prediction step. The alternation of prediction and filter steps is represented graphically in Fig. 3.5.

The filtered estimate $\mathbf{p}_{k|k}$ has all the optimum properties of a linear LS estimator. As in every linear model, there are two important test statistics that accompany the filtered estimate, the residuals and the filtered χ^2:

$$\mathbf{r}_{k|k} = \mathbf{m}_k - \mathbf{H}_k\mathbf{p}_{k|k} \tag{3.41a}$$

$$\mathbf{R}_{k|k} = \mathbf{C}(\mathbf{r}_{k|k}) = \mathbf{V}_k - \mathbf{H}_k\mathbf{C}_{k|k}\mathbf{H}_k^T \tag{3.41b}$$

$$\chi^2_{k,F} = \mathbf{r}_{k|k}^T\mathbf{R}_{k|k}^{-1}\mathbf{r}_{k|k} \tag{3.41c}$$

In the normal or Gaussian case, $\mathbf{p}_{k|k}$ is the estimator that attains the minimum variance bound, and $\chi^2_{k,F}$ is χ^2-distributed with $\dim(\mathbf{m}_k)$ degrees of freedom. The 'total chi-square' χ^2_k is the sum of the individual chi-squares of all filter steps:

$$\chi^2_k = \chi^2_{k-1} + \chi^2_{k,F} \tag{3.42}$$

As the $\chi^2_{k,F}$ are independent, the numbers of degrees of freedom have to be added as well.

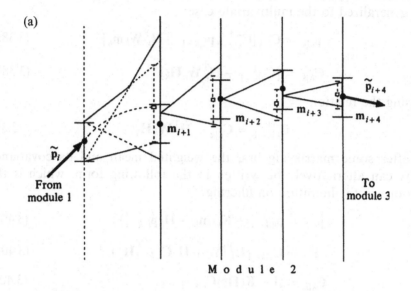

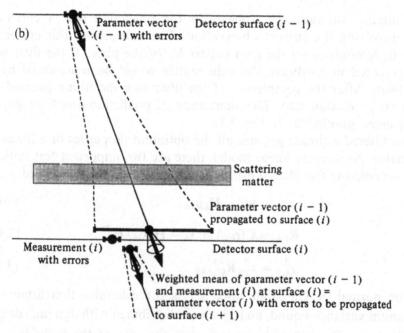

Fig. 3.5. (a) Progressive adjustment of the fitted parameters by propagation ('predictions') and the stepwise inclusion of further measurements. (b) A detailed outline of one step of the filtering procedure.

In order to get the filter going, an estimate $\mathbf{p}_{1|1}$ of the initial state of the system is required, along with its covariance matrix $\mathbf{C}_{1|1}$. If the first observation $\mathbf{m}_1$ is a full track element with a measurement of position, direction, and momentum or curvature, one takes the observation and its covariance matrix as the initial state. In general, however, the first observation will not be sufficient to determine completely the initial state. In this case the remaining components of the state vector are given arbitrary but large errors, in order not to bias the subsequent filter steps. This may, however, lead to numerical problems in the smoother (see below).

The Kalman filter can also be written down in terms of weight or information matrices instead of covariance matrices. This is called an *information filter*. The prediction step now reads:

$$\mathbf{p}_{k|k-1} = \mathbf{F}_k \mathbf{p}_{k-1|k-1} \tag{3.43a}$$

$$\mathbf{G}_{k|k-1} = (\mathbf{F}_k \mathbf{G}_{k-1|k-1}^{-1} \mathbf{F}_k^{\mathrm{T}} + \mathbf{P}_k \mathbf{Q}_k \mathbf{P}_k^{\mathrm{T}})^{-1} \tag{3.43b}$$

In the absence of process noise this can be simplified to

$$\mathbf{G}_{k|k-1} = (\mathbf{F}_k^{\mathrm{T}})^{-1} \mathbf{G}_{k-1|k-1} \mathbf{F}_k^{-1} \tag{3.44}$$

The filter step in the information filter is given by

$$\mathbf{p}_{k|k} = \mathbf{G}_{k|k}^{-1} [\mathbf{G}_{k|k-1} \mathbf{p}_{k|k-1} + \mathbf{H}_k^{\mathrm{T}} \mathbf{W}_k \mathbf{m}_k] \tag{3.45a}$$

$$\mathbf{G}_{k|k} = \mathbf{G}_{k|k-1} + \mathbf{H}_k^{\mathrm{T}} \mathbf{W}_k \mathbf{H}_k \tag{3.45b}$$

One of the advantages of the information filter is that unknown components of the initial state can be given zero weight corresponding to infinitely large errors.

When the filter has reached the last (nth) measurement, the final estimate $\mathbf{p}_{n|n}$ contains the full information from all measurements $\{\mathbf{m}_1, \ldots, \mathbf{m}_n\}$. The full information can then be passed back to all previous estimates by the *smoother*, updating all filtered estimates:

$$\mathbf{p}_{k|n} = \mathbf{p}_{k|k} - \mathbf{A}_k(\mathbf{p}_{k+1|k} - \mathbf{p}_{k+1|n}) \tag{3.46a}$$

$$\mathbf{C}_{k|n} = \mathbf{C}_{k|k} - \mathbf{A}_k(\mathbf{C}_{k+1|k} - \mathbf{C}_{k+1|n})\mathbf{A}_k^{\mathrm{T}} \tag{3.46b}$$

$$\mathbf{A}_k = \mathbf{C}_{k|k} \mathbf{F}_{k+1}^{\mathrm{T}} \mathbf{C}_{k+1|k}^{-1} \tag{3.46c}$$

As $\mathbf{p}_{k|n}$ contains more information than $\mathbf{p}_{k|k}$, its covariance matrix is smaller. The smoothed estimate is again accompanied by residuals and a χ^2:

$$\mathbf{r}_{k|n} = \mathbf{m}_k - \mathbf{H}_k \mathbf{p}_{k|n} \tag{3.47a}$$

$$\mathbf{R}_{k|n} = \mathbf{C}(\mathbf{r}_{k|n}) = \mathbf{V}_k - \mathbf{H}_k\mathbf{C}_{k|n}\mathbf{H}_k^T \qquad (3.47b)$$

$$\chi^2_{k,S} = \mathbf{r}_{k|n}^T\mathbf{R}_{k|n}^{-1}\mathbf{r}_{k|n} \qquad (3.47c)$$

In the normal case, $\chi^2_{k,S}$ is χ^2-distributed with dim $(\mathbf{m}_k)$ degrees of freedom. As all estimates $\mathbf{p}_{k|n}$ contain the same information, the $\chi^2_{k,S}$ are no longer independent.

In this version of the smoother, the covariance matrices $\mathbf{C}_{k|n}$ are computed by forming the difference of two positive definite matrices. This difference should again turn out to be positive definite, but may fail to do so because of rounding errors. Most of the time this happens in the starting detector of the filter, where the state vector is usually not fully defined and its covariance matrix contains some large elements reflecting the uncertainty of the corresponding elements of the state vector.

Alternatively, the smoother can be implemented by running a second filter backwards and by combining the filtered estimates of one filter with the predicted estimates of the other filter by a weighted mean. If the filters are implemented as information filters, the computation of the weighted mean of the estimates is very fast. This approach is known as the 'two-filter formula' for smoothing. Experience shows that it is numerically more stable than the covariance matrix version of the smoother.

In special cases also the two-filter smoother may be plagued by numerical problems. In this case, an attempt can be made to implement the filters in square-root form working only with the Cholesky factors of the covariance or weight matrices (Anderson and Moore 1979). There is empirical evidence indicating that the square-root form of the information filter is numerically the most stable one.

Track fitting with the Kalman filter/smoother has many advantages.

- It is suitable for simultaneous track finding and track fitting (see Subsection 2.4.2.2), possibly after some prefiltering by other pattern recognition methods.

- No large matrices have to be inverted, and the computational cost is proportional to the number of surfaces crossed by the track. It is therefore fast even when there is frequent multiple scattering and a large number of measurements.

- The estimated track parameters closely follow the real path of the particle, and not just the extrapolation of the initial conditions.

- The linear approximation of the track model need not be valid over the entire length of the track, but only between adjacent surfaces.

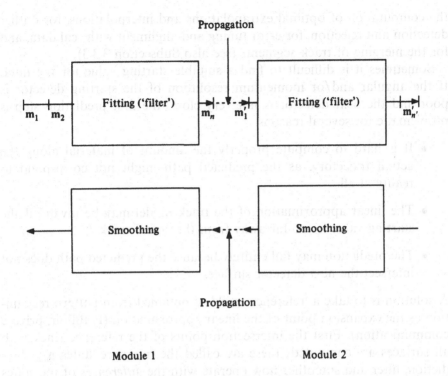

Fig. 3.6. After the complete filtering procedure (track fit) the *full information* is transmitted back to all intermediate points by the smoothing algorithm.

- Optimal predictions into other detectors (vertex detectors, calorimeters, muon chambers) can be computed from both ends of a track segment. Optimal interpolations, for instance into RICH detectors, can be computed as well.

- It is easy to remove the information of measurement m_k from the smoothed estimate $p_{k|n}$. This yields an optimal estimate of the track parameters p_k that is based on all measurements with the exception of m_k. Since this estimate normally gives a better definition of the track than the filtered estimate, it should be used for the final decision on outliers and ambiguities; a similar, but less restrictive, procedure can be applied in the filter stage. This new estimate can also be used for the checking and tuning of the detector alignment and resolution.

- The smoother can also be used for efficient track segment merging (see Fig. 3.6).

It can be concluded that the combined Kalman filter/smoother is a powerful, flexible, and efficient tool not only for track fitting, but also for the computation of optimal extrapolations and interpolations, for outlier detection and rejection, for error tuning and alignment with real data, and for the merging of track segments (see also Subsection 3.3.3).

Sometimes it is difficult to find a suitable starting value for the filter. If the angular and/or momentum resolution of the starting detector is poor and the next detector is not very close, the first prediction step is problematic for several reasons.

- It is hard to compute properly the amount of material along the actual trajectory, as the predicted path might not correspond to reality at all.

- The linear approximation of the track model may be invalid if the starting value is too far away from the actual track.

- The prediction may fail entirely because the predicted path does not intersect the next detector surface.

A solution is to take a 'reference track' as obtained from pattern recognition as the expansion point of the linear approximation (P. Billoir, private communication). First the intersection points of the reference track with all surfaces are computed; these are called the reference states $\mathbf{p}_{k,r}$. Prediction, filter and smoother now operate with the *differences* of the states $\mathbf{p}_k$ and the reference states $\mathbf{p}_{k,r}$, and all derivative matrices are computed at the reference states. In particular:

- the state vector $\mathbf{p}_k$ is replaced by $\Delta\mathbf{p}_k = \mathbf{p}_k - \mathbf{p}_{k,r}$;
- the estimates $\mathbf{p}_{k|j}$ are replaced by $\Delta\mathbf{p}_{k|j} = \mathbf{p}_{k|j} - \mathbf{p}_{k,r}$;
- the measurement $\mathbf{m}_k$ is replaced by $\Delta\mathbf{m}_k = \mathbf{m}_k - \mathbf{H}_k\mathbf{p}_{k,r}$.

With these substitutions both the filter and the smoother formulae can be used without modifications. It is strongly recommended that a reference track be used whenever possible.

3.2.6 Robust filtering

All estimators that have been presented in this section are based on the least-squares principle. They have optimal properties if two conditions are met.

- The track model, if not linear, can be approximated by a linear function in a sufficiently large neighbourhood of the actual track.

- The distribution of the measurement errors and of the process noise (mainly multiple scattering) is Gaussian or very close to a Gaussian.

The first condition can usually be met by a suitable choice of the track parameters and by a careful choice of the expansion point. The second condition, however, may not always be fulfilled. In this case a nonlinear estimator may do better than the linear least-squares estimator. It will be shown below how the linear Kalman filter can be generalized in order to deal with non-Gaussian noise, provided that all distributions involved are mixtures of Gaussians or Gaussian sums. The resulting algorithm, in which several Kalman filters run in parallel, is called the Gaussian-sum filter (Kitagawa 1989; Frühwirth 1997).

There are several cases in which the application of the Gaussian-sum filter seems promising.

- In real detectors the measurement errors are not always Gaussian, and there is virtually always a tail of outlying observations. In principle there are two approaches to the treatment of these outliers. In the first approach one tries to identify and to eliminate the outliers, based on the estimates obtained by the smoother. This results in an iterative and lengthy procedure, particularly if there are several outliers. One can avoid this by accommodating the outliers by a robustified estimator. For example, the least-squares estimator can be generalized to an M-estimator by downweighting outlying measurements (see Section 3.5). More sophisticated methods such as the Gaussian-sum filter make use of prior information or assumptions on the distribution of the outliers. In most cases it is sufficient to model the distribution of the measurement errors by a mixture of two Gaussians: the first component describes the 'good' measurements or the 'core' of the distribution, the second one describes the outliers or the 'tails'.

- The distribution of the process noise can be approximated sufficiently well by a Gaussian if it is dominated by multiple scattering and if the scatterer is homogeneous. If a scatterer is inhomogeneous it is the usual practice to smear the material and to work with an average thickness and radiation length. In this case the distribution of multiple scattering can again be described in more detail by a Gaussian sum.

- For minimum ionizing particles the process noise is indeed dominated by multiple scattering; although energy loss by ionization is a random process, its standard deviation is small compared to its mean. It is therefore usually approximated by a constant value. For electrons, however, energy loss is dominated by bremsstrahlung above a certain threshold and requires a stochastic model with a strongly non-Gaussian distribution (Stampfer *et al.* 1994). In order to apply the Gaussian-sum filter this distribution has to be approximated by a Gaussian sum (Frühwirth and Frühwirth-Schnatter 1998).

The Gaussian-sum filter is closely related to the Kalman filter; actually, it consists of several Kalman filters running in parallel. It can be derived most easily by recurring to Bayesian principles. Consider first the filter step at measurement surface k. The predicted distribution of the state vector $\mathbf{p}_k$ conditional on the observations $\mathbf{M}_{k-1} = \{\mathbf{m}_1, \ldots, \mathbf{m}_{k-1}\}$ can be assumed to be a Gaussian sum with N_{k-1} components:

$$g(\mathbf{p}_k | \mathbf{M}_{k-1}) = \sum_{j=1}^{N_{k-1}} \pi_k^j \varphi(\mathbf{p}_k ; \mathbf{p}_{k|k-1}^j, \mathbf{C}_{k|k-1}^j), \quad \sum_{j=1}^{N_{k-1}} \pi_k^j = 1 \qquad (3.48)$$

where $\varphi(\mathbf{x}; \boldsymbol{\mu}, \mathbf{V})$ is a multivariate Gaussian density with mean $\boldsymbol{\mu}$ and covariance matrix $\mathbf{V}$. This is the prior distribution of the state vector $\mathbf{p}_k$.

The distribution of the observation error $\boldsymbol{\varepsilon}_k$ is modeled by a Gaussian sum with M_k components:

$$g(\boldsymbol{\varepsilon}_k) = \sum_{i=1}^{M_k} p_k^i \varphi(\boldsymbol{\varepsilon}_k ; \mathbf{0}, \mathbf{V}_k^i), \quad \sum_{i=1}^{M_k} p_k^i = 1 \qquad (3.49)$$

Normally a mixture of two components is sufficient to model long-tailed measurement errors. The density of the observation $\mathbf{m}_k$ conditional on the state $\mathbf{p}_k$ is given by

$$g(\mathbf{m}_k | \mathbf{p}_k) = \sum_{i=1}^{M_k} p_k^i \varphi(\mathbf{m}_k ; \mathbf{H}_k \mathbf{p}_k, \mathbf{V}_k^i) \qquad (3.50)$$

Application of Bayes' theorem then leads to the following posterior density which is the density of the filtered estimate:

$$g(\mathbf{p}_k | \mathbf{M}_k) \equiv g(\mathbf{p}_k | \mathbf{m}_k, \mathbf{M}_{k-1}) = \frac{g(\mathbf{m}_k | \mathbf{p}_k) g(\mathbf{p}_k | \mathbf{M}_{k-1})}{\int g(\mathbf{m}_k | \mathbf{p}_k) g(\mathbf{p}_k | \mathbf{M}_{k-1}) \, d\mathbf{p}_k} \qquad (3.51)$$

The posterior density can be written in the following way:

$$g(\mathbf{p}_k | \mathbf{M}_k) = \sum_{i=1}^{M_k} \sum_{j=1}^{N_{k-1}} q_k^{ij} \varphi(\mathbf{p}_k ; \mathbf{p}_{k|k}^{ij}, \mathbf{C}_{k|k}^{ij}) \qquad (3.52)$$

Thus the posterior density is a sum of $n_k = M_k N_{k-1}$ Gaussian components with the following posterior weights:

$$q_k^{ij} \propto p_k^i \pi_k^j \varphi(\mathbf{m}_k ; \mathbf{H}_k \mathbf{p}_{k|k-1}^j, \mathbf{V}_k^i + \mathbf{H}_k \mathbf{C}_{k|k-1}^j \mathbf{H}_k^{\mathrm{T}}) \qquad (3.53)$$

The constant of proportionality is determined by the requirement that the sum of all q_k^{ij} is equal to 1. The posterior weights are functions of the observation and of the prediction. Their size depends on the (weighted) distance of the observation from the prediction. If this distance is large, the posterior weight is small, and the corresponding component does not contribute very much to the total mixture. Thus outlying measurements are automatically downweighted.

The mean and the covariance matrix of each component is obtained by a Kalman filter

$$\mathbf{p}_{k|k}^{ij} = \mathbf{p}_{k|k-1}^{j} + \mathbf{C}_{k|k}^{ij} \mathbf{H}_k^T \mathbf{W}_k^i (\mathbf{m}_k - \mathbf{H}_k \mathbf{p}_{k|k-1}^{j}) \tag{3.54a}$$

$$\mathbf{C}_{k|k}^{ij} = [(\mathbf{C}_{k|k-1}^{j})^{-1} + \mathbf{H}_k^T \mathbf{W}_k^i \mathbf{H}_k]^{-1} \tag{3.54b}$$

There is a 'chi-square' statistic associated to each Kalman filter:

$$(\chi_F^2)_k^{ij} = (\mathbf{m}_k - \mathbf{H}_k \mathbf{p}_{k|k}^{ij})^T (\mathbf{V}_k^i - \mathbf{H}_k \mathbf{C}_{k|k}^{ij} \mathbf{H}_k^T)^{-1} (\mathbf{m}_k - \mathbf{H}_k \mathbf{p}_{k|k}^{ij}) \tag{3.55}$$

The final Bayesian estimate $\mathbf{p}_{k|k}$ and its covariance matrix $\mathbf{C}_{k|k}$ are obtained as the mean and the covariance matrix of the posterior distribution $g(\mathbf{p}_k|\mathbf{M}_k)$. Normally it is computed only in the last detector in the fit. After renumbering the components with a single index the posterior can be written as

$$g(\mathbf{p}_k|\mathbf{M}_k) = \sum_{l=1}^{n_k} q_k^l \varphi(\mathbf{p}_k; \mathbf{p}_{k|k}^l, \mathbf{C}_{k|k}^l) \tag{3.56}$$

Then the mean and the covariance matrix of the mixture are equal to

$$\mathbf{p}_{k|k} = \sum_{l=1}^{n_k} q_k^l \mathbf{p}_{k|k}^l \tag{3.57a}$$

$$\mathbf{C}_{k|k} = \sum_{l=1}^{n_k} q_k^l (\mathbf{C}_{k|k}^l + \mathbf{p}_{k|k}^l \mathbf{p}_{k|k}^{lT}) - \mathbf{p}_{k|k} \mathbf{p}_{k|k}^T \tag{3.57b}$$

The 'chi-square' statistic of the filter step is computed by the appropriate weighted sum:

$$(\chi_F^2)_k = \sum_{l=1}^{n_k} q_k^l (\chi_F^2)_k^l \tag{3.58}$$

Summing all 'chi-squares' yields a 'total chi-square' statistic of the track. It is, of course, not actually chi-square distributed.

The posterior distribution of filter step k has $n_k = M_k N_{k-1}$ components. In the subsequent prediction step each component is propagated separately. If the process noise in the prediction step is a sum of L_k Gaussians, the prior of filter step k is a sum of $N_k = L_k M_k N_{k-1}$ Gaussians. The number of components thus rises very rapidly and has to be reduced to a manageable number. This is best done by clustering components which are close to each other as measured by the Kullback–Leibler distance.

The Gaussian-sum smoother is implemented most easily by the two-filter formula (Kitagawa 1994; Frühwirth 1997). The smoothed estimate at measurement surface k is computed according to the following prescription. The prior of the forward filter at step k is given by

$$g(\mathbf{p}_k|\mathbf{M}_{k-1}) = \sum_{j=1}^{N_{k-1}} \pi_k^j \varphi(\mathbf{p}_k; \mathbf{p}_{k|k-1}^j, \mathbf{C}_{k|k-1}^j) \tag{3.59}$$

where $\mathbf{M}_{k-1}$ denotes the set of observations $\{\mathbf{m}_1, \ldots, \mathbf{m}_{k-1}\}$. The posterior of the backward filter at step k has a similar form:

$$g(\mathbf{p}_k|\mathbf{M}^k) = \sum_{l=1}^{N_k'} \beta_k^l \varphi(\mathbf{p}_k; \mathbf{p}_{k|k\to n}^l, \mathbf{C}_{k|k\to n}^l) \tag{3.60}$$

where $\mathbf{M}^k$ denotes the set of observations $\{\mathbf{m}_k, \ldots, \mathbf{m}_n\}$. The posterior estimate of the backward filter can be interpreted as a virtual observation of the state $\mathbf{p}_k$, the distribution of the observation error being given by the posterior density. Application of Bayes' theorem immediately yields the smoother density

$$g(\mathbf{p}_k|\mathbf{M}_n) \propto \sum_{j=1}^{N_{k-1}} \sum_{l=1}^{N_k'} \pi_k^j \beta_k^l \varphi(\mathbf{p}_k; \mathbf{p}_{k|k-1}^j, \mathbf{C}_{k|k-1}^j) \varphi(\mathbf{p}_k; \mathbf{p}_{k|k\to n}^l, \mathbf{C}_{k|k\to n}^l) \tag{3.61}$$

The smoother density can be written in the following form:

$$g(\mathbf{p}_k|\mathbf{M}_n) = \sum_{j=1}^{N_{k-1}} \sum_{l=1}^{N_k'} \gamma_k^{jl} \varphi(\mathbf{p}_k; \mathbf{p}_{k|n}^{jl}, \mathbf{C}_{k|n}^{jl}) \tag{3.62}$$

the weights γ_k^{jl} being given by

$$\gamma_k^{jl} \propto \pi_k^j \beta_k^l \varphi(\mathbf{p}_{k|k-1}^j; \mathbf{p}_{k|k\to n}^l, \mathbf{C}_{k|k-1}^j + \mathbf{C}_{k|k\to n}^l) \tag{3.63}$$

The constant of proportionality is determined by the requirement that the sum of all γ_k^{jl} is equal to 1. The mean $\mathbf{p}_{k|n}^{jl}$ and the covariance matrix $\mathbf{C}_{k|n}^{jl}$ of each component are the result of a Kalman filter, written here as a weighted mean

$$\mathbf{p}_{k|n}^{jl} = \mathbf{C}_{k|n}^{jl} [(\mathbf{C}_{k|k-1}^j)^{-1} \mathbf{p}_{k|k-1}^j + (\mathbf{C}_{k|k\to n}^l)^{-1} \mathbf{p}_{k|k\to n}^l] \tag{3.64a}$$

$$\mathbf{C}_{k|n}^{jl} = [(\mathbf{C}_{k|k-1}^j)^{-1} + (\mathbf{C}_{k|k\to n}^l)^{-1}]^{-1} \tag{3.64b}$$

The estimate $\mathbf{p}_{k|n}$ of the Bayesian smoother and its covariance matrix $\mathbf{C}_{k|n}$ are again obtained as the first and second moment of the smoother density. They need to be computed only in those places where optimal interpolations are required.

The properties of the Gaussian-sum filter/smoother in the context of track fitting have been investigated in a number of simulation experiments, both for long-tailed measurement errors (Frühwirth 1997) and for the energy loss of electrons (Frühwirth and Frühwirth-Schnatter 1998). In both cases it has been shown that frequently the Gaussian-sum filter gives substantially better estimates than the linear least-squares filter.

3.3 Fitting the tracks of charged particles

In high-energy physics it is important to determine the momentum of a charged particle. The most common way to achieve this is to deflect it with a magnetic field and to measure the deflection by position-sensitive detectors. Such a device is called a *'magnetic particle spectrometer'*. In some special cases the information from the tracking detectors for charged particles is also complemented by measurements from calorimeters. A typical example is the energy measurement of very-high-momentum electrons from a Z^0 decay. However, at least the sign of the curvature must still be measured by the tracking devices.

3.3.1 The track model

3.3.1.1 The equations of motion. The trajectory of a particle in a *static* magnetic field $\mathbf{B}(\mathbf{x})$ must satisfy the equations of motion given by the *Lorentz force*. It is assumed that there is no electric field. Neglecting bremsstrahlung and material effects, this force $\mathbf{f}$ is derived from Maxwell's equations to be

$$\mathbf{f} \sim q\mathbf{v} \times \mathbf{B} \qquad (3.65a)$$

where $\mathbf{v}$ is the velocity of the particle, $v = |\mathbf{v}|(\mathbf{v} \equiv d\mathbf{x}/dt)$ and q is the (signed) charge. This gives the equation of motion in a vaccum:

$$d\mathbf{p}/dt = d(m\gamma\, d\mathbf{x}/dt)/dt = c^2\kappa\, q\, \mathbf{v}(t) \times \mathbf{B}(\mathbf{x}(t)) \qquad (3.65b)$$

where κ is a proportionality factor, dependent on the choice of units (see below), $\mathbf{B}(\mathbf{x})$ is the static magnetic field (defined by its flux density), m is the rest mass and $\mathbf{x}$ is the position (a space point) of the particle, c is the velocity of light, and t is the time in the laboratory frame. The relativistic Lorentz factor γ is given by

$$\gamma = (1 - \beta^2)^{-\frac{1}{2}}$$

where

$$\beta = |\boldsymbol{\beta}| = v/c, \quad \boldsymbol{\beta} = \mathbf{v}/c = d\mathbf{x}/d(ct)$$

This equation can be rewritten in the form of geometrical quantities only:

$$d^2\mathbf{x}/ds^2 = (\kappa q/P)(d\mathbf{x}/ds) \times \boldsymbol{B}(\mathbf{x}(s)) \tag{3.66a}$$

where $s(t)$ is the distance along trajectory (path length), with $ds/dt = v$, and $\mathbf{p} = m\gamma\boldsymbol{\beta}c$ is the momentum of the particle, $P = |\mathbf{p}| = m\gamma\beta c$ (laboratory frame).

Proof:

$$d(m\gamma\, d\mathbf{x}/dt)/dt \cdot (d\mathbf{x}/dt) \sim \mathbf{v}(t) \cdot (\mathbf{v}(t) \times \mathbf{B}(\mathbf{x}(t))) \equiv 0 \quad \text{(see Equation (3.65b))}$$

and therefore:

$$|d\mathbf{x}/dt| = v = \beta c = \text{constant} \ (\gamma \equiv \text{constant})$$
$$d\mathbf{x}/dt = (d\mathbf{x}/ds)(ds/dt) = (d\mathbf{x}/ds)\beta c$$
$$d^2\mathbf{x}/dt^2 = (d^2\mathbf{x}/ds^2)\beta^2 c^2$$

∎

In particle physics, the following standard units are used (for more details on units see e.g. Jackson (1998), Bock *et al.* (1984)):

q in multiples of the positive elementary charge (dimensionless),
$\mathbf{x}$ and s in metres (sometimes still in centimetres)
P in GeV/c
$\boldsymbol{B}$ in tesla
κ is proportional to the velocity of light and is therefore defined as 0.299 792 458 (GeV/c)T^{-1} m^{-1}.

Sometimes, mainly in experiments with fixed targets, and hence comparatively small P_T/P_L at least in the forward region (P_T and P_L being respectively the transverse and longitudinal components of $\mathbf{p}$ with respect to the beam), it might be advantageous to rewrite Equation (3.66a) choosing the coordinate along the beam (say z) as a variable (the primes denote derivatives with respect to z), giving

$$\left. \begin{array}{l} x'' = (\kappa q/P)(ds/dz)[x'y'B_x - (1 + x'^2)B_y + y'B_z] \\ y'' = (\kappa q/P)(ds/dz)[(1 + y'^2)B_x - x'y'B_y - x'B_z] \end{array} \right\} \tag{3.66b}$$

with

$$ds^2 = dx^2 + dy^2 + dz^2$$
$$ds/dz = (1 + x'^2 + y'^2)^{\frac{1}{2}}$$

(The third equation becomes a trivial identity.)

When integrating Equation (3.66a) there are six integration constants (the unknown momentum P follows from the curvature), but with the identity

$$(dx/ds)^2 + (dy/ds)^2 + (dz/ds)^2 \equiv 1$$

and an arbitrary choice of one coordinate (the 'reference surface'), there are, in fact, (for known mass) only *five free parameters* (see also Section 3.2) defining the track (e.g. two for the impact with a reference surface at $z_r =$ const., two for the direction at that point, and one for the momentum – this is more transparent in Equation (3.66b)). Note that s can be given an arbitrary value on the reference surface.

In order to become familiar with these equations, the special case of a *homogeneous magnetic field* will first be discussed. Without loss of generality, **B** is chosen to be parallel to the z axis, $\mathbf{B} = B\mathbf{e}_z$, with $\mathbf{e}_z^T = (0, 0, 1)$.

Equation (3.66a) then takes the form

$$\left. \begin{array}{l} d^2x/ds^2 = (\kappa q/P)(dy/ds)B \\ d^2y/ds^2 = -(\kappa q/P)(dx/ds)B \\ d^2z/ds^2 = 0 \end{array} \right\} \tag{3.66c}$$

and the solution is a *helix* with an axis parallel to z (Fig. 3.7):

$$\left. \begin{array}{l} x(s) = \overset{0}{x} + R_H[\cos(\Phi_0 + hs\cos\lambda/R_H) - \cos\Phi_0] \\ y(s) = \overset{0}{y} + R_H[\sin(\Phi_0 + hs\cos\lambda/R_H) - \sin\Phi_0] \\ z(s) = \overset{0}{z} + s\sin\lambda \end{array} \right\} \tag{3.67a}$$

or with $\varphi(s) = \varphi_0 + hs\cos\lambda/R_H$:

$$\left. \begin{array}{l} x(\varphi) = \overset{0}{x} + hR_H(\sin\varphi - \sin\varphi_0) \\ y(\varphi) = \overset{0}{y} - hR_H(\cos\varphi - \cos\varphi_0) \\ z(\varphi) = \overset{0}{z} + hR_H \cdot \tan\lambda \cdot (\varphi - \varphi_0) \end{array} \right\} \tag{3.67b}$$

where s is the path length along the helix, which increases when moving in the particle's direction, $\overset{0}{x}$ is the starting point at $s = \overset{0}{s} = 0$, λ is the slope ('dip') angle $(= \arcsin(dz/ds), -\pi/2 \leqslant \lambda \leqslant \pi/2)$, and R_H is the radius of the helix $(= (P\cos\lambda)/(|\kappa q B|))$. The sense of rotation of the *projected helix* in the xy plane, h, is given by $h = -\text{sign}(qB_z) = \pm 1 (= \text{sign}(d\varphi/ds))$, where φ is the track direction, and z is the polar axis parallel to the helix axis); Φ_0 is the azimuth angle of the starting point (in cylindrical coordinates) with respect to the helix axis (the azimuth angle of the track direction) and $\varphi = \Phi + h\pi/2$. For many applications, the assumptions for an explicit solution, namely a homogeneous **B** field and the absence of matter, are

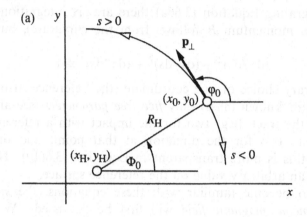

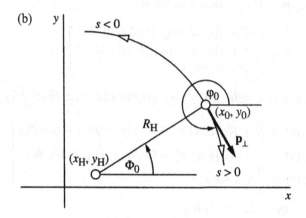

Fig. 3.7. (a) For a constant magnetic field parallel to the z axis, the solution of the equation of motion is a helix. The projection on the xy plane is a circle, which is chosen here with a positive sense of rotation ($h = +1$). (b) Projected helix with negative sense of rotation ($h = -1$). The projected curvature is $1/R_H$ and the curvature in space ($\equiv |d^2\mathbf{x}/ds^2|$) is $\cos^2 \lambda/R_H$.

approximately fulfilled so that Equation (3.67a) is precise enough to serve as a model for the track fit (Mitaroff 1987). This is the case in traditional bubble chambers and in the central region of track detectors at many storage ring experiments. In other cases, the approximation may still be good enough for pattern recognition.

   In storage ring experiments, the form of the magnetic field can be a dipole field (as in the UA1 detector at the CERN $p\bar{p}$ collider or the Double Arm Spectrometer (DASP) detector at DESY) or a solenoidal field (like most of the existing e^+e^- storage ring detectors, where synchrotron radiation makes a strong field perpendicular to the beam direction prohibitive). In such a detector the azimuth angle of a track intersecting with a cylindrical detector plays an important role. Substantial computing effort can be saved by using the variable $\varphi - \Phi$, thus eliminating temporarily one track parameter due to rotational invariance. The treatment of **B** fields that occur in practice will be discussed in more detail later.

3.3.1.2 The choice of track parameters. Using the helix solution (Equation (3.67a)), the choice of parameters to be fitted for a track model will now be discussed. Considering the usual set-up of a central detector consisting of coaxial cylinder surfaces with fixed radii R_i, one measures for each track impact two coordinates $(R\Phi)_i$ (Fig. 3.7) and z_i. The main magnetic field component is assumed to be parallel to the axis of the cylinder. The reference surface for defining the parameters may be chosen to be the cylindrical beam tube itself. Then the following choice for the parameters to be fitted will give a track model which is locally not too far from linear (Equations (3.1) and (3.4)):

$$\left.\begin{array}{ll} p_1 = (R\Phi)_r, & p_2 = z_r \\ p_3 = \varphi_r(\text{or } \varphi_r - \Phi_r), & p_4 = \tan \lambda_r \\ p_5 = (1/R_H)_r \times \text{sense of rotation} \end{array}\right\} \tag{3.68}$$

where the subscript 'r' denotes the actual value of the track parameters at the intersection with the reference cylinder $R = R_r$ (Fig 3.8(a)). Note that only p_3 depends on s for the helix track model (Equation (3.67b)).

For the *downstream spectrometer arm* of a *fixed-target detector* (Fig. 3.8(b)), the choice of parameters will be inspired by the Equations (3.66b). Such a spectrometer often consists of a first 'lever arm' with position-sensitive detectors in a field-free region (where the solution of the equation of motion simply gives a straight line), followed by a bending magnet with a strong magnetic field perpendicular to the main direction, covering only a small region as compared to the total length of the spectrometer (z was chosen in Equation (3.66b)), finally followed by a second lever arm of position-sensitive detectors. (For different kinds of detector arrangements see also Chapter 1.)

If the magnetic field is quite homogeneous, $\mathbf{B} = Be_y$ (the beam direction is again e_z), and the length of the magnet is L, the following approximate solution of Equation (3.66b) can be given:

$$\left.\begin{array}{l} \Delta x' = x'_2 - x'_1 \cong -\kappa qBL/P \\ \Delta y' = y'_2 - y'_1 \cong 0 \end{array}\right\} \tag{3.69a}$$

with

$$B_x, B_z \ll B_y$$
$$x', y' \ll 1, |ds/dz| \cong 1$$

From $p_x = P \, dx/ds \cong P \, dx/dz$ it follows that the change $|\Delta P_T|$ of the transverse momentum $P_T(= (p_x^2 + p_y^2)^{\frac{1}{2}})$ is approximately (Equation (3.69a))

$$|\Delta P_T| \cong \kappa |q B| L \qquad (3.69b)$$

and the deflection $\Delta \alpha$

$$|\Delta \alpha| \cong \kappa |q B| L / P \qquad (3.69c)$$

In practice, this formula gives an error of 10–20%, due to the fringing field.

The natural choice for the track parameters is now

$$\left. \begin{aligned} p_1 &= x_r, \quad p_2 = y_r \\ p_3 &= (dx/dz)_r, \quad p_4 = (dy/dz)_r \\ p_5 &= (1/P) \times \text{sense of rotation} \end{aligned} \right\} \qquad (3.70)$$

at a reference plane $z = z_r$ (see also Fig. 3.8).

3.3.1.3 Several types of track models. In the presence of an *inhomogeneous* magnetic field, appropriate algorithms are needed to allow a particle to be followed efficiently through a given detector set-up. The algorithms discussed in this subsection deal only with the *deterministic* part of the *track model*, i.e. with the solution of the equation of motion (Equations (3.66)).

Material effects (Subsection 3.3.1.5) are nondeterministic, and are treated in two different ways. Energy loss is added in a *deterministic approximation* as an average with the possible exception of electrons or, at LHC, even muons. Multiple scattering is treated either as a *stochastic perturbation* of the trajectory (noise contribution) or added to the measurement errors. The track model can be formulated either as the full trajectory from an analytical or numerical solution of the equation of motion (e.g. Equation (3.67)), or as the functional relation between the impact points on specific detectors and some initial parameters **p** (integration parameters), thus defining one single point in the measurement space for one set of initial parameters (Equation (3.1) and Fig. 3.2). The set of all possible *undisturbed tracks* defines a *five-dimensional hyper-surface* in the measurement space.

To account for multiple scattering in obstacles between the detectors, one may also need the derivatives $\partial \mathbf{p}(s_j)/\partial \mathbf{p}(s_i)$. If there is no multiple scattering, it is sufficient to compute the functional dependence $\mathbf{f}(\mathbf{p})$ of the measurement vector on the initial parameters. Usually $\mathbf{f}(\mathbf{p})$ is approximated by a linear expansion which again requires the evaluation of $\partial \mathbf{f}/\partial \mathbf{p}$ in Equation (3.4): $\mathbf{f}(\mathbf{p}) = \mathbf{f}(\overset{0}{\mathbf{p}}) + \mathbf{A} \cdot (\mathbf{p} - \overset{0}{\mathbf{p}})$.

Track models which do not exactly fulfill the equations of motion can still give quite valuable parameter estimations. However, in such cases

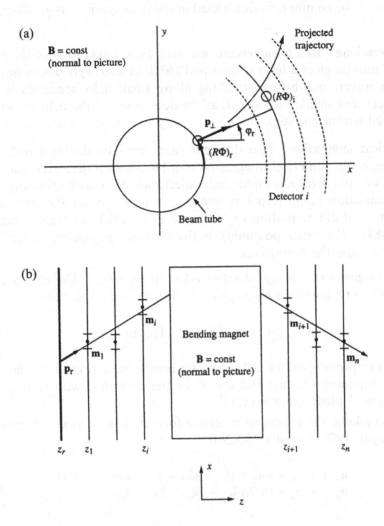

Fig. 3.8. (a) In many storage ring experiments, the magnetic field is rotationally invariant about the beam axis. In this case, the reference surface (at least for the barrel region) is often chosen to be a cylinder. (b) In the forward arm of a fixed-target spectrometer, the natural choice of parameters is: two Cartesian coordinates at the reference plane, two directions (tangents or direction cosines) and $1/P$ (or the deflection angle). Note that m_i can either be a single coordinate or a set of coordinates (in the latter case the error bars are symbolic).

the statistical properties of the LSM are no longer those described above, and the loss of information depends on whether the deviation of the approximate track model from the ideal one is significant compared to the measurement errors. Special care is then advised to avoid a systematic bias.

A few of the common methods used in today's experiments are discussed below.

Helix tracking   In a *homogeneous magnetic field*, integration of the equation of motion gives a helix (Equation (3.67)). In a *strongly inhomogeneous field*, however, step-by-step tracking along small helix segments is very inefficient and *should be avoided*, as the derivatives of the field are totally neglected within one step.

Numerical integration   Two different cases must be distinguished. The first one is the numerical integration of a bundle of trajectories, one 'zero trajectory' and five accompanying trajectories each corresponding to a slight variation of an initial parameter, in order to get the derivatives by numerical differentiation (e.g. Metcalf *et al.* 1973; Eichinger and Regler 1981). The other possibility is the parallel integration of the 'zero trajectory' and the derivatives.

(a) The fourth-order Runge–Kutta method of Nyström.   The equations of motion can be rewritten more generally for a given momentum:

$$\mathbf{u}'' = \mathbf{g}(\mathbf{u}', \mathbf{B}(\mathbf{u})) = \mathbf{f}(\mathbf{u}', \mathbf{u}) \tag{3.71}$$

where the prime denotes either the derivative with respect to the path length (Equation (3.65a)) and $\mathbf{u} = \mathbf{x}$, or the derivative with respect to z (Equation (3.14b)) and $\mathbf{u} = (x, y)^{\mathrm{T}}$.

For tracking, the following recursive formula is used (e.g. Abramowitz and Stegun 1970 (formula 25.5.20))

$$\left.\begin{aligned}
\mathbf{u}_{n+1} &= \mathbf{u}_n + h\mathbf{u}'_n + (h^2/6)(\mathbf{k}_1 + \mathbf{k}_2 + \mathbf{k}_3) + O(h^5) \\
\mathbf{u}'_{n+1} &= \mathbf{u}'_n + (h/6)(\mathbf{k}_1 + 2\mathbf{k}_2 + 2\mathbf{k}_3 + \mathbf{k}_4)
\end{aligned}\right\} \tag{3.72}$$

where

$$\begin{aligned}
\mathbf{k}_1 &= \mathbf{f}(\mathbf{u}'_n, \mathbf{u}_n) = \mathbf{g}(\mathbf{u}'_n, \mathbf{B}(\mathbf{u}_n)) \\
\mathbf{k}_2 &= \mathbf{f}(\mathbf{u}'_n + (h/2)\mathbf{k}_1, \mathbf{u}_n + (h/2)\mathbf{u}'_n + (h^2/8)\mathbf{k}_1) \\
\mathbf{k}_3 &= \mathbf{f}(\mathbf{u}'_n + (h/2)\mathbf{k}_2, \mathbf{u}_n + (h/2)\mathbf{u}'_n + (h^2/8)\mathbf{k}_2) \\
\mathbf{k}_4 &= \mathbf{f}(\mathbf{u}_n + h\mathbf{k}_3, \mathbf{u}_n + h\mathbf{u}'_n + (h^2/2)\mathbf{k}_3)
\end{aligned}$$

In practice, it is quite often sufficient to take the same field values when evaluating $\mathbf{k}_2$ and $\mathbf{k}_3$ (this is Nyström's advantage), which corresponds to replacing $\mathbf{k}_2$ by $\mathbf{k}_1$ in the second argument of $\mathbf{k}_3$. Another frequent approximation is to use the field value in $\mathbf{k}_4$ of one step to evaluate the $\mathbf{k}_1$ of the subsequent one (see below).

(b) *The Runge–Kutta method of Simpson.* A slightly different approach is to transform the equation of motion into a system of two first order, simultaneous differential equations:

$$\left.\begin{array}{l} \mathbf{u'} = \mathbf{v} \\ \mathbf{v'} = \mathbf{g}(\mathbf{v}, \mathbf{B}) \end{array}\right\} \tag{3.73}$$

Results of this approach are very similar – quite independent of the field inhomogeneities – to those obtained with the Nyström algorithm, depending slightly on the different algorithms applied to solve Equation (3.73) (e.g. Abramowitz and Stegun (1970)).

(c) *Parallel integration of the derivatives.* In order to minimize the 'χ^2 ansatz' (Equation (3.7)), the functional dependence of the track interceptions with the detectors from the track parameters (Equation (3.1)) is needed. In practice, this is achieved either by numerical differentiation (see below) or by integrating the derivatives $\mathbf{A}(s)$ (Equation (3.4)) together with the 'zero trajectory'. However, it should be mentioned that the exact algorithm also requires knowledge of the field derivatives, except when the influence of the field gradient transverse to the trajectory integrated over one step length can be neglected for the derivatives. In practice this approach saves about a factor of 2 in computing time spent in the tracking module (Myrheim and Bugge 1979; Bugge and Myrheim 1981). It was successfully applied in several experiments, e.g. for the recoil particle in the WA6 experiment at CERN ($pp\uparrow \to pp$) measuring elastic proton–proton scattering on a polarized target (Fidecaro *et al.* 1980).

(d) *'Numerical differentiation'.* If the gradient of the magnetic field transverse to the trajectory cannot be neglected and if it is not easy to obtain the field derivatives from the 'field model', numerical differentiation can be used:

$$(\mathbf{A})_{ik} = \frac{\partial f_i}{\partial p_k} = \frac{f_i(\overset{0}{\mathbf{p}} + \Delta_k \mathbf{p}) - f_i(\overset{0}{\mathbf{p}})}{\Delta p_k} \tag{3.74}$$

where $\Delta_k \mathbf{p}$ is a vector with all components being zero except the kth component which is Δp_k.

The variations should be small, but they must be large enough not to run into trouble with the machine precision of the computer. A lower limit is given by the word length of the computer used. As a rule of thumb typical variations are 0.1–1.0 mm for space coordinates and 10^{-4}–10^{-3} rad for angles. As, in practice, the fifth parameter p_5 is usually chosen to be 1/momentum or 1/transverse momentum ($1/P$ or $1/P_T$), the variation should be chosen accordingly. In the case of a fixed-target spectrometer (Fig. 3.8b) for the variation of p_5 ($= 1/P_T$) – using the deflection angle $\Delta\alpha$ (Equation 3.69c) and the lever arm of a downstream spectrometer with detectors in the region from 1 m to 10 m from the centre of the magnet – a variation of the order of 10^{-3}–10^{-4} corresponds to a lateral displacement of $100\,\mu$m–1 mm. If the variation is applied to the momentum by error propagation, then $\Delta P = |\Delta p_5/p_5^2| \cong 10^{-3}/p_5^2$. A fast check of the proper choice of the variations is the stability of the derivative values when the variations are changed slightly.

To avoid discontinuities during numerical differentiation if the field is represented in separate volumes (see Subsection 3.3.1.4), the *same* 'boxes' *of the field model* (Subsection 3.3.1.4) should be used for the zero track and for the variations at the borders of a box.

Predictor–corrector methods   Multistep integration has been applied for the AFS at the CERN-ISR. This method first makes a prediction for $\mathbf{u}'_{n+1}$, using a polynomial for $\mathbf{u}''$ through the previous points, and then a prediction for $\mathbf{u}_{n+1}$, before computing the final 'corrected' $\mathbf{u}_{n+1}$, again by integrating a polynomial. The method can be tuned by several parameters, but this makes the method less transparent. Considering the fact that at the beginning of the track no polynomials are yet available, the modest gain in computing time compared to the Runge–Kutta method does not really suggest this method should be used. However, if the method is used in connection with the analytical properties of the magnetic field, a slightly larger step length than with the Runge–Kutta method can be achieved for a required precision.

Taylor expansion

(a) *General case using derivatives of the field model.*   If not only the field values but also the derivatives can be evaluated by the field representation up to order $n-2$, an exact Taylor expansion can be evaluated up to order n. The Taylor expansion is obtained by comparing the coefficients of both sides of the equations of motion at a given point:

$$
\left.
\begin{aligned}
x_i''/\text{const} &= x_j'B_k - x_k'B_j \\
x_i'''/\text{const} &= x_j''B_k - x_k''B_j + \sum_\alpha \left[x_j'x_\alpha'\frac{\partial B_k}{\partial x_\alpha} - x_k'x_\alpha'\frac{\partial B_j}{\partial x_\alpha} \right] \\
x_i''''/\text{const} &= x_j'''B_k - x_k'''B_j \\
&+ \sum_\alpha \left[2\left(x_j''x_\alpha'\frac{\partial B_k}{\partial x_\alpha} - x_k''x_\alpha'\frac{\partial B_j}{\partial x_\alpha} \right) + x_j'x_\alpha''\frac{\partial B_k}{\partial x_\alpha} - x_k'x_\alpha''\frac{\partial B_j}{\partial x_\alpha} \right] \\
&+ \sum_{\alpha\beta} \left[x_j'x_\alpha'x_\beta'\frac{\partial^2 B_k}{\partial x_\alpha \partial x_\beta} - x_k'x_\alpha'x_\beta'\frac{\partial^2 B_j}{\partial x_\alpha \partial x_\beta} \right] \\
x_i'''''/\text{const} &= x_j''''B_k - x_k''''B_j + \cdots
\end{aligned}
\right\}
\quad (3.75a)
$$

where i, j, k denote the three coordinates cyclically, and α, β are the summation indices running from 1 to 3 (x, y, z), from which, e.g.

$$\left.\begin{aligned}
\Delta x_i &= x_i's + x_i''\frac{s^2}{2!} + x_i'''\frac{s^3}{3!} + x_i''''\frac{s^4}{4!} + x_i'''''\frac{s^5}{5!} \\
\Delta x_i' &= x_i''s + x_i'''\frac{s^2}{2!} + x_i''''\frac{s^3}{3!} + x_i'''''\frac{s^4}{4!}
\end{aligned}\right\}$$ (3.75b)

The gain in precision when adding higher orders than those suggested by the field representation depends on the analytical properties of the field itself. For example in the case of a constant field all terms in Equation (3.75a) which contain derivatives of the field are equal to zero. The remaining terms with the field itself constitute the series expansion of a helix. This shows that in practice it might be meaningful to extend the series to orders higher than suggested by the field derivatives available (i.e. $> n$).

In general this method allows larger steps than the Runge–Kutta method (but needs more computer time per step for higher orders), and the step size is only limited by the volume of the box in which the local field model is valid. If the field computations are fast, and if the formulae for the derivatives are coded efficiently, the method is competitive with the other methods mentioned above, and has the advantage that the track's position and direction are known for any point along the path, *allowing the evaluation of material effects*, the *simulation of decays* etc.

The method mentioned above has been successfully applied in the SFM at the CERN-ISR after careful tuning (Metcalf and Regler 1973; Metcalf, Regler and Broll 1973). A general program performing the comparison of coefficients for a general field model not fulfilling the 'Laplace equation' turned out to be of no practical use other than for a comparison of the precision which can be achieved by this method (Regler 1968).

If another method has been chosen, e.g. because of unknown field derivatives in a very inhomogeneous field, a second-order or 'truncated' third-order expansion (constant field approximation) may still be useful for several purposes: evaluation of multiple scattering and energy loss in inhomogeneous matter, simulation of electromagnetic processes such as ionization, or simulation of secondary reactions such as particle decays.

(b) Expansion in an axially symmetric magnetic field. An interesting special case arises if the field expansion is still more restrictive, i.e. for a field of *axial symmetry* and *mirror symmetry about the median plane*:

$$\left.\begin{aligned}
B_z(R, z) &= B_z(R, -z) \\
B_R(R, z) &= -B_R(R, -z) \\
B_\Phi &\equiv 0
\end{aligned}\right\}$$ (3.76a)

The vector potential $\mathbf{A}$ of the magnetic field (with $\mathbf{B} = \text{rot}\,\mathbf{A}$) in this special case has only one nonvanishing component – the azimuthal one $A_\Phi(R, z)$ – which can be obtained from

$$A_\Phi(R, z) = (1/R) \int_0^R r B_z(r, 0)\,\mathrm{d}r - \int_0^z B_R(R, \zeta)\,\mathrm{d}\zeta \qquad (3.76b)$$

(A_R or A_z would also give contributions to B_Φ.) The motion of a particle is then completely described by Equation (3.66)

$$\left. \begin{array}{l} P R \sin\beta \cos\lambda + \kappa q R A_\Phi(R, z) = \text{const} \\ P\,\mathrm{d}(\sin\lambda)/\mathrm{d}R - \kappa q \tan\beta\,\partial A_\Phi/\partial z = 0 \end{array} \right\} \qquad (3.76c)$$

where β is the angle between the radius vector and the direction of the particle projected onto the median plane. An efficient power series expansion can be obtained, providing the coordinates along the trajectory in terms of the inverse momentum of the particle, the distance of the particle from the symmetry axis, and the initial inclination with the median plane. For a small dip angle λ (Equation (3.67)), the series converges rapidly, and the procedure of evaluating the intercepts with the detectors is very efficient when the detectors are cylindrical in shape and concentric to the symmetry axis of the field (Birsa *et al.* 1977) (Fig. 3.9). By taking higher orders into account, any desired precision can be obtained.

For tracks going primarily along the direction of the symmetry axis, corrections to a helix are given in Billoir (1987a), which handles the inhomogeneities $\Delta\mathbf{B}$ of the field as a perturbation and computes the first order of this perturbation with integrals depending linearly on $\Delta\mathbf{B}$; the integration path is the helix approximation. (For the derivatives see Billoir (1986; 1987b).)

(c) *Magnetic quadrupole fields.* In some experiments, when measuring very forward particles, i.e. those at a small angle with respect to the beam, it may be necessary to place tracking detectors behind the last 'magnetic quadrupole lens' of the beam line. This must be taken into account in the analysis program. The vector potential $\mathbf{A}$ of the magnetic field of a pure quadrupole is particularly simple (G is the field gradient):

$$\mathbf{A}(\mathbf{x}) = \pm \frac{G}{2} \begin{pmatrix} 0 \\ 0 \\ x^2 - y^2 \end{pmatrix} \qquad (3.77a)$$

and

$$\mathbf{B}(\mathbf{x}) = \mp G \begin{pmatrix} y \\ x \\ 0 \end{pmatrix} \qquad (3.77b)$$

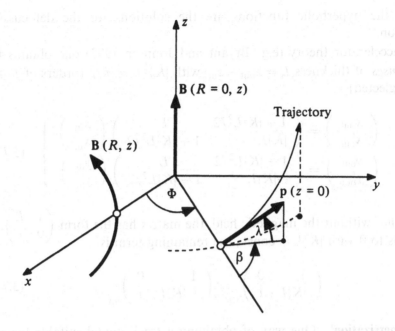

Fig. 3.9. A field of axial symmetry and mirror symmetry about the median plane. Φ is the azimuthal angle of the intersect of the trajectory with the median plane, β is the difference between the direction of the projected tangent and Φ, λ is the angle between the tangent and its projection χ, and $\mathbf{p}$ is the momentum ($\mathbf{p} \equiv \mathbf{P}$). ($\mathbf{B}$ is schematic.)

The equations of motion (Equation (3.66b)), with $dx/dz \ll 1$, $dy/dz \ll 1$, $ds/dz \cong 1$, give

$$d^2x/dz^2 = \pm(\kappa q/P)Gx = \pm Kx$$
$$d^2y/dz^2 = \mp(\kappa q/P)Gy = \mp Ky \qquad (3.77c)$$

with $K = \kappa q G/P$, where $\pm$ defines the polarity of the quadrupole and together with the sign of q, defines the focussing resp. defocussing direction: if the factor is positive, the lens is defocussing in this direction, and focussing in the other one.

If the quadrupole field is traversed along a path of length $L = z_{out} - z_{in}$, and with $k = |K|^{\frac{1}{2}}L$, the solution of the equations of motion is (a prime denotes the derivative with respect to z)

$$\left.\begin{array}{l} x(L) = x_{in}\cosh(k) + x'_{in}\sinh(k)/|K|^{\frac{1}{2}} \\ x'(L) = x_{in}|K|^{\frac{1}{2}}\sinh(k) + x'_{in}\cosh(k) \\ y(L) = y_{in}\cos(k) + y'_{in}\sin(k)/|K|^{\frac{1}{2}} \\ y'(L) = -y_{in}|K|^{\frac{1}{2}}\sin(k) + y'_{in}\cos(k) \end{array}\right\} \qquad (3.77d)$$

where the hyperbolic functions are the solutions for the defocussing direction.

In accelerator theory (e.g. Bryant and Johnsen 1993) one obtains for *thin lenses* of thickness $L = z_{out} - z_{in}$, with $|K|^{\frac{1}{2}}L \ll \pi/2$ (orders of $L \geqslant 3$ are neglected)

$$\left.\begin{array}{l}\begin{pmatrix} x_{out} \\ x'_{out} \end{pmatrix} = \begin{pmatrix} 1 + |K|L^2/2 & L \\ |K|L & 1 + |K|L^2/2 \end{pmatrix}\begin{pmatrix} x_{in} \\ x'_{in} \end{pmatrix} \\[12pt] \begin{pmatrix} y_{out} \\ y'_{out} \end{pmatrix} = \begin{pmatrix} 1 - |K|L^2/2 & L \\ -|K|L & 1 - |K|L^2/2 \end{pmatrix}\begin{pmatrix} y_{in} \\ y'_{in} \end{pmatrix}\end{array}\right\} \quad (3.77e)$$

Note that without the magnetic field, the matrix has the form $\begin{pmatrix} 1 & L \\ 0 & 1 \end{pmatrix}$. If L tends to 0 with $|K|L = \text{const.}$, the remaining term is

$$\begin{pmatrix} 1 & 0 \\ |K|L & 1 \end{pmatrix}_{x,x'} ; \begin{pmatrix} 1 & 0 \\ -|K|L & 1 \end{pmatrix}_{y,y'} \quad (3.77f)$$

'Parametrization'   One way of obtaining a track model suitable for rigorous track fitting is the evaluation of a *local linear track model*. Many attempts have been made to find a more '*global* – in general *nonlinear* – *parametrization*', and some of them have been successful, mainly if some of the five independent parameters vary over only a small range (e.g. limited interaction region in storage rings, small fixed target, limited momentum range in elastic scattering). Furthermore invariance of the magnetic field under rotation can reduce the dimension of the parameter vector (precisely speaking it leads to a trivial functional relation for one of the parameters), and symmetry planes of the magnetic field can reduce the number of possible coefficients for the track model. The technique consists of making a direct 'ansatz' for the coordinate functions of the undisturbed track as a linear combination of (if possible orthogonal) functions $\boldsymbol{\Phi}(\mathbf{p})$:

$$\mathbf{F}(\mathbf{p}) = \sum_{i_1,\ldots,i_5}^{m_1,\ldots,m_5} a_{i_1,\ldots,i_5} \times \prod_{j=1}^{5} \Phi_{i_j}(p_j) \quad (3.78)$$

The functions Φ_i may be powers of p_j, trigonometric functions, Chebyheff polynomials, etc.

The choice of the functions Φ_i determines the properties of the approximation Equation (3.78). For instance, the choice of Chebyheff polynomials guarantees the minimization of the maximal linear (absolute) residual, if the appropriate 'training sample' of tracks is chosen ($\mathbf{p}^{(k)}$), $\mathbf{f}(\mathbf{p}^{(k)})$; $k = 1,\ldots,M$, with $M = m_1 \times \cdots \times m_5$. The coefficients $\mathbf{a}$ are then evaluated by requiring the quadratic form

$$Q^2 = \sum_{k=1}^{M} (\mathbf{f}(\mathbf{p}^{(k)}) - \mathbf{F}(\mathbf{p}^{(k)}))^2 \qquad (3.79)$$

to be a minimum, leading to a set of $m_1 \times \cdots \times m_5$ linear equations.

If the functions Φ_i are *orthogonal* for the appropriate 'training sample', some advantage can be taken of an efficient choice of coefficients. First, the matrix to be inverted is *diagonal*, and second, each coefficient contributes *independently* to the total sum of Q^2, immediately allowing the selection of the most relevant coefficients. However, this requires a choice of the training sample according to the type of functions chosen, which is not always feasible in practice, owing to limitations of the field map, of nonphysical regions etc. In the latter case, an iterative procedure has to be chosen, using a '*Gram–Schmidt transformation*' for orthogonalization. An efficient program performing the choice of coefficients is available at CERN (Brun *et al.* 1979). The algorithm is described in Wind (1972). Finally an *independent* uniform 'test sample' must be generated to test the quality of the approximation, and both the averaged squared and the maximal residual must be examined carefully, in comparison with the detector resolution, for a possible distortion of the χ^2 distribution, and a possible bias due to a χ^2-cut. However, it should be mentioned that the addition of higher-order functions makes sense only if the averaged squared residual of the test sample *and* the maximal residual, together with the 'Q^2 ansatz', are still decreasing.

The method was successful in several experiments, a few of which should be mentioned. In the R401 experiment (using the SFM at CERN-ISR) the reaction $pp \rightarrow pp$ and $pp \rightarrow p(n\pi^+)$ have been measured. The result of parametrization was typically to keep 40 coefficients after having subdivided the phase space into several cells (Aubert and Broll 1974). For regions of a very inhomogeneous field, the classical numerical integration was kept as back-up. The computer time used in the tracking routine was reduced by an order of magnitude for this large detector. A sophisticated correction procedure had to be applied when a detector was remounted in a slightly different position after having been removed for maintenance. In the $pp\uparrow \rightarrow pp$ elastic scattering experiment with a fixed target (CERN WA6, see above) the forward particle's trajectory could be parametrized by 30 coefficients for the deflection plane and by 15 for the perpendicular one, for a total spectrometer length of 50 m and a variation of deflection power '$\Delta BL/BL$' $\cong 20\%$. The computer time for evaluating these functions in the track fitting program could almost be neglected.

A drawback of this method is the necessity to reevaluate the coefficients when the position of a detector inside the magnetic field is changed during the overall lifetime of the experiment (outside the field simple corrections can be applied).

Spline approximation   Another way to represent a trajectory is to connect the coordinates in the two projections by a 'spline curve'. A first approximation would be obtained by interpolating between two subsequent coordinates along the track by (quadratic) parabolas, such that the following continuity constraints hold for the resulting curve of subsequent parabola segments:

- the curve passes through the measured coordinates and is therefore continuous;

- the slope of the curve is continuous at the measured coordinates (this assumes that multiple scattering is not concentrated in the dense material in the vicinity of these measurements (detector frames) or anywhere else (walls)).

Cubic splines are also often used (Subsection 2.2.1).

Such a model allows the evaluation of the *direction* at each measured point. After computation of the *magnetic field* at these points, the *second derivatives* ('curvature') can be calculated up to an overall proportionality factor $(1/P)$ from the equations of motion. If continuity is also requested for the second derivatives, a *cubic spline* to the measurements can be used.

However, a spline is an approximation to the scattered path and also includes the measurement errors; it does not correspond to an exact solution of the equations of motion and is therefore not suitable for a rigorous track fit in the sense of this chapter. The more appropriate way of using splines for the final track fit is to assume a *cubic spline* for *the track's second derivatives*, i.e., a quintic spline model for the track as given by the equations of motion (Wind 1974, 1978); the most elegant case is when the equations of motion are written in the form of Equation (3.66b) (see also Bugge and Myrheim 1981):

$$\left.\begin{array}{l} x'' = (\kappa q/P)X''(\mathbf{x}, \mathbf{x}') \\ y'' = (\kappa q/P)Y''(\mathbf{x}, \mathbf{x}') \end{array}\right\} \tag{3.80a}$$

It is assumed that from pattern recognition (possibly combined with a cubic spline or parabolic fit to the measured coordinates) the full space point for each measured coordinate is available as well as the direction vector. Knowing the field, the right hand sides of the Equations (3.80a) can now be calculated at each space point corresponding to a coordinate:

$$\left.\begin{aligned}
X''(z_i) &= [x_i'y_i'B_x(x_i) - (1 + x_i'^2)B_y(x_i) \\
&\quad + y_i'B_z(x_i)](1 + x_i'^2 + y_i'^2)^{\frac{1}{2}} \\
Y''(z_i) &= [(1 + y_i'^2)B_x(x_i) - x_i'y_i'B_y(x_i) \\
&\quad - x_i'B_z(x_i)](1 + x_i'^2 + y_i'^2)^{\frac{1}{2}}
\end{aligned}\right\} \tag{3.80b}$$

where i denotes the value at $z = z_i$. In order to get a continuous interpolation between these points of curvature but to keep the freedom to account for measurement errors, a cubic spline interpolation is applied to the second derivatives, and then twice integrated along the preliminary reconstructed path of the particle:

$$\left.\begin{aligned}
X''(z) &= S_x^{(3)}[X''(z_1), \dots, X''(z_n)] \\
Y''(z) &= S_y^{(3)}[Y''(z_1), \dots, Y''(z_n)]
\end{aligned}\right\} \tag{3.80c}$$

where $S_x^{(3)}$, $S_y^{(3)}$ are cubic spines interpolating the second derivatives of x and y.

If energy loss has to be taken into account $X''(z_i)$, $Y''(z_i)$ have to be replaced by $X''(z_i)/(1 - \varepsilon(z_i))$, $Y''(z_i)/(1 - \varepsilon(z_i))$, with $P(z_i) = P_0(1 - \varepsilon(z_i))$, and with first guess of P_0 from pattern recognition.

The ansatz for the track model is (Equation (3.1), Fig. 3.2a):

$$\left.\begin{aligned}
f_{x,i} &= x(z_i) = a_1 + a_2z + (\kappa q/P) \int_{z_1}^{z_i} \left[\int_{z_1}^{v} X''(u)\, du \right] dv \\
f_{y,i} &= y(z_i) = b_1 + b_2z + (\kappa q/P) \int_{z_1}^{z_i} \left[\int_{z_1}^{v} Y''(u)\, du \right] dv
\end{aligned}\right\} \tag{3.81}$$

where the integration parameters a_1, b_1, a_2 and b_2, and the inverse of the momentum $1/P$ can be considered as the track parameter vector $\mathbf{p}$, giving in many cases a reasonable five-dimensional track model to be used in the LSM (Wind 1979; Zupančič 1986), and also allowing the evaluation of some kind of a χ^2, and pull quantities, and the application of error propagation.

In order to be able to construct these cubic splines passing through $X''(z_i)$, $Y''(z_i)$, one needs continuity constraints and appropriate boundary conditions.

(a) Ansatz: a cubic spline is a polynomial expression of order 3 (at most) for each interval (z_i, z_{i+1}).

(b) Continuity condition: if the spline is of order k, $k - 1$ derivatives must be continuous. This also basically holds at the points z_i, the 'knots'.

(c) Boundary conditions: from the continuity condition, $k - 1$ free parameters are left open, i.e. two for a cubic spline, giving one degree of freedom for each boundary. The boundary conditions should be chosen according to the spectrometer set-up, assuming either a constant field or a field-free region at the end points. This can be taken into consideration by requiring no change in the curvature at the end points $(S_x''(z_1) = S_y''(z_1) = S_x''(z_n) = S_y''(z_n) = 0)$.

Having performed the integration interval-by-interval, a particle's trajectory is now represented by a *doubly integrated cubic spline* (with the continuity conditions of a quintic spline but without a strict interpolation constraint). The five parameters can then be determined from a linear least squares fit. The evaluation of the cubic spline expressions $S_x^{(3)}(z)$, $S_y^{(3)}(z)$ could also be based on more data points than have actually been measured; these 'additional knots' then, of course, have to be excluded from the least squares fit. The additional knots may be necessary if the magnetic field varies strongly between the measured coordinates.

The method has been successfully applied in many experiments, giving good results, χ^2 distributions and pull quantities (SFM and AFS detectors at the CERN-ISR, OMEGA detector at the CERN-SPS and some others). However, some drawbacks should be mentioned.

- Only the global track fitting method can be used.

- The track model depends in a hidden way on the measurements. Therefore it is difficult to determine the rigorous weight matrix of the least squares fit.

- If multiple scattering is significant compared to the detector resolution, the method follows the physical path of a particle rather than the ideal one too closely, giving in practice pull quantities and χ^2s that are too small (Subsection 3.2.3). This is because multiple scattering is a discontinuous term superimposed on the curvature as given by the equation of motion (Equation (3.66b)), and is therefore not well described by s-spline interpolation. Some preliminary smoothing is already done by the model itself before the proper fit procedure.

- The trajectory of the particles should have a common 'drift direction' (z), and the deflection angle should not be large ($\ll \pi/2$). The detectors should all be arranged perpendicular to this drift direction.

- The parameters chosen give a very simplified derivative matrix, corresponding to just a variation of a rigid curve as far as the first four parameters are concerned; the field gradient transverse to the trajectory is completely neglected.

Several attempts have been made to overcome these drawbacks (e.g. more additional points, piecewise track reconstruction, break points, iterative procedures, etc.), but – reliability and clarity being major requirements in large collaborations – it seems advisable to limit the application of this method to tracks with the following properties:

- similar order of magnitude for the resolution of the individual detectors;

- detector errors which are not too large;

- little multiple scattering;

- small transverse field gradient;

- general 'main direction' perpendicular to the detectors;

- limited deflection.

It should be mentioned that all these track models are already of some historical interest only, owing to the dramatic increase in computing capacity during the past two decades. Nowadays it is more essential that the track model in use is simple and easy to understand for all collaborators. In addition it must meet the challenge of the extreme precision of modern vertex detectors.

3.3.1.4 The field representation. When integrating the equations of motion (Equations (3.65) and (3.66)), knowledge of the magnetic field (defined by its induction **B**) is needed at some known points (Equation (3.71)). Other tracking methods require the knowledge of higher derivatives, e.g. Equation (3.75).

Only the use of the field representation as needed for tracking will be discussed here. The problem of preevaluation of the field during the design of a spectrometer magnet will not be covered, nor will the algorithms for controlling or smoothing field measurements in practice be considered.

The most straightforward way to establish a field representation, which gives a fast response when looking for the field at a certain point **x**, is a dense grid of points at each of which the field vector is stored

$$\mathbf{B}_{ijk} = \mathbf{B}(x_i, y_j, z_k) \tag{3.82}$$

where i, j, k are obtained by rounding down the values x, y, z according to the grid; for an equidistant grid: $i = \text{integer}(x/\Delta x + 0.5\,\text{sign}(x))$ etc. This method is frequently used with a possible improvement of the precision by simple interpolation formulae.

However, this field representation is only efficient if the map is stored in a fast-access memory, leading to some limitation of the number of values that can be stored. For 'virtual memory' computers, the size is not limiting, but some care must be taken to avoid excessive 'page faults', making it highly inefficient in time and limiting the portability of the map. Further optimization may be gained by using sophisticated grid search algorithms. In a very inhomogeneous field as many as 100 steps in each direction can be needed, leading to some 3×10^6 field values.

If some symmetry planes exist, the mechanical precision and homogeneity of the magnet yoke and the coils should be good enough that the dimension of the field map could be reduced by a factor 2, 4 or even 8.

A measure of the required precision is again the detector resolution, with respect to which the momentum-dependent lateral displacement along the full track due to the field approximation should be negligible. Whether this displacement, summed over all the boxes along a track, cancels on average and can be treated as a pseudostochastic process $\sigma(\Delta x) \sim$ (number of boxes)$^{\frac{1}{2}}$, or whether it corresponds to a biassed summation, must be checked by a Monte Carlo calculation. Sometimes $\Delta P / P$ originating from the detector errors versus $\Delta|\mathbf{B}|/|\mathbf{B}|$ is examined instead of the error in the lateral displacement.

A way of avoiding numerical tables that describes magnetic field components in sufficient detail for simple interpolation formulae is to subdivide the volume into sufficiently small fractions (e.g. nonoverlapping 'boxes') and also to store the derivatives. Then the box size can be enlarged by an important factor. Although the higher-order coefficients must be stored in addition (three for the field itself plus $3 + 2n$ for the n'th derivative), the number of values to be stored can be reduced considerably. Note that fulfilling the Laplace equation (magnetic fields in spectrometer magnets are constant in time) does not always give the minimum number of coefficients for a given precision, but it has the advantage of always selecting the same set for a given order of the polynomial, speeding up those tracking methods which use the field derivatives. However, an interpolation polynomial must be evaluated, limiting the usefulness of a higher-order representation due to the time needed for interpolation. This limit depends on whether tracking makes direct use of the derivatives, and is of the order $n = 1$ to $n = 3$. (For more details of this and also symmetry operations on the derivatives, see Metcalf *et al.* (1973).)

Another approach is an expansion in terms of low-order orthogonal polynomials for each field component inside each fractional volume. An implementation in terms of Chebycheff polynomials has frequently been used for experiments at CERN. The output generated by this program is a FORTRAN source code, containing a loop-free evaluation of the polynomial, with all insignificant terms eliminated. No precision is lost by the omission of the polynomials not satisfying the Laplace equation. For example, in the UA1 experiment at CERN, the entire magnetized zone was represented by these polynomials (including all three components, and both high-precision tracking zones and instrumented iron volumes). Again the highest order of polynomials was no more than 3.

If the field has rotational invariance within the precision limits discussed above, the field map grid is only *two-dimensional*, allowing more freedom for the choice of the field representation. If these symmetries are also observed when designing the detector, some additional computation can be avoided, e.g. how to reach the detector surface exactly during numerical integration, or to ease parametrization (Equation (3.76)). The field shapes for several storage ring experiments are given in Fig. 3.10.

In practice it is often convenient to establish a *global field model* in parallel, allowing a check of the measurements on the boundary versus those inside the field volume, to smooth out the field measurements and to detect outliers. Such a model should be kept as a back-up, allowing a change in the field representation of the tracking program if needed. Such a model, although slow for the field evaluation, can also be a useful tool for evaluating the training sample for parametrization purposes, avoiding discontinuities which can occur between adjacent field map boxes. These discontinuities can cause difficulties during numerical differentiation (Equation (3.74)); a solution is to choose the boxes when tracking the zero track, and to keep them fixed as the track varies, even if a slight extrapolation is needed, although in general, extrapolation should be avoided.

To summarize: when speaking about the field representation one must distinguish between the *general field map*, which allows an evaluation of a field value at any point with low speed, but high precision, and the usually *boxwise field representation for the tracking program*, where the kind of field model must be chosen in conjunction with the tracking algorithm. A trade-off is necessary between memory utilization, speed, and precision.

3.3.1.5 The effects of matter on the trajectory. So far it has been assumed that the trajectory of the charged particle is not affected by any material. In reality, however, several types of secondary interaction between particle and material may occur: multiple scattering and energy loss due to electromagnetic interactions (also, the external 'bremsstrahlung' may be not negligible for high-energy electrons deflected by strong magnetic fields), elastic nuclear scattering, and processes of any kind that create additional new particles.

In this subsection, only *multiple scattering and energy loss due to electromagnetic interactions* are discussed. The theory of these processes is well

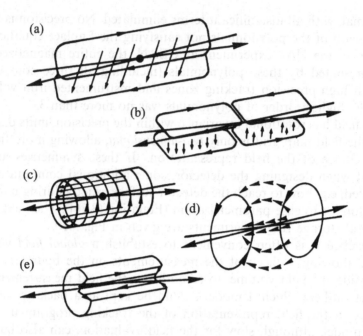

Fig. 3.10. Several field shapes from storage ring experiments (from C. Fabjan). (a) Dipole (UA1 CERN, DASP DESY); dipoles (*H*-magnets, *C*-magnets) are also used in the forward spectrometer of fixed-target experiments; (b) split field magnet (SFM CERN-ISR; (c) solenoid (DELPHI CERN-LEP); (d) axial field magnet (AFS CERN-ISR); (e) toroid (MARK II SLAC).

understood, and good descriptions of these phenomena are given in text books, e.g. Allison and Wright (1987) and Fernow (1986); for numerical values see also Particle Data Group (1998). A standard text book on this subject is by Rossi (1965). For multiple scattering see also Gluckstern (1963).

Multiple scattering When an electrically charged particle traverses a layer of matter, it is deflected from the path determined by the equations of motion in a vacuum by scattering on the electrons or the nuclei of that matter. For educational reasons it is convenient to assume that the projectile is a muon, because a muon

- is heavy compared to an electron,

- but light compared to each nucleus,

- and is not subject to the strong interaction.

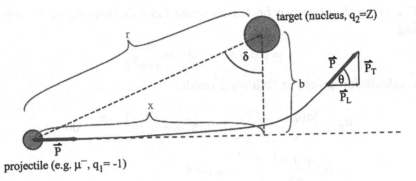

Fig. 3.11. Scattering of an incident particle with momentum P and charge q_1 on a target of charge q_2.

The transverse momentum transfer can be calculated by integrating the transverse components of the force acting between two charged particles (e.g. Jackson 1998), one moving (a *projectile*) and the other a *target* at rest:

$$P_T = \int_{-\infty}^{\infty} f_T \, dt$$

$$= q_1 e \int_{-\infty}^{\infty} \varepsilon_T \, dt$$

$$= \frac{q_1 q_2 e^2}{v} \int_{-\infty}^{\infty} \frac{b}{(x^2 + b^2)^{\frac{3}{2}}} \, dx \qquad (3.83a)$$

with

$$\varepsilon_T = \frac{q_2 e}{x^2 + b^2} \cos \delta, \qquad \cos \delta = \frac{b}{\sqrt{(x^2 + b^2)}} \qquad (3.83b)$$

where t is the time (i.e. integration over the interval of interaction), f_T the transverse force component between the particles, ε_T the transverse component of the electric field of the target, e the positive elementary charge, q_1 and q_2 the charges (in multiples of e) of the projectile and target respectively, v the velocity along the path of the projectile, b the impact parameter of the undisturbed path with respect to the target, P the momentum of the projectile and P_T the transverse momentum transfer caused by the interaction. Although the electric field used here is based on a quasi-static approximation, the final result for P_T is also correct for relativistic projectiles. In the proper electrodynamic treatment one obtains the relativistic factor γ in the formula for ε_T, but considering that one must now use rather the time in the rest frame of the projectile than the laboratory time, this factor is compensated again in the formula for P_T.

The integral above can be solved using Gauss's theorem or by substituting

$$x = b \tan \delta, \qquad dx = \frac{b \, d\delta}{\cos^2 \delta}$$

The calculation gives the following result:

$$
\begin{aligned}
P_T &= \frac{|q_1 q_2| e^2}{v} \int_{-\pi/2}^{\pi/2} \left(b^2 \tan^2 \delta + b^2\right)^{-\frac{3}{2}} \frac{b^2}{\cos^2 \delta} \, d\delta \\
&= \frac{|q_1 q_2| e^2}{vb} \int_{-\pi/2}^{\pi/2} \cos \delta \, d\delta \\
&= \frac{2|q_1 q_2| e^2}{vb}
\end{aligned}
\tag{3.83c}
$$

Note that the longitudinal momentum transfer of the scattering process vanishes because of symmetry:

$$\int_{-\infty}^{0} f_L \, dt = - \int_{0}^{\infty} f_L \, dt$$

The deflection angle θ is therefore given by

$$\theta \cong \sin \theta \equiv P_T/P = 2|q_1 q_2| e^2 / Pvb = \text{const}/b \tag{3.84}$$

The *differential cross section* can be calculated from Equation (3.84) for small deflection angles θ (note that $\cos \theta$ runs from $+1$ to -1):

$$\frac{d\sigma(\cos \theta, \varphi)}{d\Omega} = \frac{d\sigma(\cos \theta, \varphi)}{d \cos \theta \, d\varphi} = \frac{d\sigma(\theta)}{\sin \theta \, d\theta} \cdot \frac{1}{2\pi} \cong \frac{1}{2\pi\theta} \frac{d\sigma}{d\theta} = \frac{1}{2\pi\theta} \frac{d\sigma}{db} \frac{db}{d\theta} \tag{3.85a}$$

The derivative $d\sigma/db$ can be computed from a purely geometrical interpretation of the cross section. With axial symmetry the relevant area for impact parameters between b and $b + db$ has the form of an annulus with the size $d\sigma = 2\pi b \, db$. With

$$d\sigma/db = 2\pi b$$

and with (see Equation (3.84))

$$b = \text{const}/\theta, \quad |db/d\theta| = |\text{const}/\theta^2|$$

one obtains

$$\frac{d\sigma}{d\Omega} = \left(\frac{2q_1 q_2 e^2}{Pv}\right)^2 \frac{1}{16} \left(\frac{2}{\theta}\right)^4 = \frac{\text{const}^2}{\theta^4} \tag{3.85b}$$

This is an approximation of the more general *Rutherford formula* for the scattering of *point-like spinless particles*:

$$\frac{d\sigma_R}{d\Omega} = \frac{1}{4} q_1^2 q_2^2 \left(\frac{e^2}{Pv}\right)^2 \frac{1}{\sin^4(\theta/2)} \tag{3.85c}$$

If scattering occurs on an electron, then $q_2 = -1$, and the cross section per atom increases linearly with Z (the atomic number, which is also the number of electrons in the atom). On the other hand, for scattering on a nucleus, $q_2 = Z$ and the cross section per atom increases with Z^2 (Fig. 3.11). Therefore, setting the charge of the projectile $q_1 \equiv z$, one obtains

$$\frac{d\sigma_R}{d\theta} d\theta = \frac{d\sigma_R}{d\Omega} 2\pi \sin\theta \, d\theta \cong 8\pi z^2 (Z + Z^2) \left(\frac{e^2}{Pv}\right)^2 \frac{d\theta}{\theta^3} \tag{3.85d}$$

Note that because the scattering process is independent of φ the transition to the marginal distribution of θ (by integration over φ, see the left part of Equation (3.85d)) just yields a constant factor of 2π. Note further that

$$\int_{(\cos\theta)_{min}}^{(\cos\theta)_{max}} d\cos\theta = \int_{\theta_{max}}^{\theta_{min}} -\sin\theta \, d\theta = \int_{\theta_{min}}^{\theta_{max}} \sin\theta \, d\theta$$

It is very instructive to factor $d\sigma_R/d\theta$ into two components:

$$\frac{d\sigma_R}{d\theta} = \sigma_R f(\theta)$$

with the integrated cross section:

$$\sigma_R \cong 8\pi z^2 (Z + Z^2) \left(\frac{e^2}{Pv}\right)^2 \int_{\theta_{min}}^{\theta_{max}} \frac{d\theta}{\theta^3} \tag{3.85e}$$

and the normalised density function (i.e. the conditional density function for one scattering process) used for calculating $\langle\theta^2\rangle$

$$f(\theta) = \begin{cases} (1/\theta^3)/\int_{\theta_{min}}^{\theta_{max}} (1/\theta^3) \, d\theta & \text{if } \theta \in (\theta_{min}, \theta_{max}) \\ 0 & \text{otherwise} \end{cases}$$

The *second moment about zero of θ per scattering process*, where θ is the polar angle in space, can then be evaluated from $f(\theta)$ by choosing suitable bounds θ_{min} and θ_{max} (see below):

$$\langle\theta^2\rangle = \int_{\theta_{min}}^{\theta_{max}} \theta^2 f(\theta) \, d\theta = \frac{\int_{\theta_{min}}^{\theta_{max}} \theta^2 (1/\theta^3) \, d\theta}{\int_{\theta_{min}}^{\theta_{max}} (1/\theta^3) \, d\theta} = \frac{\ln(\theta_{max}/\theta_{min})}{\int_{\theta_{min}}^{\theta_{max}} (1/\theta^3) \, d\theta} \tag{3.86}$$

Considering the *second moment of the projections of* θ in a local Cartesian coordinate system with one axis pointing in the projectile's direction ($\theta^2 = \theta_1^2 + \theta_2^2$, $\theta\, d\theta\, d\varphi = d\theta_1\, d\theta_2$ – this is true as long as $\theta^2 = \theta_x^2 + \theta_y^2$, i.e. $\sin\theta \cong \theta$, $\cos\theta \cong 1$) yields an extra factor of $1/2$ for the second moment per unit length, since the process is isotropic in φ. This means that the variance of the projected scattering angle is one half of the second moment of θ in space, i.e.

$$\langle \theta_1^2 \rangle = \langle \theta_2^2 \rangle = \tfrac{1}{2} \langle \theta^2 \rangle$$

Note that $\langle \theta_1 \rangle = \langle \theta_2 \rangle = 0$ and therefore $\langle \theta_{1,2}^2 \rangle$, the second moment about zero, coincides with the variance. This holds also for multiple scattering as long as the condition mentioned above is fulfilled. It follows that for independent consecutive scattering processes $\langle \theta^2 \rangle$ is also additive.

Using the normalized density function one could write Equation (3.85d) also in the following way (with $\int (1/\theta^3)\, d\theta \cong 1/\theta_{min}^2$):

$$\frac{d\sigma_R}{d\theta}\, d\theta \cong 8\pi z^2 (Z + Z^2) \left(\frac{e^2}{Pv} \right)^2 \underbrace{\frac{d\theta}{\theta^3} \frac{1}{\int_{\theta_{min}}^{\theta_{max}} (1/\theta^3)\, d\theta} \int_{\theta_{min}}^{\theta_{max}} (1/\theta^3)\, d\theta}_{f(\theta)\, d\theta}$$

(3.87a)

The average number of scattering processes in a thin layer of material with thickness Δx is equal to:

$$n_{scatt} = \sigma_R n_A \Delta x$$

where n_A is the density of atoms per unit volume: $n_A = N\rho/A$, N is Avogadro's constant ($\cong 6.022 \times 10^{23}$ atoms per mol), ρ is the density in grams per unit volume of matter and A is the number of nucleons (protons and neutrons) in the nucleus. We assume that the scattering material is 'amorphous', i.e. that the individual scattering processes are independent. In this case the variances are additive and therefore the total variance can be calculated by multiplying the variance per scattering with the number of scattering processes.

The *differential cross section per unit length* calculated from

$$\frac{d^2\sigma}{d\theta\, dx}\, d\theta\, dx = n_A 8\pi z^2 Z(Z + 1) \left(\frac{e^2}{Pv} \right)^2 \frac{d\theta}{\theta^3}\, dx \qquad (3.87b)$$

From Equation (3.87b) one obtains

$$\frac{d\sigma}{dx} \cong n_A 8\pi z^2 Z(Z + 1) \left(\frac{e^2}{Pv} \right)^2 \cdot \frac{1}{\theta_{min}^2}, \qquad (3.87c)$$

with $\theta_{min} = \alpha Z^{\frac{1}{3}} m_e c / P$ (for α and m_e see below). Note that the dominant contribution results from the scattering on the nuclei ($\sim Z^2(!)$) and that the consideration of sophisticated corrections for excitation processes can be postponed to the treatment of energy loss.

The second *moment about zero of θ per unit length* is therefore

$$\langle \theta^2 \rangle_s = \frac{d\sigma}{dx} \int_{\theta_{min}}^{\theta_{max}} \theta^2 f(\theta)\, d\theta = (N\rho/A) 8\pi z^2 Z(Z+1)(e^2/Pv)^2 \ln(\theta_{max}/\theta_{min})$$

(3.88a)

θ_{min} can be derived from screening effects when b is large, and θ_{max} from a more sophisticated model which takes into account the projectile's Compton wavelength (see e.g. Fernow 1986, pp. 66–7):

$$\theta_{max}/\theta_{min} \cong \frac{2}{\alpha^2 (AZ)^{\frac{1}{3}}} \cong (173 Z^{-\frac{1}{3}})^2$$

To obtain this simple relation, one must use the approximation $A \cong 2Z$, so it is not valid for very heavy nuclei.

Inserting $\theta_{max}/\theta_{min}$ into Equation (3.88a) and in consideration of the factor $1/2$ one obtains for the variance of the projected scattering angle per unit length:

$$\langle \theta_{proj}^2 \rangle = \frac{1}{2}(N\rho/A) 8\pi z^2 Z(Z+1) \left(\frac{e^2}{Pv}\right)^2 2\ln(183 Z^{-\frac{1}{3}})$$

(3.88b)

Note that the use of $(183 Z^{-\frac{1}{3}})$ instead of $(173 Z^{-\frac{1}{3}})$ is just for convenience: it allows us to simplify the expression above by using the radiation length as used for bremsstrahlung, taking advantage of the similar dependence on all material related quantities (ρ, A, Z; see remarks on bremsstrahlung). Because the result shows only a logarithmic dependence of this term (and the term itself is furthermore just an approximation) the error made by this simplification is not too large (it is in fact closer to reality).

The result can also be written as

$$\langle \theta_{proj}^2 \rangle = \frac{1}{2}\left(\frac{E_S}{\beta Pc}\right)^2 z^2 \frac{\rho}{X_0}$$

(3.88c)

where

$$E_S \equiv m_e c^2 (4\pi/\alpha)^{\frac{1}{2}} \cong 0.0212\,\text{GeV}, \quad E_S/\sqrt{2} = 0.015\,\text{GeV}$$
$$\beta = v/c \quad \text{(see Subsection 3.3.1.1)}$$

m_e is the electron mass ($m_e \cong 0.511 \times 10^{-3}\,\text{GeV} \cong 9.11 \times 10^{-28}\,\text{g}$) and α is *Sommerfeld's fine structure constant* ($\alpha = e^2/(\hbar c) \cong 1/137$). Absorbing

$1/\alpha \sim 1/e^2$ in the constant E_S^2 leads for the remaining term to the same e^6-dependence as for X_0. X_0 is a density-independent scaling variable (in grams per unit area), called the 'radiation length' (see also Equation (2.60)). It is given by

$$X_0 = [(4N\alpha/A)Z(Z+1)r_e^2 \ln(183Z^{-\frac{1}{3}})]^{-1}$$

where r_e is the classical (electromagnetic) electron radius ($r_e \equiv e^2/(m_e c^2) \cong 2.818 \times 10^{-13}$ cm) and N is Avogadro's constant. Note that $1/X_0$ is proportional to the density of atoms n_A and to Z^2, as multiple scattering is dominated by the scattering on the nuclei.

If scattering occurs sufficiently often in a layer of thickness x, the joint probability density function of the projected scattering angles (θ_1, θ_2) can be approximated by a Gaussian distribution (see also Equation (3.100a))

$$f(\theta_1, \theta_2) = \frac{1}{2\pi x \cdot \langle \theta_{\text{proj}}^2 \rangle} \cdot \exp\left[-(\theta_1^2 + \theta_2^2)/(2x \cdot \langle \theta_{\text{proj}}^2 \rangle)\right] \qquad (3.88d)$$

with

$$x \cdot \langle \theta_{\text{proj}}^2 \rangle = \text{var}(\theta_{\text{proj}})$$

However, since the differential cross section for individual scattering has a large tail, convergence to a Gaussian distribution is slow, and for thin layers of matter these tails have to be taken into account (e.g. Fernow 1986; Particle Data Group 1998). This can be done by modeling the distribution as a mixture of two Gaussians, one for the 'core' and one for the 'tails' (see Subsection 3.2.6 for the handling of this mixture in the track fit). Note also that nuclear scattering can contribute a non-negligible share of the tails, the variance of the nuclear scattering angle being rather large.

A fit of a Gaussian distribution to the central 98% of the projected angular distribution yields an additional 'logarithmic correction':

$$\langle \theta_{\text{proj}}^2 \rangle \cdot x = \frac{1}{2}\left(\frac{0.91E_S}{\beta P c}\right)^2 z^2 \frac{\rho x}{X_0}[1 + 0.04\ln(x/L_0)]^2 \qquad (3.88e)$$

This is known as *Highland's formula* (with $E_S/\sqrt{2} = 0.015$ GeV, $0.91E_S/\sqrt{2} = 0.0136$ GeV). $L_0 \equiv X_0/\rho$ is a geometrical (density-dependent) scaling variable of dimension unit length. This formula is valid for a wide range of the thickness x. However, it violates the additivity of variances of independent scattering processes, and, for very small x, the 'variance' might even get negative!

In high-energy physics momenta are defined in units of GeV (i.e. $c = 1$), $\beta P c$ (originating from the Rutherford formula) can be replaced by βP, with $1/\beta = \sqrt{(m^2 + P^2)}/P$.

The stochastic nature of multiple scattering of a particle passing through matter causes a random deviation from the trajectory as evaluated by the equations of motion (Equations (3.66a, b)), and *should be added* to these equations.

If only a few thin layers act as scatterers, the effects of multiple scattering can be included in the track model by a pair of extra parameters (the two scattering angles) for each layer ('break points'). These can be considered as an unbiassed measurement with mean $\langle \theta_{1,2} \rangle = 0$, and with variance $x \cdot \langle \theta_{1,2}^2 \rangle = x \cdot \langle \theta_{proj}^2 \rangle$ given by *Highland's* formula (Equation (3.88e)). For more detailed calculations see Scott (1963) and Bichsel (1970).

The general treatment of multiple scattering in the context of the LSM will be discussed in Subsection 3.3.2.2.

Energy loss In the equations of motion (Equations (3.66)), the absolute value of the momentum of the particle is assumed to be constant. In reality, when a charged particle passes through matter, some energy is transmitted to this medium, and the 'constant' must be readjusted as a function of the path length parameter.

The energy transmitted to the medium can be calculated from the transverse momentum transfer of a single scattering process; for a non-relativistic recoil electron it is

$$\Delta E = \frac{(\Delta P)^2}{2m_2} \cong \frac{P_T^2}{2m_2} \sim \frac{q_2^2}{m_2} \tag{3.89a}$$

where ΔE and P_T are respectively energy and momentum transmitted to the recoil particle ($\Delta P = P_T$, see previous subsection on multiple scattering), and m_2 is the mass of the recoil particle. q_2 is the charge of the recoil particle in multiples of the elementary charge e, e.g. $q_2 = 1$ if the target particle is an electron and $q_2 = Z$ if it is a nucleus. It follows that the energy loss due to elastic scattering on electrons dominates by orders of magnitude the energy loss induced by the heavier nucleus, so the latter can be neglected (and $m_2 = m_e$). However, the ratio $m_e/m_{nucleus} = m_e/(Am_u)$ has to be multiplied by Z for the Z electrons per atom, leading to

$$\frac{\Delta E_{nucleus}}{\Delta E_{electrons}} \cong \frac{m_e}{2m_u} \cong \frac{1}{4000}$$

where we have set $m_u \cong m_p \cong m_n$.

For a given impact parameter b, the energy loss in one scattering process follows from Equation (3.83c) with $P_T^2 = (\Delta P)^2$ and $q_1 = z$, $q_2 = -1$:

$$\Delta E(b) = \frac{P_T^2}{2m_e} = \frac{2z^2 e^4}{b^2 v^2 m_e} \tag{3.89b}$$

Using $d\sigma = 2\pi b\, db$ (see above), and with the electron density per unit area equal to $dN = n_e dx$, and therefore $d\sigma\, dN = 2\pi b n_e\, db\, dx$, and further with $n_e = Z(N\rho/A)$ (see above) the *energy loss per unit length* in matter is given by

$$
\begin{aligned}
\frac{dE}{dx} &= n_e \int_\sigma \Delta E(b)\, d\sigma \\
&= Z(N\rho/A) \frac{2z^2 e^4}{v^2 m_e} \int_0^{2\pi} d\varphi \int_{b_{\min}}^{b_{\max}} \frac{1}{b^2} b\, db \\
&= Z(N\rho/A) \frac{2z^2 e^4}{\beta^2 m_e c^2} 2\pi \ln\left(\frac{b_{\max}}{b_{\min}}\right)
\end{aligned} \qquad (3.90a)
$$

or, with the classical electron radius $r_e = e^2/(m_e c^2)$:

$$
\frac{dE}{dx} = Z(N\rho/A)\, 2(2\pi r_e^2)\, m_e c^2 z^2 \frac{1}{\beta^2} \ln\left(\frac{b_{\max}}{b_{\min}}\right) \qquad (3.90b)
$$

Note that in the literature the differential energy loss sometimes appears with a minus sign.

If the integration variable is E rather than b, we have (with $\Delta E \Leftarrow E$)

$$
\frac{db}{b} = -\frac{1}{2}\frac{dE}{E}
$$

and a factor of 2 'disappears' and a more rigorous treatment is needed

$$
\frac{dE}{dx} = Z(N\rho/A)(2\pi r_e^2)\, m_e c^2 z^2 \frac{1}{\beta^2} \int \frac{dE}{E} \qquad (3.90c)
$$

If the energy transfer is large (i.e. much greater than the ionization potential) the classical treatment is a valid approach and yields a very good approximation for the real energy loss. For small energy transfers (distant collisions) a more rigorous treatment is needed. The total energy loss (including both close and distant collisions) can therefore be written loosely as follows:

$$
Z \int \frac{dE}{E} \Longleftarrow \sum \text{contributions from distant collisions} + Z \int_\eta^{E_{\max}} \frac{dE}{E}
$$

$$(3.91)$$

with $E(b_{\min}) = E_{\max}$, $E(b_{\max}) = E_{\min}$.

The semiclassical treatment which considers the electrons as non-relativistic harmonic oscillators yields for the total expression in Equation (3.91)

$$
2Z \ln \frac{E_{\max}}{\langle I \rangle}
$$

and for the total energy loss the *Bethe–Bloch formula* (Ahlen 1980; Jackson 1998; Rossi 1965)

$$\frac{dE}{dx} = 4\pi Z (N\rho/A) e^4 \frac{z^2}{\beta^2 m_e c^2} \left[\ln \frac{2m_e c^2 \beta^2 \gamma^2}{\langle I \rangle} \right] \tag{3.92a}$$

or, using again the classical electron radius

$$\frac{dE}{dx} = 2Z (N\rho/A)(2\pi r_e^2) m_e c^2 \frac{z^2}{\beta^2} \left[\ln \frac{2m_e c^2 \beta^2 \gamma^2}{\langle I \rangle} \right] \tag{3.92b}$$

where $\langle I \rangle$ is the *mean ionization potential*.

For historical reasons, the expression 'mean ionization potential' is retained in the Bethe–Bloch formula, although it is in fact the *geometric mean*. All oscillators with frequency ω contribute in an additional way (*arithmetic mean*) to the energy loss via the *logarithmic term* $\ln I (= -\ln(1/I))$, and therefore:

$$\underbrace{\langle I \rangle}_{\text{arithmetic mean}} = \sum_j (m_j/Z) \ln(\hbar \omega_j) = \underbrace{\ln(\hbar \langle \omega_j \rangle)}_{\text{geometric mean}}$$

$\hbar$ is *Planck's constant* ($\cong 6.582 \times 10^{-25}$ GeV s), m_j is the number of bound electrons with oscillation frequency ω_j (with $\sum_j m_j = Z$), and $\gamma \equiv 1/\sqrt{(1 - \beta^2)} = E/m_1 c^2$, where m_1 is the mass of the projectile.

For a relativistic treatment of the energy loss one has to add another term to the expression within the squared brackets of Equation (3.92a):

$$\frac{dE}{dx} = 4\pi Z (N\rho/A) e^4 \frac{z^2}{\beta^2 m_e c^2} \left[\ln \frac{2m_e c^2 \beta^2 \gamma^2}{\langle I \rangle} - \beta^2 \right] \tag{3.92c}$$

The energy loss depends on β (or $\beta\gamma$), but not explicitly on the mass m_1 of the incident particle. Therefore the behaviour of the Bethe–Bloch formula is discussed in terms of the variable β.

- For small β (but still large compared to the velocity of the orbital electrons $v(\omega)/c$), the energy loss decreases proportionally to $1/\beta^2$.

- All incident particles have a region of minimum ionization with $dE/dx \cong 0.002$ GeV/(g cm^{-2}) around $\beta\gamma \equiv Pc/(m_1 c) \cong 3$.

- For very high incident particle energies ($\beta\gamma > 10$), energy loss increases proportionally with $\ln(\beta\gamma) \cong \ln(E/m_1 c^2)$ (the region of *relativistic rise*, with $\beta \cong 1$) (Fig. 3.12).

For a projectile mass is large compared to the electron mass, $E_{max} = 2m_e c^2 \beta^2 \gamma^2$.

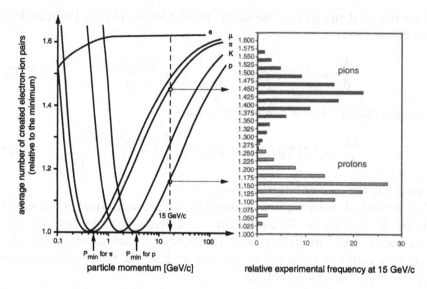

Fig. 3.12. A typical spectrometer measures the momentum of charged particles. With the measurement of a second kinetic property which does not depend in the same way on β, one can determine the mass of the particle. A sufficiently high primary ionization is required to discriminate particles with different masses with good statistic reliability. Note that the relative width decreases with $1/\sqrt{n}$ and is therefore approximately equal to the inverse square root of the detector length. Special care is needed if different kinds of particles appear with very different frequencies. (From Lucha and Regler 1997.)

For the purpose of track reconstruction, energy loss can often be approximated by a constant for $\beta\gamma > 1$. Even for a more precise treatment it is usually sufficient to calculate β and γ only once per track. However, the constant must be readjusted for each change of the scattering medium because of the different average ionization potential I (Particle Data Group 1998; ICRU 1984):

$$
I/Z \cong \begin{cases}
19.2\,\text{eV} & \text{for } H_2 \text{ gas} \\
21.7\,\text{eV} & \text{for liquid } H_2 \\
21\,\text{eV} & \text{for helium} \\
12\text{--}16\,\text{eV} & \text{for light atoms} \\
10\,\text{eV} & \text{for atoms with } Z > 16
\end{cases}
$$

Equation (3.92c) describes the energy loss in matter composed of isolated atoms. For real media the interatomic space is small and dielectric screening reduces the energy loss for collisions with large impact parameters (Sternheimer 1952)

$$
\frac{dE}{dx} \sim \frac{1}{\beta^2} \cdot \left[2\ln\left(\frac{2m_e c^2 \beta^2 \gamma^2}{\langle I \rangle}\right) - 2\beta^2 - \delta(\gamma) \right] \tag{3.92d}
$$

The correction term $\delta(\gamma)$ (Fernow 1986) cancels the factor of two in front of the logarithm for large values of γ. This is the correct formula for describing the energy loss in the track model of energetic heavy particles.

When considering the visible energy loss in a detector, however, the term $2\ln(\ldots)$ in Equation (3.92d) has to be replaced by

$$\ln\left(\frac{2m_ec^2\beta^2\gamma^2}{\langle I\rangle}\right) + \ln\frac{T_{\text{upper}}}{\langle I\rangle} = \ln\left(\frac{2m_ec^2\beta^2\gamma^2}{\langle I\rangle^2}\cdot T_{\text{upper}}\right)$$

and $-2\beta^2 \Leftarrow -\beta^2(1 + T_{\text{upper}}/E_{\text{max}})$, where T_{upper} is the limit energy of δ-electrons fully registered in the detector. This follows from the fact that the integral in the right hand side of Equation (3.91) now runs only from η to T_{upper}. In a TPC with dE/dx measurement this cutoff has contributions from the electrode geometry and/or by cuts in the pattern recognition. In a thin silicon wafer it is caused by the escape of energetic δ-electrons. The effect of the energy cutoff is that the relativistic rise is suppressed and that the energy loss approaches a constant value for $\beta\gamma > 100$ (the *Fermi plateau*).

The ratio between the energy loss at the Fermi plateau and at minimum ionization depends upon the medium. In a typical TPC the ratio is sufficiently high to allow particle identification on both sides of the ionization minimum. For instance, in argon at normal pressure, where the energy loss is negligible and may be ignored in the reconstruction of particle tracks, the ratio is $\cong 1.6$.

For crystalline silicon as used for strip and pixel detectors the ratio is less than 1.1, thus limiting the dE/dx method to the low-energy domain where dE/dx is proportional to $1/\beta^2$. This has recently been applied very successfully in the DELPHI experiment at LEP (Fig. 3.13).

The energy loss of incident electrons in material (see e.g. Allison and Wright 1987; Fernow 1986) is given by

$$\frac{dE}{dx} = 2\pi Z(N\rho/A)e^4\frac{1}{m_ec^2}\left[2\ln\left(\frac{2m_ec^2}{\langle I\rangle}\right) + 3\ln\gamma - 1.95\right] \tag{3.92e}$$

This has to be compared with the formula for heavy charged particles (see Equation (3.92c)):

$$\frac{dE}{dx} = 2\pi Z(N\rho/A)e^4\frac{z^2}{\beta^2m_ec^2}\left[2\ln\frac{2m_ec^2\beta^2}{\langle I\rangle} + 2\ln\gamma - 2\beta^2\right]$$

For electrons with 1 MeV or more, the stochastic emission of synchrotron radiation due to the bending force of the magnetic field in a spectrometer must also be taken into account (Jackson 1998).

An even more significant effect for high-energy electrons is the energy loss by the emission of photons in the electric field of an atomic nucleus,

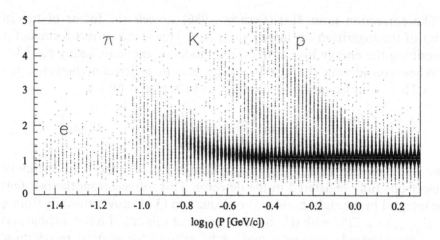

Fig. 3.13.   Results for particle identification via dE/dx with the DELPHI vertex detector from the 1997 data taking. After suitable corrections the mass separation in the $1/\beta^2$ region is in full agreement with the Monte Carlo simulation. (Courtesy of the DELPHI Silicon Tracker group.)

called *bremsstrahlung*. For an electron passing through matter the energy loss by bremsstrahlung is nearly proportional to its energy, while the ionization loss – for large energies – depends only logarithmically on it. The energy loss per unit length due to bremsstrahlung can be approximated by const $\cdot E/X_0$, where X_0 is the radiation length as used in multiple scattering (see above). Note that $X_0^{-1} \sim \alpha r_e^2 \sim e^6$, as is to be expected from the cross section of bremsstrahlung. A crude approximation for X_0 is given in Subsection 2.5.2.1:

$$X_0 = 180 \, \mathrm{g\,cm}^{-2} \, A/Z^2$$

A more precise fit to the experimental data can be obtained by the following formula (Particle Data Group 1998):

$$X_0 = \frac{716.4 \, \mathrm{g\,cm}^{-2} \, A}{Z(Z+1)\ln(287 \, Z^{-\frac{1}{2}})} \tag{3.93}$$

This formula, however, hides somewhat the functional dependence on Z, which is $\sim \ln(\mathrm{const} \cdot Z^{-\frac{1}{3}})$ as derived from physics.

The *critical energy* E_c is frequently defined as the energy at which the ionization loss rate and the loss rate due to bremsstrahlung are equal. An approximative formula for E_c, due to Bethe and Heitler (see e.g. Bethe and Ashkin 1953), is (in the same units as $m_e c^2$)

$$E_c \cong \frac{1600 \, m_e c^2}{Z}$$

An alternative definition of E_c is given by Rossi (1965), who defines the critical energy as the energy at which the ionization loss per radiation length is equal to the electron energy. The accuracy of the approximative formulas for E_c is limited by their failure to distinguish between gases and solids or liquids, where the density effect causes a significant difference in the ionization loss at the relevant energies.

As mentioned above, energy loss must be included in the track model (see Subsection 3.3.1.3) when solving the equations of motion (Equation (3.66)). In this context, however, for particles other than electrons it is treated as a deterministic process, the variance of the energy loss being small. In this way energy loss is easily implemented into the Runge–Kutta and the spline methods, but it is less obvious for explicit track models like a helix. Some care is also necessary when propagating fitted track parameters and errors by numerical calculation of the derivatives.

The effects of matter on heavy charged particles can be treated as a continuous process: white noise (see Subsection 3.3.2.2) in the case of multiple scattering, and a continuous deterministic function for the energy loss (see Subsection 3.3.1.2). The situation is somewhat different for electrons with an energy similar to or above the critical energy. In this case the probability of hard interactions with the field of the nuclei (bremsstrahlung) of the detector material requires a more sophisticated treatment. Bremsstrahlung actually dominates the energy loss and leads to large discrete jumps in the energy of the incident electron. This results in discontinuities in the curvature of the track. For the treatment of multiple scattering the problem is attenuated by the fact that the angular distribution of the emitted bremsstrahlung photons peaks in the forward direction. Note that the emission probability of bremsstrahlung photons is proportional to $1/m^2$ and is therefore, for the energy range accessible up to now, only relevant for electrons (and positrons). At the energies available at LHC bremsstrahlung starts to play a role also for muons. For track fitting with energy loss for electrons see also Subsection 3.2.6 and the references therein.

As already mentioned, measuring the energy loss of particles is a standard method of particle identification. Its result is an observable quantity that is a different function of the mass than the one obtained from the momentum measurement in the spectrometer, where the relation $p = m\gamma\beta c$ holds. Therefore particles of different masses (types) but equal momenta can be distinguished by the different amount of ionization or energy loss in a gaseous detector. The effect, however, is statistically significant only if the gas volume is sufficiently large, for instance in a time projection chamber. Because of the asymmetry of the distribution of the energy loss the discrimination power is worse than it would be in the Gaussian case. Another important aspect of the required discrimination power is the relative frequency of the different types of particles. Energy loss measurements have also been performed in the silicon vertex detectors of LEP experiments, although in silicon the height of the plateau in E/E_{min} is much smaller than in gaseous detectors. Recently, the energy loss has been studied also in diamond detectors. In such industrial diamonds a marked effect of the grain size can be observed.

3.3.1.6 The Landau distribution. The frequency distribution of the energy transfer in close collisions is proportional to $1/E^2$, with $d\theta/\theta^3 \sim dE/E^2$; this is an immediate consequence of the Molière scattering formula (Equation (3.85b)). The normalizing constant and the moments of this distribution, in particular the mean value, can be obtained only by deriving reasonable limits E_{min} and E_{max}, the latter from kinematical considerations.

In practice the data from multiple collisions are described quite well by a Landau distribution. The Landau distribution is a stable distribution, which means that it is closed under convolution. More precisely, if X, X_1 and X_2 are Landau distributed random variables, then for any positive numbers a_1, a_2 there are numbers $c > 0$ and d such that the convolution of $a_1 X_1$ and $a_2 X_2$ is distributed like $cX + d$:

$$a_1 X_1 + a_2 X_2 \sim cX + d$$

where $\sim$ denotes equality in distribution. The property of stability is obviously required by the fact that if a gas volume is divided into several compartments, the total energy loss is the sum of the energy loss in all compartments. Other examples of stable distributions are the Cauchy distribution, which is the symmetric counterpart of the Landau distribution, and the normal distribution. Like the Cauchy distribution, but unlike the normal distribution, the Landau distribution has no moments of any order, in particular no mean value. This is the source of many practical problems in dealing with the Landau distribution.

There is no analytical formula for the probability density function $f(x)$ of the Landau distribution; there is one, however, for its Laplace transform. In the simplest case the Laplace transform $\varphi(s)$ reads:

$$\varphi(s) = s^s = \exp(s \ln s) \tag{3.94a}$$

A Landau distribution with this Laplace transform can be called a 'standard' Landau distribution. The distribution is very skew with a long tail to the right. To the left it approaches 0 very fast. The maximum (mode) is at $M = -0.2228$, and the half-maximum coordinates are at $x_1 = -1.5865$ and $x_2 = 2.4319$, respectively. The quantile q_α of level α is defined by

$$\alpha = \int_{-\infty}^{q_\alpha} f(x)\,dx$$

The quantile of level $\alpha = 0.5$ is called the median m; it is at $m = 1.3558$. The mode of the Landau distribution is the quantile of level $\alpha_M = 0.2467$ and thus is very close to the first quartile $q_{0.25}$.

An affine transformation $y = ax + b\,(a > 0)$ of a 'standard' Landau distribution yields the general Landau distribution with *scale* parameter a and *location* parameter b. Its Laplace transform reads

$$\varphi(s) = \exp\left[-bs + as\ln(as)\right] \qquad (3.94b)$$

The convolution of two Landau distributions with scales a_i and locations b_i has the following Laplace transform:

$$\psi(s) = \exp\left[-(b_1 + b_2)s + a_1s\ln(a_1s) + a_2s\ln(a_2s)\right] \qquad (3.94c)$$

This is the Laplace transform of a Landau distribution with the following scale and location:

$$a = a_1 + a_2, \quad b = b_1 + b_2 + a\ln a - a_1\ln a_1 - a_2\ln a_2 \qquad (3.94d)$$

This shows that the Landau distribution is indeed a stable distribution.

The properties of the scale and location parameters are somewhat unusual, being different from those used with most other distributions. In the case of the Landau distribution the scale a is additive under convolution, whereas for all distributions with finite variances the square of the usual scale parameter, the standard deviation, is additive. The location b, on the other hand, is not additive, whereas the mean value, which is the usual location parameter, is additive under convolution, if it exists. Probably the most intuitive property of the location b is the fact that it is always the quantile of level $\alpha_b = 0.2868$, i.e. it divides the mass of the distribution approximately in the ratio $29:71$. It is somewhat larger than the mode, which is very close to the first quartile $q_{0.25}$. The scale parameter can easily be related to the width of the distribution: the full width at half maximum is equal to $4.02a$, almost precisely four times the scale parameter. This width is totally insensitive to the observations in the tail of the distribution and can therefore be used for a very robust estimate of the scale.

Since the measured energy loss is necessarily smaller than the total energy of the particle it can be argued that the actual distribution of the observations is a truncated Landau distribution and thus has moments of all orders. As the total energy is however usually larger by several orders of magnitude than the measured energy loss, this argument is not very compelling. Therefore the Landau distribution is usually considered as a suitable model of the experimental observations. The problem is then the estimation of location and scale from the observations.

If the observations follow a Landau distribution with scale a and location b, the average of the observations in a sample of size n is again Landau distributed with scale a and location $b + a \ln n$. It is therefore not particularly useful for estimating the location. In addition, it is sensitive to a truncation of the observations in the tail. Frequently the observations are censored by discarding for instance the larger half of the sample or, more generally, a fixed percentage of the observations in the tail. The relation of the censored average to the location and to the scale is unclear. On the other hand, the location and the scale can be easily determined by fitting a general Landau distribution with free parameters a and b to the observed distribution. If required, it can be convoluted with a Gaussian, in order to take into account the measurement errors of the individual energy loss measurements. In order to estimate the location with as little bias as possible, the data should not be censored or truncated, with the exception of the unavoidable truncation by the measuring device.

After estimates of a and b have been obtained from a fit to the data, they can be cross-checked with the mode, which is at $M = b - 0.22a$, with the median, which is at $m = b + 1.36a$, and with the full width at half maximum, which is very close to $4a$. Since the quantile of level $\alpha = 10^{-10}$ is close to $b - 4a$, there should not be any observations smaller than this value.

The CERN Program Library contains a set of subroutines (entry G110; see also Kölbig and Schorr 1984a, b) for computing the probability density function, the cumulative distribution function and other functions related to the Landau distribution. This is very useful for fitting the distribution to experimental observations.

3.3.2 The weight matrix

3.3.2.1 The measurement error of a detector. When a particle traverses a position-sensitive detector, the *measured position, m,* will, in practice, be different from the position of the real crossing point, $\overset{t}{c}$. This deviation can be described, in terms of statistics, by a conditional probability density function (Equation (3.13)), the *resolution function:*

$$\frac{d}{d\overset{t}{c}} P(m < \overset{t}{c} \mid \overset{t}{c}) = d(m; \overset{t}{c}) \tag{3.95a}$$

In practice, the dominant variable will be the difference between the measurement and the crossing point; however, this function quite often also depends on the crossing point itself

$$d'(\varepsilon; \overset{t}{c}) \equiv d(m; \overset{t}{c}) \tag{3.95b}$$

where $\varepsilon = m - \overset{t}{c}$; ε is the *experimental error,* although the word *error* will also be used for the *standard deviation* of ε, $\sigma(\varepsilon)$, namely the square root of the variance of ε, $\sigma^2(\varepsilon)$, or var(ε), which for an unbiassed measurement is the *expectation value* of ε^2 : $\langle \varepsilon^2 \rangle$.

Sometimes also the word 'resolution' (also experimental or detector resolution) is used for the standard deviation. This has historical reasons, when in multichannel analysis the σ of a peak was also a measure of the two-peak separation (e.g. in the dE/dx method, see Fig. 3.13). However, in some tracking detectors the two-particle separation (two-particle resolution) is often determined by quantities other than σ (e.g. electronics dead time in drift chambers).

If the word resolution is used at all, it should not be used with confusing quantitative attributes (a tiny resolution would certainly not suggest a small σ; it is better to use the terms 'good resolution' or 'small error'). In the context of separation of the signals from two different particles, 'two-particle resolution' should be used.

An estimate of the error is the RMS:

$$\text{RMS} = \left(\frac{1}{N} \sum_{\alpha=1}^{N} \varepsilon_\alpha^2 \right)^{\frac{1}{2}} \tag{3.96a}$$

If $\overset{t}{c}$ is constant but unknown, for instance in an ideal test beam, the variance should be estimated by the formula

$$\text{RMS}_N^2 = \frac{1}{N-1} \sum_{\alpha=1}^{N} \left(m_\alpha - \frac{1}{N} \sum_{\beta=1}^{N} m_\beta \right)^2 \tag{3.96b}$$

which is an *unbiassed estimator* of the variance even for finite N. (The gain of information (Equation (3.19)) is proportional to the inverse of the variance, $1/\sigma^2$, when the model related to this measurement is linear.)

A tracking detector can measure a single coordinate (a space point), a space point and a direction, or the full track parameter vector including the momentum. Here, measurement of only a coordinate will be discussed, but most of what is said can be trivially extended to the more general cases.

As examples, three types of detector will be discussed.

- An idealized detector with a constant and genuine random error.

- The drift chamber with a genuine but variable random error.

- The MWPC with a quasi-random error.

An idealized detector In an idealized detector the resolution function (Equation 3.95b) depends only on the difference between the measurement and the particle impact point (the primes for different functional relations of d have been omitted):

$$d(m - \overset{t}{c}; \overset{t}{c}) = d(m - \overset{t}{c}) = d(\varepsilon) \qquad (3.97a)$$

In this case the error does not depend on the location of the impact point itself:

$$\sigma^2 = \int \varepsilon^2 d(\varepsilon)\, d\varepsilon \qquad (3.97b)$$

Furthermore, it is assumed that the measurement is unbiassed. If it were not, the bias would be constant and one could easily correct for it by replacing ε by $\varepsilon - \langle \varepsilon \rangle$. If the errors are Gaussian (Equation (3.16)), the LSM applied to the track fit would coincide with the MLM (Subsection 3.2.3.1).

The resolution of an MWPC One of the milestones in the history of particle detectors and experimental particle physics was the invention of the MWPC (Charpak 1978; Sauli 1978; see also Subsection 2.1.1.1). It features reasonable spatial resolution, good two-track separation and a good efficiency even for large multiplicities. It has only local dead time, thus allowing high event rates.

A crude estimate of the measurement error perpendicular to the wires is straightforward: it will be seen to follow from Equation (3.99) that, for a flat distribution,

$$\sigma = d/\sqrt{12} \qquad (3.98)$$

where d is the distance between two adjacent wires (0.5–3.0 mm in practice). A *cluster* is defined as a set of adjacent fired wires belonging to one coordinate measurement. By considering odd and even wire cluster sizes for particles crossing in the neighbourhood of a wire or in a region near to the midpoint between two wires, respectively, this value could be reduced theoretically by a factor of 2, but in practice only a factor of 1.1 (i.e. about 10%) has been achieved for a particle path normal to the chamber surface.

For slightly inclined tracks the error goes through a minimum ($\cong 0.8d/\sqrt{12}$ in practice), becoming larger than the value in Equation (3.98) for angles in the region of 45°. This is due to effects such as 'cluster decays' (a missing wire in the set of wires covered by the projection of the particle's path in the chamber) (Fig. 3.14).

Using only information from single-wire clusters, the resolution function is an *ad hoc* deterministic step function when firing the wire nearest

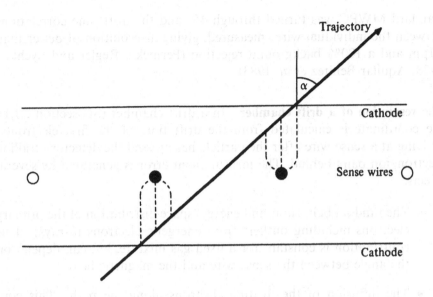

Fig. 3.14. Particles crossing the MWPC with a projected angle α can hit several wires, increasing the resolution for small α. The electrons and ions follow the field lines around the anode wires.

to the impact point of the particle's trajectory. This function is randomized by the stochastic character of the position of the impact point, and correlations between different detectors are smeared out by the different directions and curvatures due to the phase space population, deflection in the magnetic field, and multiple scattering. Therefore, the standard LSM can be applied. In a dense stack of chambers, however, the correlations between measurement errors cannot be neglected, but so far no attempt to take them into account in a rigorous way has been successful. Replacing the LSM by a 'Chebycheff norm' did not result in a significant improvement (James 1983).

A parametrized formula for the resolution is $\sigma = C \cdot (d/\sqrt{12}) f(\alpha)$ (with $C = 0.8$–0.9), where α is the angle between the track projected onto a plane perpendicular to the wires; $f(\alpha) = 1$ for $\alpha = 0$, has a minimum around 15–20°, and rises sharply for larger angles. In practice, however, the dependence of the measurement errors on the track parameters (mainly position and angle) is sufficiently smooth that the errors used for the fit be evaluated from the less precise parameters resulting from the previous pattern recognition (usually after a suitable parametrization). Only in exceptional cases must the error be reevaluated during the track fit. For the estimation of the variances see e.g. Frühwirth (1986).

An interesting application of a standard MWPC for very high resolution was performed in the European Hybrid Spectrometer (EHS) at CERN. A standard MWPC was turned through 45° and the drift time correlations between the individual wires measured, giving a resolution of better than 50 μm and a 100% background rejection (Pernicka, Regler and Sychkov 1978; Aquilar-Benitez *et al.* 1983).

The resolution of a drift chamber   In a drift chamber (Subsection 2.1.1), the coordinate is calculated from the drift time of the first electron(s) arriving at a sense wire after the particle has crossed the detector – trailing electron/ion pairs behind. The measurement error is generated by several effects.

- The random behaviour and energy/space distribution of the primary electrons including outliers from energetic electrons (δ-rays). This contribution is constant for a fixed gas mixture, but can depend on the angle between the sense wire and the magnetic field.

- The diffusion of the drifting electrons along the path. This contribution to the variance is proportional to the drift length. It is the dominant effect for large drift distances. However diffusion is largely reduced if the E- and the B-field are parallel – as is the case in many storage ring detectors.

- Different path lengths for different primary electrons due to inhomogeneities in the drift field and the 'avalanche field' near the sense wire. Special attention has to be paid to electrons coming from the region near the drift cell boundary where the field changes its sign at the adjacent cell.

- *Jitter* effects in connection with the detector and readout electronics. The TDC clock is stopped when the signal is higher than a threshold given by a discriminator. Because of the variable shapes and variable integrated charges of the avalanches in a wire chamber, the threshold is passed at different relative times for different particles crossing the detector at the same position. This jitter depends on the rise time of the pulse, the relative variation of the pulse height, and the frequency of occurrence and shape of the noise signals, all putting constraints on the choice of the threshold. *Pulse shaping* at an early time may reduce jitter effects to their *intrinsic minimum*, while the integrating effect of long cables contributes to an increase of jitter. The jitter effects can be considerably reduced if the *pulse shape* is measured by a FADC.

- The error due to the binning of the TDC. This contribution to the resolution function is of rectangular, triangular, or trapezoidal shape according to the bin sizes for the 'start' and 'stop signals', but is a real stochastic contribution due to the random relation between the arrival time and the clock status.

A typical time bin is in the order of a few nanoseconds; this is to be compared with a typical drift velocity of $50\,\mu\text{m ns}^{-1}$. The variance for a time bin of width Δt is:

$$\text{var}(c^2) = v^2 \int_{-\Delta t/2}^{+\Delta t/2} (t - \overset{t}{t})^2 f(t)\, \text{d}t = (v\Delta t)^2/12 \qquad (3.99)$$

or

$$\sigma(c) = (v\Delta t)/\sqrt{12}$$

where $f(t) = 1/\Delta t$ for t in the interval $(-\Delta t/2, +\Delta t/2)$, $f(t) = 0$ otherwise, and v is the drift velocity.

- Special behaviour must be expected for the pulse produced at the anode wire when the particle crosses in the avalanche region. In addition, the problem of how to decide on which side of the wire the particle has crossed the detector ('left–right ambiguity') is more difficult to resolve in this case.

- For typical resolutions of a few tens to a few hundreds of micrometres, the mechanical tolerances from machining and mounting, as well as from gravitational sagging and electrostatic deflections must also be considered. Some of these effects can be corrected systematically (see below), but sometimes they must be considered 'on average' by a 'pseudostochastical treatment', smeared out for different trajectories of different particles due to the distances between detectors. However, as much as possible should be done at the systematic level, in order to avoid an unnecessary loss of information.

Special problems arise inside a magnetic spectrometer, where the drift direction is determined both by the electric and the magnetic fields. If the magnetic field is constant, its effect can be compensated by inclining the direction of the electric field. Otherwise, quite complicated 'isochronous lines' (space curves of equal drift times) have to be expected. The magnetic field acts also as a focussing or defocussing force for the avalanche electrons, significantly influencing the measurement error.

Systematic effects can be due to mechanical tolerances, but also to changes in the drift velocity. Special care is needed when working in the nonsaturated region of drift velocity: the gas mixture, impurities, temperature, and pressure all influence the drift velocity and must be permanently monitored. This, of course, requires a reasonable number of calibration tracks for each monitoring period. For details on drift velocity see e.g. Fernow (1986), Leo (1987), and Sauli (1978).

3.3.2.2 Weight matrix formalism for multiple scattering.

Multiple Coulomb scattering acting on a particle sums up to relatively small but random change of the direction of flight. The resulting effect is a *stochastic process*, in particular a *Markov process*, and one can only evaluate the 'probable' amount of influence of a scatterer on a particle trajectory (Equation (3.88e)).

Multiple scattering can be described in two perpendicular planes by two random variables; the probability that a particle of momentum P travelling along the z axis leaves a scatterer of a length L in the interval of the projected angle $(\theta, \theta + d\theta)$ with a lateral displacement $(\varepsilon, \varepsilon + d\varepsilon)$ can be approximately described by the distribution function (Equation (3.88d))

$$dF(\varepsilon, \theta; L) = \frac{1}{2\pi} 4\sqrt{3} \frac{1}{\theta_s^2 L^2}$$

$$\times \exp\left\{-\left[\frac{4}{\theta_s^2}\left(\frac{\theta^2}{L} - \frac{3\theta\varepsilon}{L^2} + \frac{3\varepsilon^2}{L^3}\right)\right]\right\} d\varepsilon \, d\theta \qquad (3.100a)$$

where $\theta_s^2/2$ is the projected mean squared angle of the scattering per unit length as calculated by Equation (3.88e). (Note that in Subsection 3.3.2.1, following the notation in the classical text books, x was the particle direction.)

Proof: (with the particle's direction parallel to the z axis):

$$\langle \theta^2 \rangle = \left\langle \left(\int_{z_r}^{z} \frac{\partial\theta}{\partial z'} dz'\right)\left(\int_{z_r}^{z} \frac{\partial\theta}{\partial z''} dz''\right)\right\rangle \qquad (3.100b)$$

θ is a random quantity, and so is $\partial\theta/\partial z$. If multiple scattering is regarded as 'white noise' one can write with $L = z - z_r$:

$$\left. \begin{array}{c} \langle(\partial\theta/\partial z')(\partial\theta/\partial z'')\rangle \, dz' \, dz'' = (\theta_s^2/2)\,\delta(z' - z'') \, dz' \, dz'' \\ \langle\theta^2\rangle = (\theta_s^2/2)\,L \end{array} \right\} \qquad (3.100c)$$

Furthermore

$$\langle\theta\varepsilon\rangle = \left\langle \left(\int_{z_r}^{z} \frac{\partial\theta}{\partial z'} dz'\right)\left(\int_{z_r}^{z} (z'' - z_r)\frac{\partial\theta}{\partial z''} dz''\right)\right\rangle$$

$$= (\theta_s^2/2)\,L^2/2 \qquad (3.100d)$$

$$\langle\varepsilon^2\rangle = \left\langle \left(\int_{z_r}^{z} (z' - z_r)\frac{\partial\theta}{\partial z'} dz'\right)\left(\int_{z_r}^{z} (z'' - z_r)\frac{\partial\theta}{\partial z''} dz''\right)\right\rangle$$

$$= (\theta_s^2/2)\,L^3/3 \qquad (3.100e)$$

or in matrix notation

$$\left\langle \begin{pmatrix} \theta^2 & \theta\varepsilon \\ \varepsilon\theta & \varepsilon^2 \end{pmatrix} \right\rangle = \theta_s^2/2 \begin{pmatrix} L & L^2/2 \\ L^2/2 & L^3/3 \end{pmatrix} \qquad (3.100\text{f})$$

with the inverse

$$(8/\theta_s^2) \begin{pmatrix} 1/L & -(3/2)/L^2 \\ -(3/2)/L^2 & 3/L^3 \end{pmatrix} \qquad (3.100\text{g})$$

■

For the LSM it is necessary to know the covariance matrix of the measurement error vector $\varepsilon^T = (\varepsilon_1,\ldots,\varepsilon_n)$ (Subsection 3.2.3) or Equation (3.32). It can easily be evaluated from Equation (3.100e) (with $k < l$):

$$(\text{cov}\,(\varepsilon))_{kl} = \langle \varepsilon_k \varepsilon_l \rangle = (\theta_s^2/2)[(z_k - z_r)^2(z_l - z_r)/2$$
$$-(z_k - z_r)^3/6] \qquad (3.101)$$

For the general case of a coordinate vector **c** one can define a generalized geometry factor I_{kl}

$$\text{cov}\,(c_k c_l) = (\theta_{s,r}^2/2)I_{kl}$$

$$I_{kl} = \int_{s_r}^{\min(s_k,s_l)} g(s) \left(\frac{\partial f_k}{\partial \theta_1(s)} \frac{\partial f_l}{\partial \theta_1(s)} + \frac{\partial f_k}{\partial \theta_2(s)} \frac{\partial f_l}{\partial \theta_2(s)} \right) ds \qquad (3.102\text{a})$$

where

$$\iint \frac{df_k}{d\theta(s')} \frac{d\theta(s')}{ds'} \frac{df_l}{d\theta(s'')} \frac{d\theta(s'')}{ds''} ds' \, ds''$$

$$= \iint \frac{\partial f_k}{\partial \theta(s')} \frac{\partial f_l}{\partial \theta(s')} \frac{\theta_s^2}{2} \delta(s' - s'') \, ds' \, ds''$$

f is the model corresponding to **c** (see Equation (3.1)) and θ_1, θ_2 are the two uncorrelated and orthogonal scattering angles from Equation (3.88d), see Fig. 3.15 (note that the parameters θ_1, θ_2 have to be taken at s). The function $g(s)$ takes account of a possible change of medium and of energy loss. It is 0 in vacuum, usually 1 when entering the first medium (reference medium) and increases with energy loss (Equation (3.92b)) according to Molière's formula (Equation (3.88c)), where a change in radiation length must also be built in for the evaluation of $g(s)$. The particle's path is usually determined from pattern recognition information. If $g(s)$ only differs significantly from 0 when the particle passes through discrete detector layers, Equation (3.102a) can be approximated by a sum (Fig. 3.16):

$$\text{cov}\,(c_k c_l) = \sum_i^{\min(k-1,l-1)} \langle \theta_i^2/2 \rangle \left(\frac{\partial f_k}{\partial \theta_{1,i}} \frac{\partial f_l}{\partial \theta_{1,i}} + \frac{\partial f_k}{\partial \theta_{2,i}} \frac{\partial f_l}{\partial \theta_{2,i}} \right) \qquad (3.102\text{b})$$

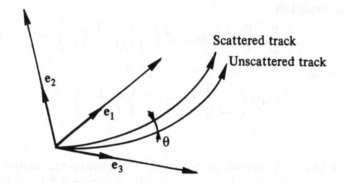

Fig. 3.15. The coordinate system with the axis e_3 parallel to the tangent of the trajectory. If multiple scattering is small it is sufficient to take into account only first order effects in the projected scattering angles θ_1 and θ_2 describing the change in the direction unit vector.

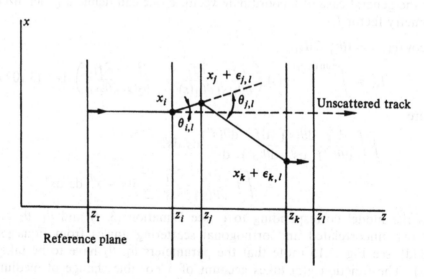

Fig. 3.16. The path of a track nearly parallel to the z axis traversing discrete scatterers.

where $\langle \theta_i^2 \rangle = (\theta_s^2/2)[(\Delta L)_i/\cos\alpha]$ is the variance of the projected scattering angle at detector layer i. It is proportional to the length of material along the path and is therefore direction dependent. Each layer is counted once, regardless of the number of coordinates measured at this layer. Here α is the angle between the particle direction and the normal to the layer. For an efficient evaluation of the geometry factors using the standard track parameters see Regler (1977).

For the direction cosines the following relation holds (with $p_3 = dx/ds$, $p_4 = dy/ds$):

$$\left.\begin{array}{l} \langle \delta p_3 \delta p_3 \rangle = (1 - p_3^2)\langle \theta_{\text{proj}}^2 \rangle \\ \langle \delta p_3 \delta p_4 \rangle = -p_3 p_4 \langle \theta_{\text{proj}}^2 \rangle \\ \langle \delta p_4 \delta p_4 \rangle = (1 - p_4^2)\langle \theta_{\text{proj}}^2 \rangle \end{array}\right\} \quad (3.102c)$$

and for the direction tangents (with $p_3 = dx/dz$, $p_4 = dy/dz$) (Eichinger and Regler 1981):

$$\left.\begin{array}{l} \langle \delta p_3 \delta p_3 \rangle = (1 + p_3^2 + p_4^2)(1 + p_3^2)\langle \theta_{\text{proj}}^2 \rangle \\ \langle \delta p_3 \delta p_4 \rangle = (1 + p_3^2 + p_4^2)p_3 p_4 \langle \theta_{\text{proj}}^2 \rangle \\ \langle \delta p_4 \delta p_4 \rangle = (1 + p_3^2 + p_4^2)(1 + p_4^2)\langle \theta_{\text{proj}}^2 \rangle \end{array}\right\} \quad (3.102d)$$

Several approaches have been invented for efficient use of the LSM. They can roughly be summarized in three classes:

- the global method;

- the break point method;

- the Kalman filter.

The '*global method*' has been discussed in this chapter to illustrate the weight matrix formalism. It is the appropriate method in the absence of multiple scattering. In the presence of multiple scattering, its contribution to the track uncertainty (Equations (3.102a, b)) is added to the error matrix (Equation (3.5)). The following ansatz has to be minimized (Equation (3.7)):

$$M(\mathbf{p}) = [f(\mathbf{p}) - \mathbf{m}]^{\text{T}}(\text{cov}(\mathbf{c}))^{-1}(f(\mathbf{p}) - \mathbf{m})] \quad (3.103a)$$

The advantage of this method is the fact that all the information is used at once. Its disadvantages are as follows.

- It is not suitable for pattern recognition.

- Inversion of an $n \times n$ matrix (number of operations $\sim n^3$).

- The fitted track follows the ideal extrapolation of the starting vector and not the scattered path of the real track. The residuals are dominated by multiple scattering, and therefore the pull quantities show the real meaning of multiple scattering rather than of the measurement errors.

The '*break point*' method (e.g. Billoir 1984) is adequate in the presence of a limited number of strong scatterers, e.g. plates and frames. An additional term is added to the least squares ansatz for each scatterer:

$$M(\mathbf{p}, \theta_{\text{bp}}) = [\mathbf{f}(\mathbf{p}, \theta_{\text{bp}}) - \mathbf{m}]^{\text{T}}[\text{cov}(\mathbf{c})]^{-1}[\mathbf{f}(\mathbf{p}, \theta_{\text{bp}}) - \mathbf{m}]$$
$$+ \theta_{\text{bp}}^{\text{T}}[\text{cov}(\theta_{\text{bp}})]^{-1}\theta_{\text{bp}} \tag{3.103b}$$

The main advantage of this method is not only that all the information is used at once, but that the real path of the particle is closely followed. This is of great importance for error checking and tuning where optimal interpolation is needed, and for extrapolation.

For a large number of break points the method is computer intensive because the derivatives $\partial \mathbf{f}/\partial \theta$ at each break point have to be calculated. It can be shown that the global method and the break point method are equivalent as far as the estimate of the initial parameter $\tilde{\mathbf{p}}_{\text{r}}$ is concerned. All formulae are given in Billoir et al. (1985).

'Recursive track fitting' by the LSM (the Kalman filter) is described in Subsection 3.2.5 (Frühwirth 1987). The contribution of multiple scattering between two subsequent steps can now be obtained from Equation (3.102a), and from discrete layers from Equation (3.102b), including inherently the break points at detector surfaces.

The advantage of this method is an efficient combination of pattern recognition and track fitting when large numbers of measurements closely follow each other.

3.3.2.3 Resolution of magnet spectrometer.
In most cases the errors in track reconstruction, as given by the LSM theory, are quite representative for the optimal precision achievable. In order to arrive at a formula for a first guess of the resolution of a spectrometer, the two main types of spectrometer must be treated separately.

(a) A spectrometer consisting of a central 'bending magnet' and two position detectors 'lever arms'. This set-up is typical in fixed-target experiments (Fig. 3.8(b)).

(b) A compact spectrometer consisting of a set of equidistant detectors, all inside a magnetic field. This is typical for detectors in colliding beam experiments (Fig. 3.8(a)).

Case (a) If one considers a charged particle moving parallel to the z axis (i.e. $1 - (dz/ds) \ll 1$), being subject to a small deflection by the spectrometer magnet's quasi-homogeneous field with only one significant component B_y, the deflection angle $\Delta\alpha$ can be obtained by integrating the equations of motion (Equation (3.66b)):

$$\Delta\alpha \cong \Delta x' \cong -(\kappa q/P)(1 + x'^2)\int_L B_y \, dz \tag{3.104a}$$

where x' is the derivative of x with respect to z ($dx/dz \ll 1$), $\Delta\alpha$ is the deflection angle, and $\int_L B_y\,dz \equiv \bar{B}L$ where $\bar{B}$ is the average field along the z axis; $\bar{B}L$ is often called the bending power (in T m). Neglecting x'^2, Equation (3.104a) can be rewritten

$$(P\Delta\alpha)/P \cong \Delta P_T/P = -(\kappa q/P)\bar{B}L$$
$$|\Delta P_T| \cong |\kappa q\bar{B}L| \tag{3.104b}$$

with P, P_T, κ and q defined as in Subsection 3.3.1.1. Differentiation of the logarithm of Equation (3.104a) yields

$$\left|\frac{\delta\Delta\alpha}{\Delta\alpha}\right| = \left|\frac{\delta P}{P}\right| = \left|\delta\Delta\alpha\frac{P}{\kappa q\bar{B}L}\right| \tag{3.104c}$$

With M position detectors available and a *symmetric spectrometer of length, l,* the best theoretical angular resolution is obtained by placing $M/4$ detectors at the end of each arm and $M/2$ detectors at the centre (i.e. near the spectrometer magnet). Note that the detectors arranged in the centre contribute in a correlated way to the direction measurement in front of and behind the magnet. The resolution obtainable is then

$$\sigma(P)/P = P \cdot 8\sigma \cdot (|\kappa q\bar{B}Ll|)^{-1} \cdot M^{-\frac{1}{2}} \tag{3.104d}$$

where $\kappa = 0.299\,792\,458$ (GeV/c)T^{-1} m^{-1} and σ is the error of an individual detector measuring the x coordinate (same units as l). This configuration of detectors, whilst optimizing geometrical precision, is a particularly unsuitable arrangement for correctly associating measured points into tracks and, therefore, can be used only in experiments with trivial track recognition problems (e.g. elastic scattering). At low energies, a limit on the precision is set by multiple scattering. Whereas the relative error in the momentum from position measurements is given by $\sigma(P)/P \sim P/l$, i.e. is proportional to P, another term arising from multiple scattering contributes $\sigma_{MS}(P)/P \sim 1/(\beta\bar{B}L)$, which is *large for small* β and *constant for high momenta* ($\beta \cong 1$). These two contributions must be added quadratically in order to obtain the error on the momentum. With Equations (3.88c) and (3.104c) one obtains:

$$|\sigma_{MS}(\Delta\alpha)/\Delta\alpha| = |q\,0.015(M/2)^{\frac{1}{2}}(d/L_0)^{\frac{1}{2}}(1/\beta Pc)(P/\kappa q\bar{B}L)|$$
$$= 0.015(M/2)^{\frac{1}{2}}(d/L_0)^{\frac{1}{2}}(1/\beta c\kappa\bar{B}L) \tag{3.105}$$

where P is the momentum (GeV/c), $\bar{B}L$ is the deflection power (in T m), $M/2$ is the number of detectors placed around the magnet, and d/L_0 is the thickness of one individual detector layer in units of radiation length (note that one detector layer can measure either one or several coordinates; the latter case can be taken into account by reducing σ). The contribution to $\sigma(P)/P$ from multiple scattering does not depend on the mass of the incident particle if β has been chosen as variable.

Case (b)   In central spectrometers, as used in most of the storage ring experiments, all tracking detectors are located inside a quasi-homogeneous magnetic field. This set-up was extensively studied for the first time for bubble chambers (Gluckstern 1963), both with and without multiple scattering.

The error in the momentum reconstructed in any projection is *inversely proportional to the field* normal to this projection, B_p, and to the square of the projected track length, l_p^2. In many storage ring experiments the field is of cylindrical symmetry, with its main component parallel to the beams (z axis). Therefore, the central detectors are designed such that they can measure the azimuthal coordinate $R\Phi$ precisely and therefore determine $P_T \equiv P_p (= P \cos \lambda = P \sin \theta)$ with ultimate precision (Fig. 3.8(a)): for the polar angle, θ, and the 'dip angle' $\lambda = \pi/2 - \theta$, see Fig. 3.9 and Equations (3.67).

Assuming a set-up consisting of M equidistant surfaces (concentric cylinders) for measurements of the coordinate $R\Phi$, and $M \gg 3$, one obtains asymptotically

$$\sigma(P_p)/P_p = P_p(|\kappa q B l_p|)^{-1}(\sigma/l_p)[720/(M+6)]^{\frac{1}{2}} \qquad (3.106a)$$

Note that comparing Equation (3.106a) with Equation (3.104d), and with $L = l = l_p$, the relative error $\sigma(P_p)/P_p$ is proportional to $1/l_p^2$. This is an obvious disadvantage for the design of compact storage ring detectors (and the reason why highly energetic particles often need an additional determination of energy by calorimeters surrounding the tracking detector; Section 2.6). For a given relative error, the spectrometer must grow in size with $P^{\frac{1}{2}}$ if M is fixed. If M also grows linearly l_p, $\sigma(P)/P$ is asymptotically proportional to $P_p l_p^{-\frac{5}{2}}$.

If the set-up is not equidistant, but such that half of the measurements are at the centre of the track and one quarter at each of the ends, the momentum error is substantially improved to

$$\sigma(P_p)/P_p = P_p(|\kappa q B l_p|)^{-1}(\sigma/l_p)[256/(M+2)]^{\frac{1}{2}} \qquad (3.106b)$$

This can easily be shown by evaluating the sagitta S:

$$S \cong l_p^2/(8R_p), \quad R_p = P \cos \lambda/(|\kappa q B|),$$

$$\sigma(S) \cong \sigma/(M/4)^{\frac{1}{2}},$$

$$\sigma(S)/S = \sigma(P_p)/P_p = [\sigma/(M/4)^{\frac{1}{2}}] \times [8P_p/(l_p^2\kappa|qB|)],$$

with $\sqrt{256} = 16$, to be compared with $\sqrt{720} \cong 27$ (from Equation (3.106a)) or, on the other hand, with 8 (from Equation (3.104d)). This is, again, only a hypothetical arrangement of detectors, as it is unsuitable for recognizing tracks and difficult to install. Note that even this formula gives – for fixed $\bar{B}L$ and fixed M – a precision worse than that of the two lever arm spectrometer by a factor of 2.

For the precision of the initial direction of the track one obtains (again in the asymptotic limit) for equidistant detectors (Gluckstern 1963):

$$\sigma(\phi_0) = (\sigma/l_p)[192/(M + 6)]^{\frac{1}{2}} \qquad (3.106c)$$

and for the correlation between ϕ_0 and $1/P_p$ a constant i.e.

$$|\rho(\phi_0, 1/P_p)| = 0.968 \qquad (3.106d)$$

Note that ρ *vanishes in the centre* of the track segment. To evaluate the error in P, the error in θ must also be taken into account

$$P = P_T / \sin\theta$$

$$\sigma(P) = [\sigma^2(P_T)/\sin^2\theta + \sigma^2(\theta)P_T^2\cos^2\theta/\sin^4\theta]^{\frac{1}{2}} \qquad (3.107a)$$

the covariance between P_T and θ being negligible in practice. For $\sigma(P)/P = P\sigma(1/P)$ one obtains

$$\sigma(P)/P = [\sigma^2(P_T)/P_T^2 + \sigma^2(\theta)\cot^2\theta]^{\frac{1}{2}} \qquad (3.107b)$$

At *low energies*, a limit of the precision is again set by *multiple scattering* and the optimization becomes definitely more complicated, as the final resolution depends not only on the number of coordinates measured, but also on the amount of matter and on the momentum spectrum of the particles. This limitation is also important for high precision vertex detectors. For an S-shaped track (as in the CMS detector at LHC (CMS collaboration 1998)) the relative momentum error increases by a factor between 2 and 4 (note that M is small). Substantial help can be obtained if the origin of the particle is known (beam position or vertex reconstruction). For the ATLAS detector at LHC a toroidal field shape has been chosen (ATLAS collaboration 1997).

Modern storage ring spectrometers try to minimize the amount of matter inside the detectors themselves, and scattering occurs mainly in the supporting frames and in the beam tube. For such discrete layers the formula derived for homogeneously scattering matter can only be used as a guideline.

The contribution of *multiple scattering* to the *relative error* in the momentum measured by a set of equidistant detectors after correction of the track length for the dip angle λ (subscript p' in contrast to p for the projection), is:

$$\sigma_{MS}(P_p)/P_p = 1.2 P_p/(\kappa |qB|)(\langle\theta_s^2/2\rangle_{p'} l_p)^{\frac{1}{2}} \qquad (3.108a)$$

where $P_p = P\cos\lambda (\text{GeV}/c)$ is the momentum component in the azimuthal plane perpendicular to **B**, $l_p = l\cos\lambda$ is the track length projected to azimuthal plane (l = track length in space in the detector module), $\lambda = \pi/2 - \theta$ is the dip angle (θ = polar angle), see Fig. 3.9, $\langle\theta_s^2/2\rangle_{p'}$ is the variance per projected unit length by multiple scattering with $\langle\theta_s^2/2\rangle$ as defined in Equation (3.100b), and

$$\langle\theta_s^2/2\rangle_p = \langle\theta_s^2/2\rangle/\cos^2\lambda, \quad \langle\theta_s^2/2\rangle_{p'} = \langle\theta_s^2/2\rangle_p/\cos\lambda$$

The term $1/\cos^2\lambda$ accounts for the projection of the scattering angle, while the term $1/\cos\lambda$ accounts for the additional matter traversed in space.

Remember that $\langle\theta_s^2/2\rangle_p \sim (m^2 + P^2)/P^4$ gives a similar behaviour as Equation (3.105).

Using the Molière formula (Equation (3.100b)) for $\langle\theta_s^2/2\rangle$ yields:

$$\sigma_{MS}(P_p)/P_p = \frac{1.2}{\kappa|B|}\frac{0.015}{\beta c}\left(\frac{l_p\rho}{X_0\cos\lambda}\right)^{\frac{1}{2}} \qquad (3.108b)$$

where $X_0/\rho = L_0$ is the geometrical radiation length (same unit as l_p) and ρ is the density. For a set-up which has coaxial cylindrical detectors parallel to **B**, any significant change of the angle β (Fig. 3.9) between the projected trajectory and the radius vector must be considered for a more accurate calculation.

It should be noted that $\sigma_{MS}(P_p)/P_p$ does not depend explicitly on the number of detectors. However, a set-up often consists of M detector layers, separated by gaps without multiple scattering. This can be accounted for by replacing ρ in Equation (3.108b) by ρ_{eff} ($= \rho_{mat}l_{mat}/l$) where ρ_{mat} is the density of the detector material, and l_{mat}/l is the fraction of track length through the layers of matter which of course depends on M.

If, in a cylindrical set-up, a total radial space $R_{max} - R_{min} = \Delta R$ is available for placing M detector layers, each with a radial thickness d, then $l_{mat}/l = Md/\Delta R$ and (Fig. 3.8(a))

$$\sigma_{MS}(P_p)/P_p = \frac{1.2}{\kappa|B|}\frac{0.015}{\beta c}\left(\frac{1}{X_0}\frac{\rho_{mat}dM}{\cos\lambda}\right)^{\frac{1}{2}} \qquad (3.108c)$$

This contribution, $\sigma_{MS}(P_p)/P_p$, has to be added quadratically to the errors in the relative momentum due to the measurements, $\sigma_{pos}(P_p)/P_p$ (see

Subsection 3.3.2.3):

$$\sigma(P_p)/P_p = [(\sigma_{pos}(P_p)/P_p)^2 + (\sigma_{MS}(P_p)/P_p)^2]^{\frac{1}{2}} \qquad (3.109a)$$

In the case of a cylindrical set-up, for a given ΔR, $\sigma_{pos}(P_p)/P_p \sim P_p/(M+6)^{\frac{1}{2}}$ (see Equation (3.106a)), and

$$\sigma(P_p)/P_p = \{[c_1 P_p/(M+6)^{\frac{1}{2}}]^2 + (c_2 M^{\frac{1}{2}})^2\}^{\frac{1}{2}} \qquad (3.109b)$$

This shows clearly that an optimal M exists for an available space ΔR, for each value of P_p. Overinstrumentation will be counterproductive, especially for low-momentum particles, unless additional measurements can be included *without further increasing the scattering matter* (e.g. by using a Time Projection Chamber (TPC)).

These considerations also have an impact on the precision of vertex reconstruction (Section 3.4). Because of the strong correlation between the track errors in the direction and the momentum (Equation (3.101d)), the error in the vertex position is largely due to the lever arm between the innermost detector (usually just outside the beam tube) and the vertex region; therefore, precise spatial vertex evaluation also requires z and θ to be measured with high precision. However, the effects of multiple scattering within the beam tube and the innermost detector (a high-precision vertex detector) can be considerable, thus also limiting the precision that can be achieved in the (x, y) projection. The only solution is a beam pipe with a minimum of material (in units of radiation lengths) at a small radius.

3.3.3 Track element merging

In a complex detector it is often necessary for track segments to be fitted separately. The problem of combining the information is discussed in this subsection. It will be assumed that there are two detector modules with two estimates ($\tilde{p}_i, i = 1, 2$) of the track parameters at the respective reference planes $z_r = z_{r,1}, z_{r,2}$. The covariance matrices of the $\tilde{p}_i$ are denoted by C_i. The reference plane of the combined track information is assumed to be equal to $z_{r,2}$ (Fig. 3.17).

The principle, which is similar to a Kalman-Filter step, is to propagate the estimate $\tilde{p}_1$ and its covariance matrix C_1 to $z_{r,2}$ (see also Equations (3.36)):

$$\tilde{p}_2^{(1)} = \tilde{p}_2(\tilde{p}_1)$$
$$C_1' = F C_1 F^T$$

with

$$F = \partial p_2 / \partial p_1$$

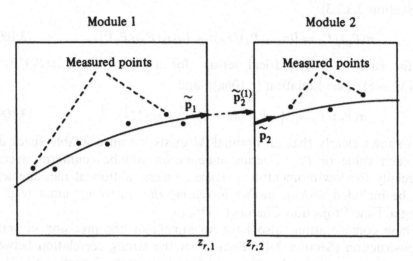

Fig. 3.17.   In track element merging, the track parameters $\tilde{\mathbf{p}}_1$ are propagated to a new reference plane. $\tilde{\mathbf{p}}_2^{(1)}$ and $\tilde{\mathbf{p}}_2$ are considered as direct measurements of the true track parameters at the reference plane $z_{r,2}$.

Contributions to multiple scattering between $z_{r,1}$ and $z_{r,2}$ (including material at $z_{r,1}$) are added to $\mathbf{C}_1'$, to give the final covariance matrix $\mathbf{C}_2^{(1)}$ of $\tilde{\mathbf{p}}_2^{(1)}$

$$\mathbf{C}_2^{(1)} = \mathbf{C}_1' + \sum_i \frac{\partial \mathbf{p}_2}{\partial(p_{i,3}, p_{i,4})} \frac{\partial(p_{i,3}, p_{i,4})}{\partial(\theta_{i,1}, \theta_{i,2})} \mathbf{C}(\theta_{i,1}, \theta_{i,2})$$

$$\times \frac{\partial(p_{i,3}, p_{i,4})^{\mathrm{T}}}{\partial(\theta_{i,1}, \theta_{i,2})} \left(\frac{\partial \mathbf{p}_2}{\partial(p_{i,3}, p_{i,4})} \right)^{\mathrm{T}} \tag{3.110}$$

$+$ contributions due to continuous scattering

(see also Equation (3.33))

where $\mathbf{C}(\theta_{i,1}, \theta_{i,2})$ is the covariance matrix of the scattering angles $\theta_{i,1}, \theta_{i,2}$, and i is the index for all scatterers in the interval $[z_{r,1}, z_{r,2})$. If one now considers $\tilde{\mathbf{p}}_2^{(1)}$ and $\tilde{\mathbf{p}}_2$ as direct measurements of $\mathbf{p}_2$, a joint least squares ansatz can be made

$$M(\mathbf{p}_2) = (\mathbf{p}_2 - \tilde{\mathbf{p}}_2)^{\mathrm{T}} \mathbf{C}_2^{-1} (\mathbf{p}_2 - \tilde{\mathbf{p}}_2) + (\mathbf{p}_2 - \tilde{\mathbf{p}}_2^{(1)})^{\mathrm{T}} (\mathbf{C}_2^{(1)})^{-1} (\mathbf{p}_2 - \tilde{\mathbf{p}}_2^{(1)})$$

$$\tag{3.111a}$$

The final estimate is a weighted mean:

$$\tilde{\tilde{\mathbf{p}}}_2 = [(\mathbf{C}_2^{(1)})^{-1} + \mathbf{C}_2^{-1}]^{-1} [(\mathbf{C}_2^{(1)})^{-1} \tilde{\mathbf{p}}_2^{(1)} + \mathbf{C}_2^{-1} \tilde{\mathbf{p}}_2] \tag{3.111b}$$

and $M(\tilde{\tilde{\mathbf{p}}}_2)$ is χ^2-distributed with five degrees of freedom.

Ansatz (3.110) contains an implicit assumption, namely that $\delta \tilde{\mathbf{p}}_2^{(1)}$ and $\delta \tilde{\mathbf{p}}_2$ are independent. This is, however, only true if the reference plane $z_{\mathrm{r},2}$ is at the near end of the module 2, as seen from module 1. In this case, the combined estimate is between the two modules and therefore not of great interest, except for a χ^2 test. In the more interesting case of $z_{\mathrm{r},2}$ being at the far end of module 2, one notes first that the difference

$$\mathbf{p}_2(\overset{t}{\tilde{\mathbf{p}}}_1) - \overset{t}{\tilde{\mathbf{p}}}_2$$

is a random quantity: it is *correlated to the contribution of multiple scattering* to the measurement errors in module 2 and hence also to $\tilde{\mathbf{p}}_2 - \overset{t}{\tilde{\mathbf{p}}}_2$. Therefore, $\tilde{\mathbf{p}}_2^{(1)} - \overset{t}{\tilde{\mathbf{p}}}_2$ is also correlated (Billoir *et al.* 1985).

It is important to note that, in the presence of multiple scattering, the following rule has to be observed: if an optimal estimate should be achieved on one side of the set of detectors, all estimates for the individual sets should be made on the corresponding side of each detector module (e.g. for vertex fitting), and vice versa (e.g. for extrapolation to the muon chambers). Otherwise a complicated smoothing algorithm must be used (Subsection 3.2.5).

If the information from individual detector modules with poor resolution is going to be merged, a 'reference track' as obtained from the global track search (pattern recognition) can be used, as described in Subsection 3.2.5.

Adding nongeometrical information In addition to the coordinates, some other information such as the 'TOF' ('Time of Flight') or 'dE/dx' is sometimes available. Usually, however, this information, although very useful for particle identification, triggering, etc., must be treated with great care when used in a geometry program. We demonstrate this with TOF measurements.

If the velocity of a particle is significantly different from the velocity of light (e.g. for a recoil particle in diffractive processes), then the TOF yields valuable information about the momentum and the mass of this particle:

$$ct \cong cL/\beta = c(L/P)(P^2 + m^2)^{\frac{1}{2}} \tag{3.112a}$$

where m and P are in GeV, and L is the path length. Using $p_5 = 1/P$ one obtains

$$ct \cong cLp_5[m^2 + (1/p_5)^2]^{\frac{1}{2}} \tag{3.112b}$$

It is usually possible to make a rough estimate of L and to consider it as a constant during the fit. Only for very slow particles and large curvature must the length L also be varied during the iteration process; but slow particles always cause several problems (important multiple scattering, energy loss, nuclear scattering).

There are now two possibilities:

- either the TOF information is used during the geometrical fit;

- or, for a better treatment of multiple scattering, this information is first used only for the mass assignment and added in an additional fit step, or even later in the kinematical fit if enough geometrical information is available for a first track fit without using additional information.

In either case, the additional term to be added to the least squares ansatz is:

$$M'(\mathbf{p}) = M(\mathbf{p}) + [t(p_5) - t_m]^2/\sigma^2(t) \qquad (3.112c)$$

3.3.4 Numerical minimization technique

Data analysis in high-energy physics is confronted with a variety of problems that require the optimization of some function with respect to a set of parameters. The MLM (where usually the negative logarithm of the likelihood function is going to be minimized) and the LSM are examples. The methods used all have in common the fact that the *optimal estimate* of the parameters is defined by the *minimum of a function*, which depends explicitly on these parameters. The measurements, together with their covariance matrix, are fixed parameters in these functions. The numerical technique will be outlined below only for the case of the LSM. The only assumption is an approximate quadratic behaviour of the least squares function in the immediate vicinity of its minimum. In the linear model, a sufficient condition for this assumption is that, within the measurement space $\{\mathbf{c}\}$ there exists a subspace with dimension $n \geqslant$ dimension of $\mathbf{p}$ for which the covariance matrix is not singular, and the corresponding rows of the matrix of derivatives, $\mathbf{A}$, are at least of the rank of the dimension of $\mathbf{p}$. If this is not the case, part of the parameter vector may still be defined (e.g. the three-dimensional projection $R\Phi$, ϕ, $1/P_T$, without the knowledge of z and θ).

   If the model for the LSM is linear, the minimization procedure is trivial: the least squares ansatz (Equation (3.7)) for the function $M(\mathbf{p})$ is of second order, therefore, the minimum condition $\partial M/\partial \mathbf{p} = 0$ yields a system of linear equations for the estimate $\tilde{\mathbf{p}}$ with the explicit solution (Equation (3.8)):

$$\tilde{\mathbf{p}} = \overset{0}{\mathbf{p}} + (\mathbf{A}^T\mathbf{W}\mathbf{A})^{-1}\mathbf{A}^T\mathbf{W} \cdot [\mathbf{m} - \mathbf{f}(\overset{0}{\mathbf{p}})]$$

where $\overset{0}{\mathbf{p}}$ is the expansion point (Equation (3.4)) of $\mathbf{f}(\mathbf{p})$ and $\mathbf{W} = \mathbf{V}^{-1} = \mathbf{cov}^{-1}(\varepsilon)$. For the least squares ansatz, in addition to the track model, the covariance matrix $\mathbf{cov}(\varepsilon)$ must also be evaluated and inverted, which, in the presence of multiple scattering causing a nondiagonal covariance matrix (Subsection 3.3.2.2), might become a time consuming procedure if the number of measurements is large.

If, however, the track model is far from being linear, the linearization will give a result, $\tilde{p}$, strongly dependent on the choice of $\overset{0}{p}$, and an *iterative procedure* has to be applied, using the 'Newton method' or the quasi-Newton method, for iteratively approximating $\overset{0}{p}$ towards $\tilde{p}$. This is usually successful for the 'least squares function' $M(p)$. First and second derivatives of M with respect to p can be expected to exist everywhere in the region of interest of p. One can then write the Taylor series expansion of M around some point p, say $\overset{0}{p}$

$$M(p) = M(\overset{0}{p}) + q \cdot (p - \overset{0}{p}) + \tfrac{1}{2}(p - \overset{0}{p})^T G \cdot (p - \overset{0}{p}) \qquad (3.113a)$$

where $\overset{0}{p}$ is an *ad hoc* estimate (e.g. from a previously performed pattern recognition). (As mentioned earlier (Subsection 3.3.2) the covariance matrix $V = \mathbf{cov}(\varepsilon)$ varies only slowly with p in most cases and can therefore be kept constant; only when $\overset{0}{p}$ must be corrected drastically, i.e. $|\overset{0}{p} - \tilde{p}_1|$ large, must $V (= W^{-1})$ be reevaluated for the second iteration, sometimes for both contributions V_{detector} and V_{MS} (Subsection 3.3.2).)

For the derivatives of the LSM ansatz, one gets for q and G from Equation (3.113a)

$$q = -2A^T W \cdot [m - f(\overset{0}{p})] \qquad (3.113b)$$

and the 'Hessian matrix' G, i.e. the matrix of second derivatives, is

$$G = 2\{A^T W A - [\partial^2 f/(\partial p)^2]^T W \cdot [m - f(\overset{0}{p})]\} \qquad (3.113c)$$

where

$$[\partial^2 f/(\partial p)^2]_{i;kl} \equiv [\partial^2 f_i(p)/(\partial p_k \partial p_l)]_{\overset{0}{p}}$$

for summation over i (the 'T' in Equation (3.113c) refers to i).

The calculation of the second derivatives of f would cause prohibitive computing time if done numerically together with step-by-step tracking. Fortunately the second term of Equation (3.113c) can be neglected as long as *both* the second derivatives of the track model *and* all $m_i - f_i(\overset{0}{p})$ are small (note that in the case of linear functions this term vanishes anyway). This leads to the same formula as for the linear case, see above and Equation (3.8). If the functions $f(p)$ are explicitly known, there is no reason to suppress the second derivatives and thus reduce the convergence properties of the iterative procedure. The *convergence point* $\tilde{p}$ remains the same, but is approximated faster with the second derivatives retained.

Neglecting the second term of Equation (3.113c) results in $\mathbf{G}$ being necessarily positive definite, but does not necessarily yield a meaningful curvature at $\overset{0}{\mathbf{p}}$ and may cause 'divergent oscillations' of the procedure (in case of unprecise initialization), whereas with the second derivatives retained in the least squares ansatz, the Newton method, may diverge when the initial parameter values, $\overset{0}{\mathbf{p}}$, are outside the domain where M is parabolic, i.e. beyond an inflection point. This is not always a disadvantage, because it may be a useful test of the *ad hoc* method for initialization and, for instance, indicate an error in the previous pattern recognition.

In practice, two or three iterations are necessary in the nonlinear case until the minimum is reached, i.e. until the estimated parameters become stable; then the *change* in 'χ_i^2' $\equiv M(\tilde{\mathbf{p}}_i)$ for the last iteration becomes negligibly small

$$0 \leqslant M(\tilde{\mathbf{p}}_{i-1}) - M(\tilde{\mathbf{p}}_i) \ll 1$$

usually less than 10^{-2}. (However, this convergence limit should not be smaller in order to avoid numerical oscillations.) Therefore, it is sufficient in most cases to reevaluate the derivatives only once. The decision criteria must be deduced from practical experience; they may depend on $\overset{0}{\mathbf{p}}$ itself. During the tuning period of the fit program, however, one extra iteration of the track fit at the expansion point $\overset{0}{\mathbf{p}}_{i+1} = \tilde{\mathbf{p}}_i$ yielding $\tilde{\mathbf{p}}_{i+1}$, may be needed. For more details see Eichinger and Regler (1981), Frodesen *et al.* (1979), James (1972) and James and Roos (1986).

3.4 Association of tracks to vertices

3.4.1 Basic concepts

The vertex fit serves two purposes. The first is to estimate the position of the point of interaction and the momentum vectors of the tracks emerging from this point (with improved precision due to the *vertex constraint*). The second is to check the association of tracks to a vertex, i.e. the decision of whether the track does indeed originate from this vertex. The following discussion is applicable both to the primary interaction vertex and to an eventual secondary vertex (a decay or secondary interaction). While in the first case the exact position of the vertex might seem a simple mathematical constraint, it is of some importance in the second case, since it determines the direction of the (possibly unseen) track connecting the two vertices.

In both cases the momentum vectors of all emerging charged tracks should be computed with the best possible precision together with their common covariance matrix, since they are the input for a subsequent kinematical fit. Here the LSM turns out to be the best method for the subsequent merging of information in the presence of additional constraints (Fig. 3.18). (The same holds for the final fit of kinematics by adding the momentum and energy conservation constraints.)

The method described below has been proposed and used successfully by the first generation experiments at the CERN-ISR (Metcalf *et al.* 1973). It involves the *inversion of a matrix of the order* $3n$ (n = number of tracks). Since the number of arithmetical operations for the inversion increases with the third power of the order, this method becomes prohibitive with a further increase in energy and the resulting higher multiplicities. Also, the events become more complex with an increasing need to eliminate tracks which do not belong to the primary interaction vertex. Therefore, a new algorithm for the computation of the estimates, their covariance and the χ^2 was developed (Metcalf, Regler and Broll 1973; Billoir, Frühwirth and Regler 1985; Regler and Frühwirth 1990), allowing the application of this method to very complex events also. In order to avoid unnecessary repetitions of the vertex fit, a *recursive method* is desirable, allowing a check to be made on the association of tracks to a common vertex. The input for the vertex fit consists of information about the tracks to be grouped together. Normally one considers the estimated track parameters at a reference surface as '*virtual measurements*'. The reference surface will in most cases be a plane, a cylinder (especially in storage ring experiments) or the beam tube.

The choice of the reference surface has a certain influence on the behaviour of the fit, since it is desirable that the virtual measurements, namely the track parameters, are to a good approximation *linear functions* of the vertex position and of the parameters determining the momentum vector at the vertex. In some cases the reference surface will coincide with a physical surface, e.g. the wall of a vacuum vessel or a vertex chamber. If multiple scattering in this wall is important, it can easily be taken into account, by augmenting the covariance matrix of the estimated track parameters (Equation (3.100a)).

If the position of the vertex is known *a priori* to some precision, as is the case for the interaction region of a storage ring, this knowledge can be considered as an *independent measurement* of the position, with its proper error matrix.

If a single track is poorly defined, its coordinates should be incorporated directly into the vertex fit, instead of a possibly biassed estimate of the track parameters. This should also be done if some *a priori* knowledge of the vertex position was used in a first individual track fit.

Some care is necessary when changing the reference surface, e.g. from a cylinder parallel to the z axis (with $R = \text{const}$) to a plane at $z = \text{const}$. Although the transformation may only concern the spatial components p_1 and p_2 of the parameter 5-vector, the *full dependence* on the old parameters *must be included* in the error propagation even if a component of the

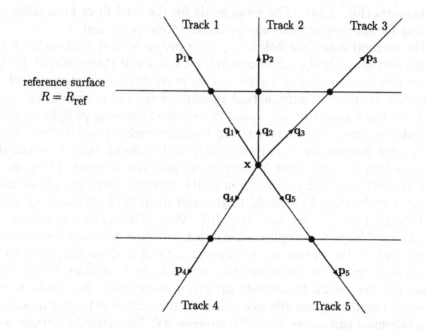

Fig. 3.18. The fitted track parameters $\mathbf{p}_i$ are available at the reference cylinder, e.g. the beam tube of the collider. For the vertex fit the parameters $\mathbf{p}_i$ are considered as 'virtual measurements' and used to estimate the vertex position $\mathbf{x}$ and the momentum vectors $\mathbf{q}_i$ at the vertex.

(virtual) measurement vector itself is not going to be propagated. That is, when changing from cylindrical coordinates with $R\Phi$, z at $R = $ const, to Cartesian coordinates x, y at $z = $ const, and when truncating higher orders in an expansion, the following arguments contribute by first order to the matrix of derivatives and therefore to the error propagation at a fixed point (R_H is the projected radius of the helix):

$$p_{1,\text{new}} = x(R\Phi, z_{\text{old}}, \theta, \phi) \text{ at } z_{\text{ref}} = \text{const}$$

$$\Delta x = -R\sin\Phi\Delta\Phi - \tan\theta\cos\phi\Delta z_{\text{old}} \text{ (with } R\sin\Phi = y)$$

$$p_{2,\text{new}} = y(R\Phi, z_{\text{old}}, \theta, \phi) \text{ at } z_{\text{ref}} = \text{const}$$

$$\Delta y = -R\cos\Phi\Delta\Phi - \tan\theta\sin\phi\Delta z_{\text{old}} \text{ (with } R\cos\Phi = x)$$

$$p_{4,\text{new}} = \phi$$

$$\Delta\phi_{\text{new}} = \Delta\phi_{\text{old}} - (\tan\theta/R_H)\Delta z_{\text{old}}$$

and p_3, p_5 remain unchanged. For the inverse transformation the corresponding derivative matrix is obtained by inverting the 3×3 non-trivial part of the derivative matrix. Note that the fixed position of the new reference surface (plane) is at $z_r = z_{old}$ but although $z_r = const$, the error in z_{old} contributes to the errors in x and y via θ and ϕ terms in the matrix of derivatives, yielding dependence on θ and ϕ. The curvature, which is proportional to $p_5 = 1/P_p$, contributes only to ϕ and the contribution to $\partial \mathbf{p}_{new}/\partial \mathbf{p}_{old}$ vanishes for $z_r \to z_{old}$.

If the direction were unknown, the matrix of derivatives would be *singular* for this special change of the reference surface, and so would the new covariance matrix. The transformation would not be a 'bijective' application, and the information about σ_z would be lost, while the correlation between x and y would be 1.

3.4.2 Global vertex fit and Kalman filter

As mentioned above, for the vertex fit the previously fitted track parameters $\mathbf{p}_k$ with their weight matrices $\mathbf{G}_k = \mathbf{V}_k^{-1}$ ($k = 1,\dots,n$) are now regarded as 'virtual measurements'; they are used to find an estimate of the vertex position $\mathbf{x}$ and of the three-dimensional new track parameters $\mathbf{q}_k$ for direction and momentum at the vertex. Prior information about the vertex position can be included via a vector $\mathbf{v}_0$ and its covariance matrix $\mathbf{V}_0$. For instance, in the fit of the primary vertex knowledge of the beam profile can be incorporated in this way. It is assumed that this prior information is stochastically independent of the virtual measurements, i.e. of the estimated track parameters.

The functional dependence of the track parameters $\mathbf{p}_k$ on the vertex parameters $\mathbf{x}$ and $\mathbf{q}_k$ is normally given by a nonlinear function $\mathbf{p}_k = \mathbf{h}_k(\mathbf{x}, \mathbf{q}_k)$. The function $\mathbf{h}_k$ is therefore Taylor-expanded to first order at some point $(\mathbf{x}_0, \mathbf{q}_{k,0})$:

$$\mathbf{h}_k(\mathbf{x}, \mathbf{q}_k) \cong \mathbf{h}_k(\mathbf{x}_0, \mathbf{q}_{k,0}) + \mathbf{A}_k(\mathbf{x} - \mathbf{x}_0) + \mathbf{B}_k(\mathbf{q}_k - \mathbf{q}_{k,0})$$
$$= \mathbf{c}_k + \mathbf{A}_k\mathbf{x} + \mathbf{B}_k\mathbf{q}_k \qquad (3.114a)$$

where $\mathbf{A}_k = \partial \mathbf{h}_k/\partial \mathbf{x}$ and $\mathbf{B}_k = \partial \mathbf{h}_k/\partial \mathbf{q}_k$ are the respective 5×3 derivative matrices computed at the expansion point. This leads to the following linear model:

$$
\begin{pmatrix} \mathbf{v}_0 \\ \mathbf{p}_1 - \mathbf{c}_1 \\ \cdot \\ \cdot \\ \cdot \\ \mathbf{p}_n - \mathbf{c}_n \end{pmatrix}
=
\begin{pmatrix}
\mathbf{I} & 0 & \cdot & \cdot & 0 \\
\mathbf{A}_1 & \mathbf{B}_1 & 0 & \cdot & \cdot \\
\mathbf{A}_2 & 0 & \mathbf{B}_2 & 0 & \cdot \\
\cdot & 0 & \cdot & \cdot & \cdot \\
\cdot & \cdot & \cdot & \cdot & \cdot \\
\mathbf{A}_n & 0 & \cdot & \cdot & \mathbf{B}_n
\end{pmatrix}
\cdot
\begin{pmatrix} \mathbf{x} \\ \mathbf{q}_1 \\ \cdot \\ \cdot \\ \cdot \\ \mathbf{q}_n \end{pmatrix}
+
\begin{pmatrix} \boldsymbol{\varepsilon}_0 \\ \boldsymbol{\varepsilon}_1 \\ \cdot \\ \cdot \\ \cdot \\ \boldsymbol{\varepsilon}_n \end{pmatrix}
$$

$$(3.114b)$$

The measurement errors ε_k can be assumed to be independent; their joint covariance matrix is therefore block-diagonal.

Estimates of $\mathbf{x}$ and all $\mathbf{q}_k$ can be determined by the LSM (see Equation (3.8)):

$$
\begin{pmatrix} \tilde{\mathbf{x}} \\ \tilde{\mathbf{q}}_1 \\ \cdot \\ \cdot \\ \cdot \\ \tilde{\mathbf{q}}_n \end{pmatrix} = \mathbf{M}^{-1} \cdot \mathbf{N} \cdot \begin{pmatrix} \mathbf{v}_0 \\ \mathbf{p}_1 - \mathbf{c}_1 \\ \cdot \\ \cdot \\ \cdot \\ \mathbf{p}_n - \mathbf{c}_n \end{pmatrix} \tag{3.114c}
$$

where $\mathbf{M}$ is a $(3n + 3) \times (3n + 3)$ matrix and $\mathbf{N}$ is a $(3n + 3) \times (5n + 3)$ matrix, both of which contain $\mathbf{A}_k$, $\mathbf{B}_k$, and $\mathbf{G}_k$. The covariance matrix of the estimate is $\mathbf{M}^{-1}$, and the number of operations required for a straightforward inversion of $\mathbf{M}$ is proportional to n^3. This leads to an unacceptably slow performance for large multiplicities.

However, it has been shown (Metcalf, Regler and Broll 1973; Billoir, Frühwirth and Regler 1985) that faster algorithms can be used. Thanks to the block structure of the model matrix (see Equation (3.114b)), the vertex fit by the LSM may be considered as a special case of the (extended) Kalman filter with a variable dimension $(3n+3)$ of the state vector $(\mathbf{x}, \mathbf{q}_1, \ldots, \mathbf{q}_n)$. Initially the state vector consists only of the prior information about the vertex position $\mathbf{v}_0$ and its covariance matrix $\mathbf{G}_0 = \mathbf{V}_0^{-1}$. The Kalman filter describes how to add a new track k with track parameters $\mathbf{p}_k$ to a vertex already fitted with $k-1$ tracks. At the same time it provides an estimate of the momentum vector $\mathbf{q}_k$ of track k at the vertex. The system equation of the filter is particularly simple:

$$
\mathbf{x}_k = \mathbf{x}_{k-1} \tag{3.115}
$$

The measurement equation is in general nonlinear

$$
\mathbf{p}_k = \mathbf{h}_k(\mathbf{x}_k, \mathbf{q}_k) + \varepsilon_k \tag{3.116a}
$$

As in the global fit, $\mathbf{h}_k$ is Taylor-expanded to first order at some point $(\mathbf{x}_{k,0}, \mathbf{q}_{k,0})$:

$$
\begin{aligned}
\mathbf{h}_k(\mathbf{x}_k, \mathbf{q}_k) &\cong \mathbf{h}_k(\mathbf{x}_{k,0}, \mathbf{q}_{k,0}) + \mathbf{A}_k(\mathbf{x}_k - \mathbf{x}_{k,0}) + \mathbf{B}_k(\mathbf{q}_k - \mathbf{q}_{k,0}) \\
&= \mathbf{c}_k + \mathbf{A}_k \mathbf{x}_k + \mathbf{B}_k \mathbf{q}_k
\end{aligned} \tag{3.116b}
$$

where $\mathbf{A}_k$ and $\mathbf{B}_k$ are the respective derivative matrices. If there is good prior knowledge of the vertex position the expansion point $\mathbf{x}_{k,0}$ can be the same for all k, just as in the global fit. The updated estimates are computed according to the following formulae:

$$\tilde{x}_k = C_k[C_{k-1}^{-1}\tilde{x}_{k-1} + A_k^T G_k^B(p_k - c_k)]$$
$$\tilde{q}_k = S_k B_k^T G_k(p_k - c_k - A_k\tilde{x}_k)$$
$$C_k = C(\tilde{x}_k) = (C_{k-1}^{-1} + A_k^T G_k^B A_k)^{-1}$$
$$D_k = C(\tilde{q}_k) = S_k + E_k^T C_k^{-1} E_k$$
$$E_k = \text{cov}\,(\tilde{x}_k, \tilde{q}_k) = -C_k A_k^T G_k B_k S_k \qquad (3.117a)$$

with

$$S_k = (B_k^T G_k B_k)^{-1}$$
$$G_k^B = G_k - G_k B_k S_k B_k^T G_k \qquad (3.117b)$$

The chi-square of filter step k has two degrees of freedom and is given by

$$\chi_{k,F}^2 = (\tilde{x}_k - \tilde{x}_{k-1})^T C_{k-1}^{-1}(\tilde{x}_k - \tilde{x}_{k-1}) + r_k^T G_k r_k \qquad (3.118a)$$

$$r_k = p_k - c_k - A_k\tilde{x}_k - B_k\tilde{q}_k \qquad (3.118b)$$

The total chi-square of the fit is equal to the sum of the chi-squares of all filter steps, which are independent random variables:

$$\chi_k^2 = \chi_{k-1}^2 + \chi_{k,F}^2 \qquad (3.119)$$

Because of the simple form of the system equation the final estimate of the vertex position based on all n tracks and its covariance matrix can actually be computed in one go:

$$\tilde{x}_n = C_n\left[G_0 v_0 + \sum_{k=1}^{n} A_k^T G_k^B(p_k - c_k)\right] \qquad (3.120a)$$

$$C_n = \left[G_0 + \sum_{k=1}^{n} A_k^T G_k^B A_k\right]^{-1} \qquad (3.120b)$$

This is the fast version of the global fit mentioned above. Smoothing is tantamount to recomputing all momentum vectors q_k and the full covariance matrix at the final vertex position $\tilde{x}_n$:

$$\tilde{q}_{k|n} = S_k B_k^T G_k(p_k - c_k - A_k\tilde{x}_n)$$
$$D_{ik|n} = \text{cov}\,(\tilde{q}_{i|n}, \tilde{q}_{k|n}) = \delta_{ik} S_k + S_i B_i^T G_i A_i C_n A_k^T G_k B_k S_k \qquad (3.120c)$$
$$E_{k|n} = \text{cov}\,(\tilde{x}_n, \tilde{q}_{k|n}) = -C_n A_k^T G_k B_k S_k$$

If the final vertex position is significantly different from the original one it is recommended to recompute the derivative matrices A_k and B_k.

The cost of computing the estimates and their respective marginal covariance matrices is proportional to n; the remaining elements of the full covariance matrix can be calculated with $O(n^2)$ additional operations. Comparing this with the straightforward global fit (see Equation (3.114c)), the break-even point should be at about $n = 4$.

3.4.3 Track association and robust vertex fitting

The separation of secondary vertices and the proper association of tracks to their respective vertices are very important in the analysis of high-energy particle reactions. Event topologies to be identified include neutral 2- and 4-prong and charged 3- and 5-prong decays. However, in order to be able to select from many tens to hundreds of tracks those which belong to a particular secondary vertex, the primary vertex must be reconstructed as precisely and with as little bias as possible. This calls for a robustification of the fit procedure which minimizes the influence of outlier tracks on the estimated vertex position. Using the beam profile is also very helpful if this information is available. In this subsection the discussion will be concentrated on decays which occur inside the beam tube. The methods, however, can also be applied to the reconstruction of V^0 decays ('vees', i.e. neutral 2-prong decays). For the treatment of 'kinks' (charged 1-prong decays) the reader is referred to the literature (Stimpfl-Abele 1991).

If both primary and secondary vertices are inside the beam tube, the reconstructed tracks have to be back extrapolated through the material surrounding the beam tube and the beam tube itself. The required precision can be achieved only by high-resolution vertex detectors which are situated as close to the beam tube as possible. The first step in the search for secondary tracks is the robust estimation of the primary vertex position. In the context of the primary vertex fit both badly measured tracks and secondary tracks are considered as outliers.

As usual, outliers can be dealt with in two different ways. In the first approach one tries to identify the outlying tracks, which are then removed from the vertex. A rough check is possible already in the filter stage, by means of the filter chi-square $\chi^2_{k,F}$ mentioned in the preceding subsection (Equation (3.118a)). This test is, however, not symmetric: track k is tested against a vertex fitted from $k-1$ tracks. Therefore the final selection very likely depends on the order in which the tracks are attached to the vertex. A symmetric test can be constructed by using the smoother chi-square. To this end each track in turn is removed from the fitted vertex by an inverse Kalman filter using a negative weight matrix $-\mathbf{G}_k$ instead of $\mathbf{G}_k$:

$$\tilde{\mathbf{x}}_k^* = \mathbf{C}_k^*[\mathbf{C}_n^{-1}\tilde{\mathbf{x}}_n - \mathbf{A}_k^{\mathsf{T}}\mathbf{G}_k^B(\mathbf{p}_k - \mathbf{c}_k)] \tag{3.121a}$$

$$\mathbf{C}_k^* = \mathbf{C}(\tilde{\mathbf{x}}_k^*) = (\mathbf{C}_n^{-1} - \mathbf{A}_k^{\mathsf{T}}\mathbf{G}_k^B\mathbf{A}_k)^{-1} \tag{3.121b}$$

The chi-square of smoother step k has two degrees of freedom and is given by

$$\chi^2_{k,S} = (\tilde{\mathbf{x}}_n - \tilde{\mathbf{x}}^*_k)^{\mathrm{T}}(\mathbf{C}^*_k)^{-1}(\tilde{\mathbf{x}}_n - \tilde{\mathbf{x}}^*_k) + \mathbf{r}^{\mathrm{T}}_{k|n}\mathbf{G}_k\mathbf{r}_{k|n} \qquad (3.122a)$$

$$\mathbf{r}_{k|n} = \mathbf{p}_k - \mathbf{c}_k - \mathbf{A}_k\tilde{\mathbf{x}}_n - \mathbf{B}_k\tilde{\mathbf{q}}_{k|n} \qquad (3.122b)$$

In contrast to the filter, the smoother chi-squares $\chi^2_{k,S}$ are not independent random variables.

Clearly, the smoother chi-square can be used as a test statistic for checking whether track k is compatible with the fitted vertex. Its usefulness, however, is rather limited, particularly when there are several outlying tracks. In this case the estimated vertex position $\tilde{\mathbf{x}}^*_k$ is biassed by the remaining outliers and the power of the test decreases. This problem can be cured by making the estimator of the vertex position less sensitive to outliers, i.e. more robust. There are robust estimators with high breakdown point that are totally insensitive to outliers as long as these are in the minority (Rousseeuw and Leroy 1987). Such estimators, however, are difficult to compute exactly. A more feasible alternative is a robustified version of the Kalman filter which is based on the M-estimator (Huber 1981).

The basic idea of the M-estimator is a modification of the objective function of the fit such that outliers have less influence on the estimate. This is achieved by downweighting measurements that have large residuals. The quadratic objective function of the LS principle is replaced by a function which rises less steeply. A typical choice is the function

$$\rho(t) = \begin{cases} -Rt - R^2/2, & t < -R \\ t^2/2, & -R \leqslant t \leqslant R \\ Rt - R^2/2, & R < t \end{cases} \qquad (3.123)$$

which is quadratic in the interval $[-R, R]$ and linear outside. R is called the constant of robustness; it is usually set to a value between 1 and 3. The M-estimator can be written down as a least-squares estimator with a modified weight matrix. It cannot be computed explicitly, however, as the weights depend on the residuals which are unknown before the estimate has been computed. Therefore one has to resort to an iterative procedure. The initial estimate is computed by the LS method; the estimate of iteration $k - 1$ is used to compute residuals, which in turn are used to compute weights for iteration k. The iteration is stopped as soon as there is no significant change in the minimum value of the objective function. The details of the implementation of the M-estimator in the vertex fit are described in Frühwirth *et al.* (1996).

The statistical analysis of the M-estimator is considerably more difficult than the analysis of the LS estimator. Under regularity conditions, the

M-estimator is consistent and asymptotically normal; little, however, can be said about the small sample properties. The test of compatibility of a track with the fitted vertex is again performed by using the smoother 'chi-squares'. With the M-estimator, these are no longer χ^2-distributed, even if there are no outliers. Therefore the critical values of the test have to be determined by extensive simulation studies. The M-estimator has been successfully used on real data of the DELPHI experiment (Frühwirth *et al.* 1996).

After all non-primary tracks have been associated to secondary vertices the event topology is fully determined. In the subsequent step of data analysis the kinematical constraint equations (three for momentum conservation and one for energy conservation) are imposed on the result of the vertex fit, using mass assignments as provided by particle identification (Cherenkov counters, muon chambers, dE/dx measurements, calorimeters, and combinations of these). A good geometrical fit of the direction between the primary and any secondary vertex will be an advantage. This is also important for measurements of the lifetime of short-lived particles. It is possible to readjust the geometrical vertex fit after the kinematics fit (see Subsection 3.4.4). However, this update is very sensitive to ambiguities in the mass assignment (Forden and Saxon 1985). If the mass assignment is doubtful only the three constraints from momentum conservation should be used.

3.4.4 Kinematical constraints

In Subsection 3.4.2 we showed how the LSM can be used to add the vertex constraint to the information obtained from the individual trajectories, and in Subsection 3.4.3 the detection of secondary vertices has been discussed. In order to add the information obtained from the four equations of energy and momentum conservation, the problem of using the LSM with linearized constraints with the help of a 'Lagrange multiplier' will be discussed in this subsection. This method is, of course, applicable to a large variety of problems.

Consider a system of n direct measurements (e.g. three virtual measurements for each particle coming from a secondary vertex), with some additional unknown parameters $p_1, \ldots, p_r$, e.g. momentum and mass (or, sometimes better, the square of the mass) of a decaying particle. The mass can be a known constant, a value to be determined or a discrete set of alternative hypotheses for particle discrimination. Note that for those vertices which should undergo a subsequent kinematical fit the three Cartesian projections of the momentum $\mathbf{p}$ should be chosen as parameters q_i. Since the errors are now smaller than for the individual track fit, the problem of the linearity of the model is less important. If because of too large errors the parameter change is only done after the vertex fit, the matrix to be transformed is of dimension $3n \times 3n$ (n is the multiplicity).

The measurement vector is then

$$\mathbf{c} = \begin{pmatrix} \mathbf{q}_1 \\ \vdots \\ \mathbf{q}_n \end{pmatrix} = \begin{pmatrix} {}^{\mathrm{t}}\mathbf{q}_1 \\ \vdots \\ {}^{\mathrm{t}}\mathbf{q}_n \end{pmatrix} + \varepsilon$$

with dimension $m = 3n$.

The covariance matrix of ε, $\mathrm{cov}(\varepsilon) = \mathbf{V}$, is built up from the 3×3 sub-matrices (see Subsection 3.4.2, Equation (3.120c))

$$\mathrm{cov}(\mathbf{q}_i, \mathbf{q}_j) = \delta_{ij}\mathbf{S}_j + \mathbf{S}_i\mathbf{B}_i^{\mathsf{T}}\mathbf{G}_i\mathbf{A}_i\mathbf{C}\mathbf{A}_j^{\mathsf{T}}\mathbf{G}_j\mathbf{B}_j\mathbf{S}_j \tag{3.124}$$

with $\mathbf{C} = \mathrm{cov}(\tilde{\mathbf{x}})$, with $\tilde{\mathbf{x}}$ from Equation (3.120a), being the 3×3 covariance matrix of the vertex corresponding to $\mathbf{C}_n$ of Equation (3.120b) or with the evident submatrix of $\mathbf{M}^{-1}$ from a global vertex fit (see Equation (3.114c)). The four constraint equations are

$$\sum_{i=1}^{n} q_{x,i} = \sum_{i=1}^{n} q_{y,i} = \sum_{i=1}^{n} q_{z,i} = 0 \tag{3.125a}$$

$$\sum_{i=1}^{n} (m_i^2 + q_{x,i}^2 + q_{y,i}^2 + q_{z,i}^2)^{\frac{1}{2}} = 0 \tag{3.125b}$$

where $i = 1, \ldots, n$ (the multiplicity) and m_i is the mass of the ith particle. All *outgoing particle terms* in the sums *have opposite signs*, and *all particles*, measured or unmeasured, *must be included*. Momenta and masses are in equivalent units, e.g. in GeV.

All the unknown quantities in Equations (3.125a, b) are contained in an additional parameter vector $\mathbf{p} = (p_1, \ldots, p_r)^{\mathsf{T}}$. Note that for the masses only *discrete mass hypotheses* will be compared by two separate fit procedures for each set. The general form of the constraint equations is

$$g_j = (c_1, \ldots, c_m, p_1, \ldots, p_r) = 0, \quad j = 1, \ldots, k$$

If the problem is to be completely determined, it follows that $r \leqslant k$, and consistency requires $k \leqslant m + r$, if the k equations are independent.

The χ^2 ansatz

$$M(\alpha) = (\mathbf{c} - \alpha)^{\mathsf{T}}\mathbf{V}^{-1} \cdot (\mathbf{c} - \alpha) = \mathbf{r}^{\mathsf{T}}\mathbf{W}\mathbf{r}$$

should now be minimized together with the constraint equations $\mathbf{g}(\tilde{\alpha}, \tilde{\mathbf{p}}) = 0$. This is done with the help of 'Lagrange multipliers' μ

$$L = \mathbf{r}^{\mathsf{T}}\mathbf{W}\mathbf{r} + 2\mu^{\mathsf{T}}\mathbf{g}(\alpha, \mathbf{p}) \tag{3.125c}$$

where α are parameters corresponding to direct measurements, and with the abbrevation $\mathbf{r} = \mathbf{c} - \alpha$ for the residual. Note that the final residual will be $\tilde{\mathbf{r}} = \mathbf{c} - \tilde{\alpha}$.

The Lagrange method requires that the variation of δL with $\delta \mathbf{r}$ must vanish.

Now the constraint equation will be linearized (care is required if $\sigma(P)/P$ is not $\ll 1$)

$$\mathbf{g}(\alpha, \mathbf{p}) = \mathbf{g}(\overset{0}{\alpha}, \overset{0}{\mathbf{p}}) + \mathbf{A} \cdot (\alpha - \overset{0}{\alpha}) + \mathbf{B} \cdot (\mathbf{p} - \overset{0}{\mathbf{p}}) \qquad (3.126a)$$

with

$$\mathbf{A} = \partial \mathbf{g}/\partial \alpha|_{\alpha = \overset{0}{\alpha}, \mathbf{p} = \overset{0}{\mathbf{p}}}$$

$$\mathbf{B} = \partial \mathbf{g}/\partial \mathbf{p}|_{\alpha = \overset{0}{\alpha}, \mathbf{p} = \overset{0}{\mathbf{p}}}$$

$$\delta = \mathbf{p} - \overset{0}{\mathbf{p}}$$

and

$$\mathbf{g}(\alpha, \mathbf{p}) = \mathbf{g}(\overset{0}{\alpha}, \overset{0}{\mathbf{p}}) + \mathbf{A} \cdot (\alpha - \mathbf{c} + \mathbf{c} - \overset{0}{\alpha}) + \mathbf{B} \cdot (\mathbf{p} - \overset{0}{\mathbf{p}})$$

$$\equiv \text{const} - \mathbf{A}\mathbf{r} + \mathbf{A} \cdot (\mathbf{c} - \overset{0}{\alpha}) + \mathbf{B}\delta \qquad (3.126b)$$

It follows

$$L = \mathbf{r}^\mathrm{T}\mathbf{W}\mathbf{r} + 2\mu^\mathrm{T}[\text{const} + \mathbf{A} \cdot (\mathbf{c} - \overset{0}{\alpha}) - \mathbf{A}\mathbf{r} + \mathbf{B}\delta]$$

$$dL/d\mathbf{r} = 2\mathbf{r}^\mathrm{T}\mathbf{W} - 2\mu^\mathrm{T}\mathbf{A} = 0 \qquad (3.127)$$

$$\mathbf{W}\mathbf{r} = \mathbf{A}^\mathrm{T}\mu$$

$$\mathbf{r} = \mathbf{V}\mathbf{A}^\mathrm{T}\mu \qquad (3.128)$$

Substituting $\mathbf{r}$ from Equation (3.128) in Equation (3.126b)

$$\text{const} - \mathbf{A}\mathbf{V}\mathbf{A}^\mathrm{T}\mu + \mathbf{A} \cdot (\mathbf{c} - \overset{0}{\alpha}) + \mathbf{B}\delta = 0 \qquad (3.129a)$$

$$\mu = (\mathbf{A}\mathbf{V}\mathbf{A}^\mathrm{T})^{-1} \cdot [\text{const} + \mathbf{A} \cdot (\mathbf{c} - \overset{0}{\alpha}) + \mathbf{B}\delta] \qquad (3.129b)$$

$$\mathbf{r} = \mathbf{V}\mathbf{A}^\mathrm{A}(\mathbf{A}\mathbf{V}\mathbf{A}^\mathrm{T})^{-1}[\text{const} + \mathbf{A} \cdot (\mathbf{c} - \overset{0}{\alpha}) + \mathbf{B}\delta] \qquad (3.130)$$

From $\partial L/\partial \delta = 0$ it follows (with $(\mathbf{A}\mathbf{V}\mathbf{A}^\mathrm{T})^{-1} \equiv \mathbf{W}'$)

$$2\mu^\mathrm{T}\mathbf{B} = 0$$

$$\mathbf{B}^\mathrm{T}\mu = 0$$

$$\mathbf{B}^\mathrm{T}\mathbf{W}' \cdot [\text{const} + \mathbf{A} \cdot (\mathbf{c} - \overset{0}{\alpha})] + \mathbf{B}^\mathrm{T}\mathbf{W}'\mathbf{B}\delta = 0$$

$$\tilde{\delta} = -(\mathbf{B}^\mathrm{T}\mathbf{W}'\mathbf{B})^{-1}\mathbf{B}^\mathrm{T}\mathbf{W}' \cdot [\text{const} + \mathbf{A} \cdot (\mathbf{c} - \overset{0}{\alpha})] \qquad (3.131)$$

This requires, however, that $\mathbf{B}^T\mathbf{W}'\mathbf{B}$ is not singular, with the necessary condition $m \geqslant k$; otherwise the complete set of equations must be solved, which would also facilitate the evaluation of the covariance matrix between $\tilde{\boldsymbol{\alpha}}$ and $\tilde{\mathbf{p}}$.

Substituting $\boldsymbol{\delta}$ in Equation (3.130) one obtains

$$\tilde{\mathbf{r}} = \mathbf{VA}^T\mathbf{W}'[1 - \mathbf{B}(\mathbf{B}^T\mathbf{W}'\mathbf{B})^{-1}\mathbf{B}^T\mathbf{W}'] \cdot [\text{const} + \mathbf{A}(\mathbf{c} - \overset{0}{\boldsymbol{\alpha}})] \tag{3.132a}$$

and finally

$$\tilde{\boldsymbol{\alpha}} = \mathbf{c} - \tilde{\mathbf{r}} \tag{3.132b}$$

$$\chi^2 = \tilde{\mathbf{r}}^T\mathbf{W}\tilde{\mathbf{r}} \tag{3.132c}$$

If necessary, Newton iteration (due to the strong nonlinearity of the fourth constraint equation) can be performed by substituting $\overset{0}{\boldsymbol{\alpha}} \leftarrow \tilde{\boldsymbol{\alpha}}, \overset{0}{\mathbf{p}} \leftarrow \tilde{\mathbf{p}}$. Note that only during the first iteration $\mathbf{c} = \overset{0}{\boldsymbol{\alpha}}$ holds, and therefore $\mathbf{A} \cdot (\mathbf{c} - \overset{0}{\boldsymbol{\alpha}}) = \mathbf{0}$.

The covariance matrices are calculated by error propagation (with $\tilde{\mathbf{p}} = \tilde{\boldsymbol{\delta}} + \overset{0}{\mathbf{p}}$):

$$\left. \begin{aligned} \text{cov}(\tilde{\mathbf{p}}) &= (\mathbf{B}^T\mathbf{W}'\mathbf{B})^{-1}\mathbf{B}^T\mathbf{W}'\mathbf{AVA}^T\mathbf{W}'\mathbf{B}(\mathbf{B}^T\mathbf{W}'\mathbf{B})^{-1} \\ &= (\mathbf{B}^T\mathbf{W}'\mathbf{B})^{-1} \end{aligned} \right\} \tag{3.133a}$$

$$\left. \begin{aligned} \tilde{\boldsymbol{\alpha}} &= \mathbf{c} - \tilde{\mathbf{r}} \\ &= \text{const}' + \{1 - \mathbf{VA}^T\mathbf{W}' \cdot [1 - \mathbf{B}(\mathbf{B}^T\mathbf{W}'\mathbf{B})^{-1}\mathbf{B}^T\mathbf{W}']\mathbf{A}\} \cdot \mathbf{c} \end{aligned} \right\} \tag{3.133b}$$

$$\begin{aligned} \text{cov}(\tilde{\boldsymbol{\alpha}}) &= \{1 - \mathbf{VA}^T\mathbf{W}' \cdot [1 - \mathbf{B}(\mathbf{B}^T\mathbf{W}'\mathbf{B})^{-1}\mathbf{B}^T\mathbf{W}']\mathbf{A}\}\mathbf{V}[\cdots]^T \\ &= \mathbf{V} - \mathbf{VA}^T\mathbf{W}'\mathbf{AV} + \mathbf{VA}^T\mathbf{W}'\mathbf{B} \cdot (\mathbf{B}^T\mathbf{W}'\mathbf{B})^{-1}\mathbf{B}^T\mathbf{W}'\mathbf{AV} \end{aligned}$$

The expectation value of the $\chi^2, \langle\chi^2\rangle$, is $k - r$, the number of degrees of freedom of the kinematical fit.

3.5 Track reconstruction: examples and final remarks

This chapter has described how to obtain the best estimates of the geometrical track parameters by fast and optimal algorithms, and how to test the hypothesis that the measurements grouped together in a track candidate actually belong to the track of one and the same particle. All problems arising in this context should be kept in mind at the design stage of the detector and of the reconstruction program. The effort necessary for this is relatively small compared to the total effort invested in an experiment, but is crucial if an irretrievable loss of information is to be avoided.

The discussion has focussed on estimators based on the least-squares principle. It allows one to perform the fit on several levels: local fit in a detector module, track fit, vertex fit, and kinematic fit. The error propagation between the subsequent fitting steps is carried out in the linear approximation. If this approximation is sufficiently good, the covariance matrix of the estimated parameters and the average value of the χ^2-statistic of the fit are correct. In the ideal case of Gaussian errors in the measured coordinates or space points, the distribution of the errors in the final estimated parameters is Gaussian as well; the χ^2-statistic of the fit is actually χ^2-distributed and independent of the estimated parameters. Therefore the losses can be precisely controlled, and they are unbiassed. However, it is essential that *all covariance matrices are carefully tuned* via the pull quantities and the χ^2 *at the earliest possible stage*, first in the local fit, then in the track fit and in the vertex fit, and finally in the kinematic fit.

Before analysing real data, the full reconstruction chain should be tested with simulated data, where the true values $\overset{t}{p}_i$ of the parameters to be estimated are also available, allowing a check on the normalized residuals $(p_i - \overset{t}{p}_i)/\sigma(p_i)$. The exact errors in the fitted parameters which might be influenced by outliers and their removal procedure as well as by misalignment of the detector should be carefully studied. Simulated data are also required to assess the efficiency and the rejection power of the track fit. In the simplest case, the efficiency can be defined as the percentage of correctly found tracks that are accepted by the fit, whereas the rejection power is the percentage of contaminated or ghost tracks rejected by the fit (see the classification tree in Subsection 2.3.6). If the rejection of outliers is enabled in the fit, the analysis is more complicated. Contaminated tracks can be rescued by finding and removing the outlier(s); on the other hand, good measurements can be erroneously rejected as outliers. It is advisable to keep track of the frequency of these events as well.

It should also be mentioned that track fitting is not always a purely geometrical task; in some cases, additional information about the mass or the type of the particle is needed in order to fully specify the trajectory and the interaction of the particle with the detector. For instance, the mass of a low-energy particle has to be known if energy loss and multiple scattering are to be treated rigorously. Similarly, if it is known that a particle is an electron its special properties with regard to energy loss can be taken into account. Sometimes measurements intended for particle identification, e.g. time-of-flight measurements, have to be fitted together with the coordinate measurements.

Many current experiments rely on the implementation of the least-squares estimator by the Kalman filter and smoother. Two examples will be given below. With the smoother the search for outliers can be performed systematically and with the best possible discrimination power. However, this requires attention to possible numerical problems, particularly if the errors of some of the measurements are very small compared to the distance between adjacent detector modules. In general, the information filter working with weight matrices seems to be less prone to numerical problems than the standard filter working with covariance matrices. In extreme cases it may be required to rewrite the filter in square-root form, involving only the Cholesky factors of the covariance or weight matrices.

There have been a few attempts to use estimators based on principles other than least squares, for instance the Chebyshev norm (James 1983; Chernov *et al.* 1993) or the maximum likelihood method (Chen and Yao 1993). However, these proposals are tied to specific detector types or to specific assumptions on the form of the measurements errors. In addition, it seems to be difficult to incorporate multiple scattering and other stochastic material effects. A very general method for coping with non-Gaussian measurement errors and/or material effects has been presented in the subsection on robust filtering (3.2.6). It has the advantage that it can be implemented by several Kalman filters running in parallel, and thus can easily be coded by using the building blocks of a standard Kalman filter. Up to now, the method has been tested only on simulated tracks; an application to real data would be a most interesting undertaking.

Of course it would be desirable to have some successful field-proven algorithms in a program library. To some extent this has been achieved, for instance in the case of helix tracking in a homogeneous field (Mitaroff 1987). Unfortunately it has turned out in practice that the bulk of a track-fitting program is usually too specific to be used in several different experiments, and too complex to allow easy extraction of just parts of it. The computing power available to the data analysis has increased enormously in the last decade, but so has the complexity of the detectors and in some cases also the number of events to be analysed. It is therefore frequently necessary to speed up the track fit by manual intervention in the critical parts. Unfortunately these are very likely those most intimately tied to the choice of the track parameters and of the track model, to the geometry of the detector, and to the shape of the magnetic field. This is an additional obstacle to the free exchange of program modules between experiments.

This has of course been recognized by many workers in the field. As a consequence, there is currently a strong trend towards object oriented implementations of the analysis software and the creation of a standard class library (see Section 4.1). It is expected that this will have many benefits: the interfaces between program modules can be defined very clearly; program modules will be easier to exchange; both the data and

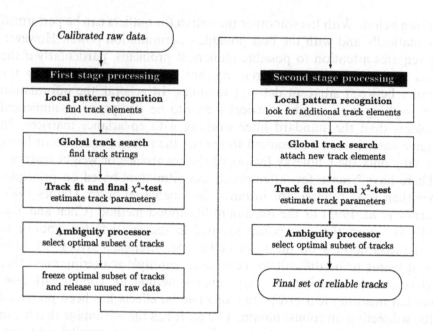

Fig. 3.19. Schematic flow chart of the track reconstruction in DELPHI.

the methods can be hidden from the user, thus ensuring the absolute integrity of the data; and both the data structures and the algorithms will be easier to maintain and to document.

This chapter concludes with two examples of track reconstruction in complex detectors. The issues will be illustrated by briefly describing track finding and track fitting in two experiments: the DELPHI experiment at the LEP electron–positron collider, and the H1 experiment at the HERA electron–proton collider.

The DELPHI tracking system is remarkable for its complexity (see Subsection 2.1.1.3). Not counting RICH counters and muon chambers, there are no less than six entirely different types of tracking detectors (Abreu *et al.* 1996). Two of these, the vertex detector and the inner detector, are in turn composed of two different subdetectors. This complexity is mirrored in the design of the track reconstruction algorithm (Fig. 3.19). Four main stages can be distinguished.

1. **Local pattern recognition** Track elements of variable dimension are found locally in all track detectors where this is possible. The method depends on the characteristics of the particular detector. For instance, in the time projection chamber (TPC), the local track finder is essentially a Kalman filter running from the outermost pad row towards the innermost one. The result of the local pattern

recognition is either a full 5-vector as in the TPC, or a subset as in most other track detectors.

2. **Global track search** Owing to the different geometries in the barrel and in the forward regions there are two track-finding modules. The track search in the barrel region relies on the TPC as the pivot detector. There is special code for locating tracks not seen by the TPC because of cracks between the sectors. The track search in the forward region is essentially combinatorial. It starts from single track elements or from pairs of track elements and looks for suitable partners either by helix extrapolation or by a conformal transformation (see Subsection 2.4.3.3). The result of the global track search is a track string, i.e. an assembly of track elements.

3. **Track fit and extrapolation** The track fit is implemented in a single module accepting track strings from both the barrel and the forward track search. It uses the Kalman filter in the information matrix formalism; smoothing is implemented by the two-filter formula (see Subsection 3.2.5). Outliers are systematically looked for and removed until all smoothed chi-squares and the total chi-square are satisfactory. After all track strings have passed through the track fit the surviving candidates are fed into the ambiguity processor which constructs an optimal subset of compatible tracks (see Subsection 2.3.5). Finally extrapolations into all detectors not taking part in the fit are computed.

4. **Second stage processing** First the raw data from track elements which have not been used in any accepted track so far are released for further use. The extrapolations from the track fit are then used in the second stage local pattern recognition to locate or create further track elements. This is followed by the second stage track search and track fit using similar methods as in the first stage. Finally the ambiguity processor is invoked again.

This is just a rough outline of the entire track reconstruction program. Many of the details depend very much on the particulars of the detectors and of their interplay. An in-depth description would be outside the scope of this book. Nevertheless, some comments reflecting the personal opinions of the authors should be made on the principal features.

- The program is modular: the interface between the modules is well defined, and the transmission of data between the modules is strictly regulated. Each module is effectively prevented from modifying data which are outside its scope of operation. This is crucial for the distributed development of the code, the local pattern recognition modules having been written by the groups who have designed and built the respective detectors. The price to be paid for this modularity is a certain overhead caused by packing the data into a standard exchange format and again by unpacking the data.

- The task is clearly split into a detector dependent part and a general part. The local pattern recognition modules need to know a lot about the internal workings of the detectors; their output is standardized. The general modules – track search, track fit, and ambiguity processor – working on this standardized output are neatly separated from the detector details. However, they still have to have knowledge about the relations between the detectors. This is particularly pertinent to the track search.

- The task is performed in two stages. In the first stage the search is concentrated on tracks which are 'easy' to find in the sense that they have enough measurements and are fairly well isolated. Only after these tracks have been found is the search extended to tracks in difficult regions of the detector and to tracks in densely populated regions.

- Local alignment and error tuning is done by the various detector groups. The global alignment and fixing, however, is performed on the level of the track elements, and the raw data cannot be involved in this process.

- The present state of the track reconstruction program has been reached after many years of experience with the detector, and after many modifications tracking the development and upgrades of the detector. Although the original design is still clearly visible, it has been overgrown by a large mass of more or less *ad hoc* modifications. This makes the code hard to read and hard to maintain. This is one of the reasons why the future LHC experiments have decided to switch to the object oriented programming paradigm (see Section 4.1).

As a second example, the track reconstruction in the Forward Track Detector (FTD) of the H1 experiment shall be briefly described (Burke *et al.* 1996). The H1 detector is very different from the DELPHI detector. Its design is conspicuously asymmetric because of the kinematics of the electron–proton collisions in the HERA collider. The FTD covers the laboratory angles between 5° and 30° with respect to the proton beam direction. The primary track multiplicity is about 10–15; there are, however, a large number of secondary tracks, about four times as many.

The detector is an assembly of three identical subdetectors. Each of these consists of three planar drift chambers with four wire planes each, oriented at $0°$, $60°$ and $-60°$ with respect to the vertical; this is followed by an MWPC, then transition radiation material, and finally a radial drift chamber with 12 planes of radial sense wires measuring the azimuth Φ; the radial coordinate is measured by charge division.

The sense wires in the forward tracker are read out using 104 MHz flash ADCs giving a history of the chamber pulses in time-slices of 9.6 ns. When an event is triggered the history is scanned for regions containing significant raw data. These regions are transferred to the next stage of the data acquisition, the QT analysis. This analysis performs a hit search on these data. For each hit found a charge (Q) and a time (T) are determined. Up to this point the analysis is done on the on-line farm of processors.

After the QT analysis and the conversion of the resultant drift times to drift distances the pattern recognition starts by finding line segments independently in the planar and radial chambers. Because of the different geometries two different procedures are used. In the planar chambers first clusters are found in each orientation separately; these are then combined to line segments. Owing to the large number of ambiguities even a small number of tracks produces a large number of false line segments. Most of these are removed by selecting a subset of line segments that do not share hits (see Subsection 2.3.5). In the radial chambers the search for line segments is based on triplets of adjacent points. These triplets are then joined to line segments.

The next step is the linking of the line segments. Since the line segments forming a track lie on a helix, an (approximate) helical track model is used to link the line segments and to verify that they are consistent with the track hypothesis.

The final track fit is done using the Kalman filter plus smoother, including outlier rejection. The smoother is implemented in a modified covariance matrix formalism. Various techniques have been used to optimize the code. The two main methods are the expansion of matrix multiplications in-line and the use of the known properties of various matrices. By the way, this is also true for the Kalman filter implementation in the DELPHI track fit. For track extrapolation the track model is approximated wherever possible, either with a straight line or with first-order approximations for changes in angle. Multiple scattering is neglected within a drift cell.

The effect of all optimization measures has been to decrease the computing time needed by roughly two orders of magnitudes, to a point where the Kalman filter accounts for about 20% of the computing time taken by the FTD reconstruction. The efficiency of the track reconstruction chain is about 80% over a wide range of the acceptance, in spite of the complexity of the events.

Part of the pattern recognition task is performed on-line, i.e. in real time. At LHC a large part of the track reconstruction, indeed of the entire event reconstruction, will have to be done in quasi real time, in order to be able to keep in step with the data taking. This requires very fast and reliable calibration and alignment procedures. A reprocessing with the ultimate precision will be confined to a small sample of particularly interesting events.

4

Tools and concepts for data analysis

In this chapter we will discuss a number of practical problems encountered in analysing data; more precisely, we will discuss how current experiments in high-energy physics confront these problems. This includes a discussion of how the somewhat abstract world of mathematical formulae is mapped onto *existing* computer and software tools, and the flow of associated data between different parts of the analysis. We will also discuss briefly how data sets can be made to reveal statistical properties, and how existing tools can help in this task.

What this amounts to in general, is a discussion of the restrictions imposed by *today's* toolkits. This contrasts somewhat with Chapters 2 and 3, where formulae were given in a classical description, in principle valid forever – were it not for the fact that detectors evolve and, therefore, new methods of analysis become necessary or possible.

Computer architectures and their performance continue to evolve rapidly, software tools come and go, are first hailed and then abandoned. As a novice, one feels a puzzling irritation faced with too many possibilities. There are no reliable textbooks that are generally recognized – whatever gets written down, simply has no time to be accepted before newer, sometimes even better, systems and tools come into being and have to be understood. And many texts are written not so much with general information in mind but for selling a specific product, be it a computer architecture, a software product, a methodology, or a service. Having found guidance somewhere, you make a choice (or a choice is forced upon you), and you then learn more about some corner of this edifice; your irritation will most likely give way to an 'expert syndrome': you cannot but be biassed by the experience you have thus acquired. And you will find out that your colleagues who may have made a different choice, also manage to solve their problems, and occasionally force you into clashes of opinion difficult to discuss in objective terms.

How will the present book be able to help? Please do not read this chapter as a *how-to* guide; the authors, at best, could be such biassed experts who pull you one way or the other – and, in fact, cannot really claim to be even that. However, we feel that not discussing these subjects at all is not a better solution: data analysis is made of so much technical detail that a text without addressing the *problems* would be unfair. Whatever *solutions* are indicated are in danger of being out of date after a few years (reading through the first edition of this book we realize that this expectation was indeed verified), but we do want to address a number of important issues, hoping to isolate some concepts with inherent stability, and thus perhaps to help the less experienced reader.

4.1 Abstracting formulae and data in the computer

This section deals with the process of rethinking a problem so that it can be brought to the computer, expressed in programming and data representation conventions. It has been mentioned above, that *formulae* used in analysis and *algorithms* that implement such formulae, are two different things. Formulae are expressed as follows. *Let x_i, y_i be a set of points measured in the plane of a Ring Image Cherenkov detector (RICH); they are due to Cherenkov radiation from a particle with velocity β. We fit a circle to these points by the least squares method, minimizing the sum of squares of shortest distances of all points to the circle. Let us call the result of this fit a circle of radius R with an error ΔR. The error on $\gamma = 1/\sqrt{(1 - \beta^2)}$ is then given by $\Delta\gamma = \gamma^3\beta^2 n\Delta R/\sqrt{(N_0 L^3)}$, where L is the depth of the radiator and N_0 its quality expressed in photons radiated per unit length and energy interval.*

Such a statement sounds perfectly meaningful to a physicist, although there are numerous assumptions and abstractions. The detector is assumed planar, Cherenkov radiation happens at a constant distance from this plane, the particle is normal to the plane (otherwise the circle becomes distorted into an ellipse), errors in x and y are the same everywhere, and Gaussian ..... There is also implied reference to pre-existing knowledge: no explanation is needed what a RICH is, how the least squares method works, or why N_0 is a sufficient description for the radiator in this context.

Let us assume that we want to write the above algorithm in a programming language like C or FORTRAN (the choice should have little influence on what follows). We then cannot escape formulations like this. *In the photosensitive chamber there are n_{xy} measured coordinates x and y, we store them in variable Nxy and the array variables Xrich, Yrich respectively.*

Well, hold it: we are forced now to declare a maximal dimension of these arrays; what is it? What will the program do if this dimension is exceeded? Or should the coordinates be stored as an image, so that there is no count, but all possible pixels are transmitted? And if so, where do we indicate the dimensionality of the pixel array, from which it is defined which pixel has which position? Let us continue.

These variables are to be used in various functions (subroutines), and should thus be declared Global and External or appear in a COMMON statement. The definition of the plane in which Xrich, Yrich are defined, is given by an (X, Y, Z) triplet for the four corners of the photosensitive chamber.

This is not quite enough yet to relate the two coordinate systems, viz. the overall detector system and the planar system in which Xrich, Yrich are given; also, the detector may well have a sensitive region that is not given by four corners, and who will check that the four corners are indeed in a plane ... ? But other questions may be more relevant: most likely, this photosensitive chamber is but one of many such chambers, and all of them together are only part of a much more complex detector; the definition of raw data and the description of all detector parts need to be carefully controlled, successive versions perhaps being managed in a data base, which is likely to put some constraints on how we store the various coordinate systems. But let us assume we have converted the original definition to the above form, and that the validity of the coordinates Xrich, Yrich has been checked already. We can then perform our pattern recognition task (*which points belong to the circle?*) and execute the least squares fit in the local system. Do we want to use a standard package for this fit? Does it fit our problem without modifications? Can it be ported to all computers on which our RICH analysis will have to run? Or should we rather code the fitting procedure ourselves? How do we evaluate the robustness of the procedure?

Here we go: without even discussing the programming language, let alone the engineering aspects of this and other pieces of software, we have touched upon a number of technical (or should we say: organizational?) problems, that at first sight seem unrelated to the problem. They deeply influence the translation of an abstract problem (*finding a circle and determining its parameters*) to the computer (*writing a procedure that reliably solves this problem in most cases*). The larger a program gets, and the longer it will stay alive, the more people will contribute to designing and implementing it, and the more the organizational aspects will be in the foreground. So much so that the algorithmic content (the lines of code that contain the procedure of finding and fitting the circle, in our example) is dwarfed by the implementation of conventions concerning data and program structuring, version control and program maintenance, documentation, etc.

It is generally accepted that optimal structuring of a program and its data should be done in function of the intended process only, including considerations concerning ease of maintenance and future adaptation, but ignore any argument of computer optimization. The underlying reason is, of course, that the process of program development, testing and, in particular, maintenance (mostly following changes in requirements) is much more costly than the execution of the code can ever be. Even so, to piece together good data and program structures is serious work, typically an order of magnitude more than finding the procedure that is at the core of the algorithm (in our example the circle finding and fitting).

Implementers have surrounded the historically dominant language for scientific programming, FORTRAN, by multiple conventions and support packages to address these problems; also, the language itself has evolved towards better data structures and more type conventions that help to avoid errors. For smaller problems or for program parts that are in need of system communication (e.g. in real time programs) or require local optimization, the C language is taking much room today; however, the idea of optimizing data and program structures in a larger context may be realized best by the comparatively recent *object-oriented* approach to programming, characterized by the programming languages C++ or (most recently) Java. In these, procedures (*methods*) and data form a single modular unit (*object*), and any information about module–internal structure is hidden from the user of a module (*information hiding* or *encapsulation*). To make objects agree with well-separated conceptual entities is a lengthy and iterative process, and clearly, there is a balance to strike between the effort of designing a program with these principles in mind, and the longevity and generality of the intended program: for a quick one-off test program the situation is different from that of a (part of a) general analysis program of an experiment.

For the longest-living programs, designed and implemented in collaboration between many persons or even several teams, the choice of programming language is, of course, not sufficient: methodologies and conventions of *software engineering* have to surround the process: software will have to be engineered. This includes foreseeing solutions for all parts of the 'life cycle' of software components: definition of requirements (including those for future changes!), analysis and general design, detailed design, module prototyping, implementation, component testing, integration, all accompanied by general and detailed documentation, etc. Don't fall for the simple buzzword solutions like *top-down* (describing a design starting from general abstract principles and moving towards finer and finer detail) or *bottom-up* design (the contrary, obviously): in one case, you will find it hard to get the details right, in the other it will be difficult to group the detailed parts into entities that still have simple relations; stepwise refinement and iterative prototyping at different levels will be unavoidable!

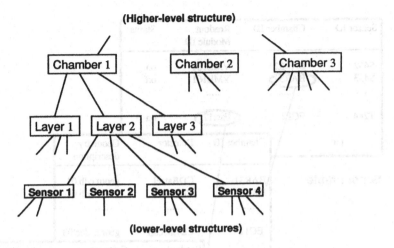

(Higher-level structure)

Chamber 1   Chamber 2   Chamber 3

Layer 1   Layer 2   Layer 3

Sensor 1   Sensor 2   Sensor 3   Sensor 4

(lower-level structures)

Fig. 4.1. Detector description data as an example for a hierarchical data structure.

The specialized literature is abundant, but not without contradictions (and commercial undertones). Around the keyword *SA/SD* (for Structured Analysis and Structured Design) multiple products are available that allow one to build and relate *process models* (flow of information and control), *data models* (see also next section), and also *class and object models*. Rumbaugh *et al.* (1991) and Booch (1994) can be used for starting, if you are inclined towards object-oriented thinking, but given the fast evolution, the best recommendation may be to search on the Internet (watching the date: there are many unburied corpses!).

4.2 Data access methods

The problem of organizing, viz. structuring, data is intimately connected with programming, and also is, independent of programs, a critical step in the decomposition of high-level concepts. Like in our thinking or memorizing, details take a meaning only *in the context* of higher-level concepts. A sense wire needs a detailed description in terms of position and orientation, and the associated electronics needs calibration constants; both take meaning as part of a chamber, which is part of some detector module, which in turn is defined for a given experiment.

This sounds like a strictly *hierarchical* structure (see Fig. 4.1), a universe in which relations are idealized like in a simple family tree, a sort of generalized 'parent–child' relation; this could also work for the hierarchy in a company: the director general 'controls' directors, who organize their respective departments into divisions, groups etc.

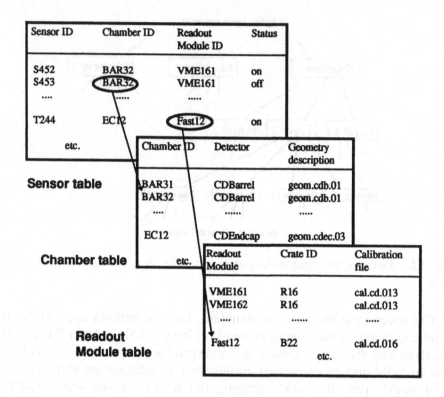

Sensor ID	Chamber ID	Readout Module ID	Status
S452	BAR32	VME161	on
S453	BAR32	VME161	off
.....	......	......	
T244	EC12	Fast12	on
	etc.		

Sensor table

Chamber ID	Detector	Geometry description
BAR31	CDBarrel	geom.cdb.01
BAR32	CDBarrel	geom.cdb.01
....	....	.....
EC12	CDEndcap	geom.cdec.03

Chamber table   etc.

Readout Module	Crate ID	Calibration file
VME161	R16	cal.cd.013
VME162	R16	cal.cd.013
....	....	.....
Fast12	B22	cal.cd.016
	etc.	

Readout Module table

Fig. 4.2. The principle of relating items (*entities*) in different tables of a data base (*entity relationship*).

The truth is that in all these examples the simple hierarchy is insufficient to represent the reality of things: detectors also have mechanical units of construction, electronics channels not only have calibration constants per wire but transmission cables which need grouping onto trays, and readout buffers which are stored in crates; the idealized family tree is perturbed by children out of wedlock or divorcees and adoption (two sets of parents); hierarchies of command have to co-exist with horizontal coordination committees etc. And, worst of all, the hierarchies keep changing, and their history has to appear in the data representation. Some of the more complicated structures can be expressed in a *relational model*: all items of the same kind are kept in a table, and identified by position inside this table along with their attributes (imagine a table as a simple spreadsheet matrix); relations between items in different tables can be expressed by 'keys', essentially pointers to items in other tables. Figure 4.2 shows a trivial example.

In short: data structures that are sufficiently local may fit a somewhat simplified data model (hierarchy, entity relationship), and may as such serve a purpose in a certain frame of locality; when used inside a program, this may be most efficient both for understanding and program performance. For applications exceeding a single program, both commercial and home-built systems have been in use in high-energy physics (often termed database management systems). Local data management, mostly using a hierarchical data model, has been grafted onto FORTRAN in the past; this has largely been replaced by data structures offered as part of programming languages. Database management systems using entity relationships exist commercially, like the products from Oracle. These *relational databases* also have a theoretical underpinning (Chen 1976).

When challenged to be the basis for many different programs, using different aspects of these data, and when addressed by users with different requirements, these simplified data models may not suffice: if data complexity is more than the simple relational database can easily handle, an *object-oriented database management system* (ODBMS) may be indicated. 'Complex' may be a combination of

- many-to-many relationships between objects;

- variable-length data;

- composite objects (e.g. structural hierarchies appear as objects);

- objects that need frequent changing of structure.

ODBMSs are comparatively new; they allow (in analogy to objects in OO programming) the association of methods and data, and experience with these systems in high-energy physics is limited at time of writing. Introductory literature does exist (Taylor 1992; Cattell 1991; Barry 1994; Halpin 1995), but this being a task for teams, it might be more useful to get in touch with the relevant working groups in experiments that come onto the floor in the late 1990s or even later. The team constraint is critical: precisely for that reason, individual knowledge is not enough; an entire group of people has to agree on a method, and be trained in using a given tool.

4.3 Graphics

Originally, graphics on computers was understood as a means of presenting data to a human, viz. a passive form of output: output on paper or on a graphics terminal. Input commands were given on a terminal, often separate from the output device. We know that the role of *interaction*, i.e. a continuous dialogue between computer and human, has taken on an overwhelming importance, usually in the form of mixed graphics and text/menu mode. Behind this type of man–machine 'conversation', there is some very clever programming, most of which is hidden to the user who uses existing graphics user interfaces (GUIs). The one-time severe limitations that could be ascribed to the limited computing and storage resources, have been largely removed by ever cheaper processors and memories, and what remains are only the problems of deciding which information to present.

We should make the side remark here that the 'reading' capabilities of computers have not nearly evolved to the same extent. Yes, acquisition of images (e.g. by scanners) and their efficient transmission (e.g. by compacting formats like JPEG) across many different operating systems (and the Internet, of course) has been largely ticked off as a problem; likewise, lower-level image processing, at the pixel level, like edge sharpening, dilation and erosion, or contrast stretching, are well-understood and often helpful operations (see the image processing literature, e.g. Jain (1989)). On the other hand, image interpretation, i.e. the understanding of the objects in an image, is still in its infancy, despite all progress made in robotics. This subject is important in the discussion in the Artificial Intelligence community (touching also neurophysiology and philosophy) about how cognitive processes function in the human brain – a fascinating field of research, but not one we can divert into. For some strong opinions in this field, see Penrose (1994), Moravec (1988), Eccles (1994), and many others.

This diversion introduces a statement that remains true despite the extraordinary progress that graphics interfaces have made over the last decade. *Graphical communication via a computer requires a substantial effort and a certain degree of specialization.* At least this is true when no intermediate package takes the burden from the programmer. It becomes relevant, then, to see in which applications graphics is to be used. What are they in high-energy physics? One usually thinks primarily of:

- data analysis and presentation;

- monitoring and control, e.g. during data acquisition;

- event display: show detector parts and the associated digitizings, be they recorded or simulated;

- visualization of physics aspects that are not normally visible; this is not strictly part of data analysis, but fascinating on the same grounds as showing hypothetical molecules in chemistry, or intergalactic interactions. The evolution of lattice calculations, or the short-range forces between particles have been shown to be applications where graphics can help, if not make a decisive difference.

The requirements for these applications are quite different; we will come back to that below. Let us first discuss what the situation is with respect to graphics packages.

At a basic level, graphics applications have to some extent converged towards a standardized interface (graphics user interface or GUI), best symbolized by the wide-spread Unix-based packages X-Windows (Nye 1992) with Xlib, and Motif (Heller and Ferguson 1994); interaction and remote application possibilities are built into them. Typically, these packages have language bindings for C++, C, and FORTRAN. Among the older products, PHIGS (Gaskins 1992) seems to survive, while GKS has been abandoned, and the Web-oriented Java is still very young. Practical choices are often determined as much by local conditions (*Which product is installed and supported? How much does it cost? What do other laboratories in the collaboration decide? Who is available to help solving problems?*) as by optimal functionality for the problem at hand and a good guess what it might develop into (the lifetime of packages in this field may be limited).

Low-level graphics tools typically allow one to manipulate the various coordinate systems needed (window coordinates, world coordinates, screen coordinates), to draw straight or curved lines, to write multiple connected lines (polylines) and possibly fill the resulting two-dimensional bounded surface with patterns or colour, to write text or symbols. The packages operate directly on a pixel map, which is the standard way of storing an image for display (a decade ago, vector-driven display hardware was still commonplace).

Higher-level tools allow three-dimensional objects with surfaces to be drawn, viz. the tool takes care not only of projecting onto a two-dimensional surface, but also of hidden-line removal, possibly also of surface shading (*rendering*) with or without ray tracing. The importance of the notion of (three-dimensional) surfaces is easily demonstrated by the limits of what wire-frame drawings (viz. points in three dimensions connected by lines) can express (Fig. 4.3): already in very simple cases, the eye needs more guidance than a low-level interface can provide.

Graphics projected onto a plane from three dimensions appeals, in fact, to quite complex pattern recognition processes in our brain. M.C. Escher has systematically exploited the possible conflict between the flat and the spatial; his 1960 lithograph (Fig. 4.4) is an example (there was a 1958 article relating to the subject of 'endless stairs' by R. Penrose, in the *British Journal of Psychology*).

Open Inventor (Wernecke 1993), Open GL (Neider *et al.* 1993; Boudreau 1997) are recent tools at higher levels of graphics; for the most advanced, Virtual Reality Markup Language (VRML) seems to offer a platform-independent Internet-oriented basis (look for it on the Web).

Even excellent tools, however, can do nothing for your problem when choices have to be made as to what objects should be drawn and in which form. Take as an example the presentation of maps: assume that all fields,

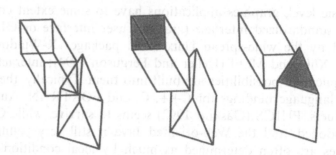

Fig. 4.3. Wire-frame drawings, as offered by the very basic graphics commands, are not satisfactory for even very simple objects; adding the notion of *surface* is essential.

forests, mountain slopes, roads, houses, etc. exist in a database, identified and in three coordinates; to make a useful map of detail at 1:5000, a road map at 1:200 000, or an overview of 1:1 000 000 will need different selection criteria and choice of symbolics, and is far from trivial.

This demonstrates that a substantial number of additional attributes are needed for graphical representation; objects for graphics require data structures separate from those describing the objects themselves (e.g. for processing). The graphics structure may be generated from the latter, but much extra information will need adding.

Choices in graphics are particularly relevant in the area of event display: one wants to display (selected parts of) observed events as recorded or simulated in the various parts of a complex detector. This is of major interest during development of programs, but equally as a constant check on the functioning of analysis. Real-time interactive control of the analysis programs is a permanent need in experiments, and requires three-dimensional representation of clusters and tracks along with the raw recorded information and the detector layout. This includes not only proper choices of the information to display, but also the representation of this information; for a selection of literature, see Drevermann and coworkers (1990, 1995, 1997), Boudreau (1997), and literature quoted there. Not solved, for the moment at least, is the linking of the physicists' event display software to the engineers' detector design packages, the latter usually commercial products for mechanical engineering (CAD systems) with excellent built-in graphics.

A similar transition problem arises in other applications: in data presentation, good commercial analysis packages exist (see Section 4.6 below); however, neither the data input in formats as recorded by physics analysis programs, nor their output towards graphics editing tools are catered for. The former is an obvious requirement, the latter should allow

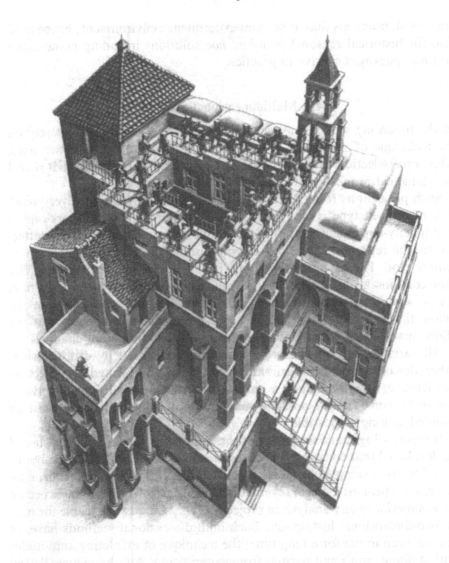

Fig. 4.4. The perception of three dimensions from a projection can be dramatically misguided; here done on purpose by M.C. Escher.

publication-quality graphs to be made in all generality (like in Powerpoint, Framemaker, or even MSword). This possibility should include combining output files coming from different products. Usually, one resorts to local data transfers in some ASCII format. More on this in the discussion of data presentation tools below.

Similarly, in the area of monitoring, commercially available and popular packages exist (e.g. LabView; Wells and Trevis (1996)), but they do not address all problems that arise in an experiment's environment; hence (and also for historical reasons) many *ad hoc* solutions including home-made graphics packages survive in practice.

4.4 Multidimensional analysis

In the preceding sections, aspects of analysis were discussed that rely on the technique of projecting the event description onto one or two axes, after some selection. It was implied that projections are chosen that reveal the statistical properties of the class of events selected.

Such projections result, of course, in a loss of information. Every reaction of a given type is characterized, by a number of parameters and is multiplicity-dependent, and only the simplest reactions like elastic scattering can be represented without information loss in a single histogram or scatter plot. The example of Fig. 4.5 is characteristic for these: a Dalitz plot contains all kinematically relevant information for three-body final states, and its analysis can go along known lines, because it is not the image that has to be analysed, but predetermined interference relations allow one to extract the relative contributions from different channels.

All variables are, in addition, potentially influenced by distortions or other detector-dependent calibration problems, or biassed by selection inefficiencies; therefore such a reduction in dimensionality nearly always has to be considered a possible source of problems, and hence must be avoided as long as possible.

It seems a logical consequence of these thoughts that techniques should be developed that consider *simultaneously* all or at least a larger number of variables for a sample of events of a given type. If the sample is sufficiently large, statistical methods in several dimensions exist which can be expected to be superior to any analysis in projection, i.e. methods applicable for one- or two-dimensional histograms. Such multidimensional methods have, of course, been in use for a long time; the technique of extracting amplitudes with different spins and parities from experimental data known as 'Partial Wave Analysis', is one of these. In this technique, limited assumptions are made about the quantum numbers of the contributing waves, and the higher contributing spin amplitudes are either ignored or calculated from some model (for a better introduction and a bibliography, Litchfield (1984) should be consulted).

Such analysis methods using *a priori* knowledge, or at least a model about reactions, are multidimensional, but do not correspond to what is to be touched upon in this section. We refer here to the attempts, popular in the 1970s and somewhat more dormant today, to extract information in several dimensions simultaneously *without* any preconceived idea about the physical process. Mostly, this amounts to looking for *structure* in

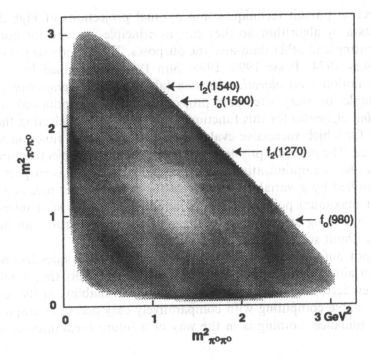

Fig. 4.5. Image-quality results from an experiment: with sufficient statistics, extraordinary detail can be extracted from data. The Crystal Barrel experiment (Landau 1996a, b) entered more than 700 000 annihilations $p + \bar{p} \rightarrow \pi^0 + \pi^0 + \pi^0$ in this Dalitz plot.

event distributions without making assumptions as to what might be at the origin of such structure. The natural consequence of detecting structure would, of course, be to *classify* events into categories, belonging to different 'clusters' of which the structure is made.

In order to define structure and clusters, great care has to be taken in choosing the *variables* that are meant to reveal the structure, and in defining a *metric* that corresponds to the space spanned by these variables (see Subsection 2.2.5). By metric, we mean the recipe that tells us how the variables are interconnected and allows us to define a measure of *closeness* or *similarity*. In Euclidean space, two points are said to be close if their distance is small, and distance is defined by $(\Delta x^2 + \Delta y^2 + \Delta z^2)^{\frac{1}{2}}$. In a space defined by variables such as the scattering angle (or rapidity), the azimuth, the energy ratio, how does one define closeness? Without this very central notion of a metric, there is no defined procedure for finding clusters.

Various attempts have been made in the past 20 years to define general clustering techniques that would produce good physics results in the context of high-energy physics data analysis.

Projection pursuit techniques find optimal projections of high-dimensional data by algorithm, so they can, in principle, be used for nonparametric fitting and other data-analytic purposes. These methods (Friedman and Tukey 1974; Posse 1990, 1995; Sun 1991) are defined by two important notions: an *objective function* (also called the *projection index*), computable for every attempted projection and to be optimized, and an *optimizing algorithm* for this function, viz. a strategy for selecting the projections for which successive evaluations of the objective function should take place. The practical problem remains: to define the objective function, possibly also the optimization algorithms, when structure is suspected but not identified by a variable. As *optimal* for the computer means a minimum or maximum point, but for the application means *most interesting*, or *best exhibiting structure*, it is apparent that one hardly can achieve results without some *a priori* assumptions.

It is not unfair to say that the pursuit of such techniques has not, so far, been able to enrich high-energy physics with results that would not have been accessible otherwise. Given the wide availability of the toolkits, viz. affordable computing with comparatively easy access to graphics for human guidance, nothing is in the way of a future breakthrough in this area.

4.5 Data selection

One of the most frequent and critical problems in preparing a data sample for final analysis is that of deciding which events to include and which to reject. Equally important is the question of how to account for the errors that will result from inevitable wrong decisions. Except in the most simple cases, this is a process that needs iterations, test runs, and extended simulation. Very often results obtained after many successive selection filters indicate the need to go back and verify that an observed effect is indeed inherent in the physics or the detector, and not one that was introduced artificially by the selection criteria.

We will refrain here from discussing, in general, the ways of deciding between different hypotheses. Textbooks of statistics deal competently with the subject. In some situations, philosophical arguments cannot be avoided, like the Bayesian argument dealing with the use of *a priori* knowledge.

It may seem appropriate and of practical use, however, to attempt to give a few rules of thumb, which would at least be applicable to the simpler cases. Let us therefore define a restricted decision problem in which we are interested. Let each event be characterized by a fixed number n of elements (an n-vector or n-tuple). Let the events be drawn randomly from a mixture of two distributions. Call the members of these distributions *good* and *background* events. They should form different, ideally separate, clusters in *n*-space (the feature space of Subsection 2.2.4). We can make assumptions about the shape of the distributions of the two classes, and about the mixing ratio. We can also assume that we can tune the performance of our algorithm on representative *training samples* of good and background events.

We seek a classification algorithm expressed as one or several test statistics, i.e. single-valued functions of the *n* elements. In order to be useful, a test statistic must separate the two samples into two clusters, which are as distinct as possible. To each test statistic we can therefore associate a cut parameter, which divides that projection of the feature space into a 'good' and a 'background' fraction. These are called the 'acceptance region' and the 'rejection region' respectively (see also Subsection 3.2.4).

In other words, our selection mechanism is meant to operate like this. We form a test statistic from a given event, and obtain a value that is a random element from one of the distributions, good or background. This element is now compared with a cut parameter. Depending on the sign of the difference between the cut parameter and the function value, the event is classified as a *good* or *background* event. A *correct decision* will then classify a good event as good, a background event as background. A false decision will classify a background event as good (*contamination*), or a good event as background (*loss*). Assuming sufficiently different distributions for good and background elements, the 'significance of the test' (the probability of rejecting good events) will be small, and the 'power of the test' (the probability of rejecting background events) will be large, i.e. the procedure will classify individual elements correctly most of the time (cf. Subsection 3.2.4). Note that various terminologies are used: significance is also denoted by 'cost', and power by 'purity'.

Plotting the acceptance of background events against the acceptance of good events (which are the complements of power and significance), we obtain what is known as the *Neyman–Pearson diagram* (or decision quality diagram). Different points on the curve correspond to different values of a cut parameter, if a single test statistic is used. As both coordinates are integrals of a probability density function, they increase monotonically. A good test statistic brings the curve close to the point of *ideal separation*, viz. the corner of the diagram that corresponds to the coordinate pair $(0, 1)$, i.e. full acceptance of good events and no acceptance of background. An example is shown in detail in Fig. 4.6 below.

If more than a single test statistic is involved, the space can be explored by fixing all but one of them at a time, and plotting multiple curves in the same diagram; if high dimensionality makes this tedious, the exploration of the available space of cuts on different test statistics can be done in some problem-oriented order; each set of applied tests will give a unique classification of events and hence result in a point in the Neyman–Pearson diagram.

An *optimal* decision function will use algorithms that minimize the losses for a given contamination. The optimal decision function will have the property that *no other decision function will have smaller losses for the same background*. In the language of statistics (Kendall and Stuart 1967) this would be a 'uniformly most powerful test'. In the Neyman–Pearson diagram, the optimal decision function results in the surface of closest possible approach to the point where both losses and contamination are zero. The cut parameter(s) can then be chosen to define a point on the curve or surface in the region of that minimal distance. As there is usually a wide choice of possible decision functions, one might be tempted to give a recipe for constructing the optimal decision function based on mathematical principles. This seems a possible goal as we have assumed that we know the statistical properties of our good and background event samples. Even more modestly, it might seem possible (and useful) to give a rule such as *if criteria of distinction exist in several independent dimensions, use them independently rather than combining them into a single-parameter criterion.*

Unfortunately, in general even such a simple rule cannot be formulated. It is comparatively easy to find background shapes which favour, for the same sample of good events, different decision functions. And it is typically the assumption about the shape of background which is the least robust, subject to numerous hypotheses. The finding of an optimal decision function invariably needs a heuristic approach, one of trial and error.

The question is, obviously, of particular importance in triggering: triggering means taking decisions in real time, and in this case the discarded events are lost, and there is no way of repeating trial and error. Before the experiment starts, the exploration of trigger decision variables is done with large samples of simulated data; when real events become available, one creates training and control samples by loosening the cuts in special runs or with some acceptable low (scaled-down) rate.

4.5.1 A simple fictitious example

In order to show the problem of optimizing decision functions in more detail, we will concentrate on a very simple case. Let each event be characterized by two elements. For the sample of good events, let these be random elements drawn from two independent χ^2 distributions, each with three degrees of freedom (see Subsection 3.2.5 for an explanation). Let the corresponding elements in the sample of background events, instead, be random variables from two identical flat distributions, extending

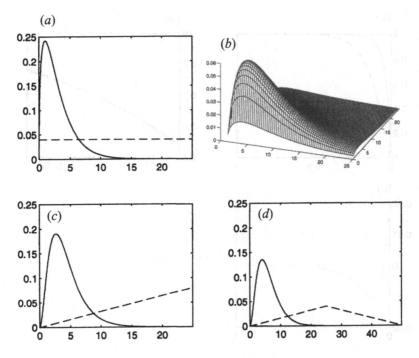

Fig. 4.6.   An example of constructing a decision quality (Neyman–Pearson) diagram. (a) A single χ^2 probability density function (pdf) with three degrees of freedom, considered to be the signal, and a flat distribution between 0 and 25, considered to represent background. (b) and (c) A two-dimensional pdf with the same function as in (a) along the two axes, and the projected pdfs for signal and background obtained by applying an identical cut along the two axes. (d) The pdfs obtained by reducing the decision to a cut in the single variable which is the sum of two variables as in (a) (see text for discussion).

(for normalization) from 0 to 25 (any background beyond that value is considered to be without impact on our study). Fig. 4.6(a) shows these distributions for a single variable.

We now limit our study to comparing two decision functions, hoping that one of them will be uniformly superior, i.e. provide a better recipe for separating events in a mixed sample into good and background events, whatever contamination or loss we are willing to tolerate.

(1) We use the two elements as independent test statistics, with the two cut parameters assumed to be equal. An event is considered good if both its elements are below the cut parameter; it is considered background otherwise. This reflects the fact that the χ^2 distributions peak at low values, whereas the background distributions are flat.

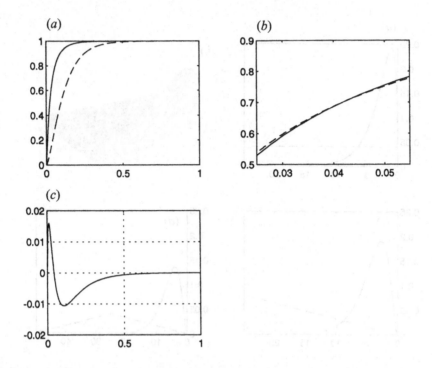

Fig. 4.7. Using Neyman–Pearson diagrams. (a) The decision quality curve for the pdfs of both 4.6(c) and (d), indistinguishable at this scale (full line), and the same obtained for a single χ^2 variable as in 4.6(a) (dashed line). (b) A blow-up of the decision quality curve for the pdf as in 4.6(c) (dashed line) and in 4.6(d) (full line). (c) The difference between the two curves magnified. (See text for discussion.)

(2) We construct a single test statistic by forming the sum of the two original test statistics, and then use a single cut parameter.

The first decision function uses two variables separately, and hence is two-dimensional. The single cut parameter used on both statistics translates into the prescription that good events have to have both χ^2s below the single cut parameter. This corresponds to cutting out a square in the plane (2-space) of the two elements, in which the probability density is a two-dimensional surface (Fig. 4.6(b)). This surface is peaked towards the origin for the good events (but with a zero at the origin), and is entirely flat for the background event sample. Figure 4.6(c) shows the probability density integrated over a square with a given side length corresponding to the cut parameter, and normalized to the integral over the entire plane.

In the second decision function, we have reduced the problem to a one-dimensional case. The test statistic will be a χ^2 with six degrees of freedom for good events, and will have a linearly rising and falling probability density for background as shown in Fig. 4.6(d). Here again, low values are more likely to be associated to good events, and the cut parameter is used to separate good from bad.

We now compare these two test statistics, cases (1) and (2), in the Neyman–Pearson diagrams of Fig. 4.7. Diagram (a) shows, as full curve, the result of test statistics (1) and (2), not distinguishable at this scale, and opposed to the dashed line, which corresponds to the decision quality of a single χ^2 variable as in Fig. 4.6(a). If we look closer (we can do this reliably, as we deal with a fictitious analytical problem), we see that the test statistics (1) and (2) show a small difference, and the curves intersect at a purity of 0.7, as shown in Fig. 4.7(b). From this curve, and more precisely when looking at the difference between the curve derived from two independent variables (two χ^2s with 3 d.o.f. each) and the single-variable curve (single χ^2 with 6 d.o.f.), we see that up to a significance of 0.04 the two-variable decision is superior, and the single test statistic dominates beyond: no approach is uniformly superior.

It is understood that the two approaches are not very different qualitatively in this particularly simple example, and our interest appears somewhat academic. The example is, however, a useful demonstration of a *genuine difficulty*.

4.5.2 An example from an experiment

The experiment UA1 at the CERN SPS $p\bar{p}$ collider isolated the reaction $p\bar{p} \rightarrow W^{\pm} + X, W^{\pm} \rightarrow \tau + \nu$, by using a test statistic optimized and invented for the purpose, called the 'τ-likelihood'. The signal of τ-decays of the charged intermediate vector boson $W^{\pm}$ is seen with a background of Quantum ChromoDynamics (QCD) jets, and for the optimal separation of the two samples some specific properties of τ-decays have been used. τs appear as very collimated jets owing to their decay into few hadrons, whereas QCD jets are caused by parton hadronization, and may radiate gluons. τs are associated, like some rare QCD events, with a large unobserved energy vector, inferred by forming the vector sum of all energies observed in the hermetically closed calorimeter (see Subsection 2.5.6). Both τ-decays and QCD jets appear on top of an event-internal background of tracks due to the remnants from the two spectator quarks and antiquarks (that do not participate in the hard interaction).

After various cuts (see the full description in Albajar *et al.* 1987), three *distinct* relevant parameters (test statistics) are evaluated: (a) the ratio F of the jet energies, contained in a narrow and a wider cone around the jet direction, (b) an angular separation r between the leading track and the jet direction, and (c) a multiplicity n of charged particles with high

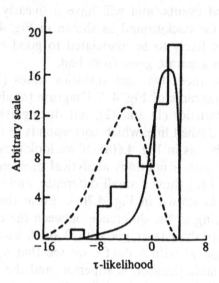

Fig. 4.8.   The test statistic 'τ-likelihood' explained in the text, plotted for UA1 events (histogram), for τ-decays from Monte Carlo (solid curve), and from observed jet events (dashed curve), from Albajar *et al.* (1987). All curves are normalized to the same number of events.

transverse momentum in a narrow cone. All of F, r, and n are expected to be larger for the signal (τ-decays) than for the background (QCD jets). A *single combined test statistic* L is therefore defined by $L = \log(nrF)$ (the 'τ-likelihood'). The probability densities for L as measured in a sample of known QCD jets, as Monte Carlo simulated for τ-decays, and as observed in the candidate events, are shown in Fig. 4.8. A cut parameter $L = 0$ results in a loss of signal of 22%, with a contamination of 11% (of QCD events!).

4.5.3 Practical conclusion

What, then, can be said in general that is of practical help? The answer is: little more than the following.

- Hypothesis deciding deserves careful study, making models for both good and background events, and trying different decision criteria on *training samples* of events, constructed typically by a Monte Carlo program with the statistical properties of the samples under study.

- The more test statistics that can be found that are not interdependent and in which probability densities are significantly different, the easier it will be to find a satisfactory cut parameter(s) and hence an acceptable decision function.

- In a systematic study, the graphical presentation of decision functions in a Neyman–Pearson diagram may be helpful, in particular to show the sensitivity to background assumptions.

- If test statistics closely relate to physical or detector properties, it may be an advantage to use them separately in order to identify 'outliers'. Outliers are elements that fail to follow the statistical properties assumed in the model, either because of a truly exceptional low probability, or, more likely, because of some gross errors, e.g. δ-rays in a chamber, readout or transmission errors, wrong association in the pattern recognition etc.

4.6 Data accumulation, projection, and presentation

We now assume that we have reduced our data sufficiently to be left with a random sample of events, each characterized by n elements which represent the detector- and physics-dependent variables of interest in further analysis. In other words, we have used successive steps of *abstraction* (to represent each event by few variables) and *selection* (to define the event sample). Whatever information has been lost in these steps has to be carefully and often independently evaluated, and the sample may have to be corrected for the losses. The corrections must also be as independent as possible with respect to the cut parameters ('robust correction'). Typically, the selection and abstraction process is iterative and tentative stepwise, i.e. samples are formed under certain assumptions, studied, and the result verified by changing some of the assumptions. Consequently, final event samples as stored on tape or disk are often not purified nor abstracted to a fixed number of variables of each event; in practice, gross event samples are often kept together, maintaining highly redundant data for each event, the selection and abstraction processes being left to what is called the 'final event analysis'.

In this final event analysis, then, the sample will be subjected to the usual statistical methods. In order to extract the physics content of quantum-mechanical nature, most of the analysis will concentrate on projecting the multidimensional space spanned by the event variables onto lower-dimensional spaces, and usually one will use probability densities in one or two dimensions, i.e. *accumulation* and *projection*. Accumulated and projected representations of data stand for observations of probability densities. Typically, they are shown as graphical presentations, i.e. as histograms in one or two dimensions (here we somewhat generalize the notion of a histogram, which is usually restricted to the one-dimensional projection). The contents n_i of bin i of a histogram allow the deduction of the probability frequency $P(x)$, from:

$$\langle n_i \rangle = \sum n_i \int P(x)\, dx \quad \text{(one-dimensional)}$$

or

$$\langle n_i \rangle = \sum n_i \iint P(x, y)\, dx\, dy \quad \text{(two-dimensional)}$$

with the sums extending over all bins, and the integrals over the limits of bin i in x and y. Statistics textbooks give more details. In these projections, comparisons with theoretically predicted or independently derived experimental probability densities become possible, unexpected structure may be observed, correlation may be seen and may lead to regression analysis, etc.

In what follows, we raise a few practical problems concerning projecting and presenting data, whose solution is not always apparent from the more general textbooks; we do this under a few unrelated headings.

4.6.1 Binning

In order to obtain probability densities, interesting data that represent typically continuous variables are usually discretized, i.e. all elements between given limits are grouped into 'bins'. The frequency of events in a given bin is a measure of the average probability density in the region spanned by the bin. Any resolution contained in the original measurements finer than the bin size will be lost; the loss is the more severe the better the experimental resolution is, compared with the bin size. The choice of bin size, therefore, is of prime importance. Bins, on the other hand, should be chosen such that the bin content should not be less than some statistically significant number (no rule of thumb exists!). In other words, there should not be too many bins. Too few bins may, of course, result in loss of information about the probability distribution.

If this dilemma presents a serious problem, the computational advantage of equal-size bins may be abandoned. Statistically optimal binning has been shown to correspond to bins of equal probability, hence not to equal-size bins, in general. For small samples, binning may be outright impossible. The use of unbinned data may instead be indicated, and methods other than standard ones (often based on the laws of large numbers) should be considered. James (1981) gives some specific hints for small samples. A method that works reliably on unbinned data, without being restricted to this application, is the Maximum Likelihood Method.

Note that a compromise between the data reduction inherent in binning and the necessity of keeping analysis options open, is frequently found

by *truncating* numerical data to some acceptable level of precision without accumulating them statistically into bins. This results in creating a
small set of (truncated) numbers, for each event (often called an '*n*-tuple').
All correlations between variables, lost when histogramming (i.e. projecting onto a lower-dimensional surface), are preserved, and later statistical
analysis for any bin size larger than the truncation precision is possible
(with the caveat, though, that periodic effects due to interferences between
truncation precision and the chosen bin size can cause unpleasant surprises). In fact, *n*-tuples are more often than not used without truncation,
which is only useful for reducing the space occupied by large data sets.

4.6.2 *Error analysis*

A general classification of errors (see also Subsection 3.3.2) is into the
following four groups: *random errors* (those caused by random processes,
decreasing by a factor $\sqrt{f}$ if the sample size is increased by a factor f),
truncation and rounding errors (arising from digital signal processing, and
decreasing with sample size only if kept carefully uncorrelated), *systematic
errors* (caused by incomplete knowledge or inadequate correction, and
causing the same bias independent of sample size), and *gross errors* (due
to wrong assumptions such as including measurements in the wrong
sample).

Most error estimates are based on random errors. Great care therefore
has to be taken to estimate the influence of non-random errors correctly,
or to ensure that their influence is negligibly small. In judging the influence
of errors on statistical estimators derived from a sample, the most general
and computer-adapted method may be the 'bootstrap' error estimate of
Efron (1982): new artificial samples, of the same size as the original
sample, are formed by randomly drawing the sample members from the
original sample (hence possibly introducing some members several times).
The estimator in question will then show a probability density function of
its own, which allows the derivation of confidence limits.

4.6.3 *Presentation*

This may seem a purely technical or aesthetical issue, but the choice
of presentation is *not* only that, but may well help you recognize what
otherwise would escape your eye. Take the example of Fig. 4.9: data
have been generated from the hypothetical function $z = (1 + \alpha \sin n\phi)$
$(1 + \sin r)$, with a small coefficient α, $r^2 = x^2 + y^2$, and x, y proportional
to $\cos \phi, \sin \phi$. The very same data are represented as two-dimensional
probability densities in three different ways with quite different results.

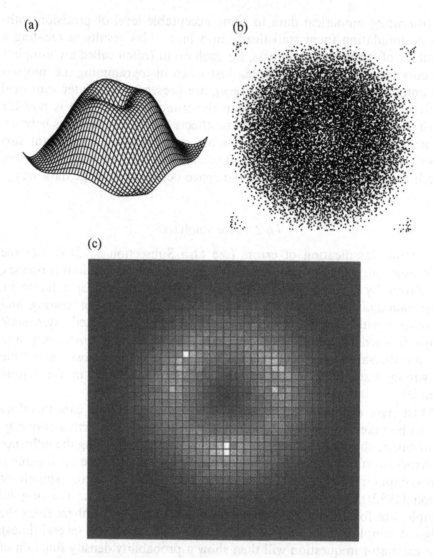

Fig. 4.9. Choosing the best representation: a fictitious two-dimensional probability distribution function is defined by pdf $= (1 + 0.03 \sin(3\phi))(1 + \sin(r))$, with ϕ and r the polar coordinates, obvious functions of x and y. (a) Shows a mesh plot of this function (with x, y extending from -4 to $+4$), without fluctuations; (b) and (c) show data with 15 000 points, (b) a scatter plot, (c) a two-dimensional histogram with the bins represented as pixels of an image, with the gray scale carefully adjusted to show the humps along the rim.

For the above examples, a commercial presentation tool has been used. In fact, there is a choice of popular commercial *interactive software packages* like Matlab (Matlab 1997) or Mathematica (Wolfram 1991). Both are language-based and cater for many statistical analysis tasks, numerical computation and visualization, for scientific and engineering applications. Other packages exist, commercial or public-domain, some of them most likely have equal merit as those we happen to mention.

Some limitations of commercial packages concerning their interfacing to other software tools in data input and final graphs were mentioned in Section 4.3 above; also price and local support may play a role in deciding which tool to use.

Before the advent of these packages, the high-energy community had created its own toolset, known under the package name PAW (PAW 1994), which at time of writing still has a faithful user community, and neither support nor development have been abandoned.

References

Aarnio, P. *et al.* (1991): The DELPHI detector at LEP. *Nuclear Instruments and Methods in Physics Research* **A303**, 233

Abachi, S. *et al.* (1994): The D0 detector. *Nuclear Instruments and Methods in Physics Research* **A338**, 185

Abbott, B. *et al.* (1997): *Fixed Cone Jet Definition in D0 and R_{sep}*. Fermilab, Batavia, Fermilab-PUB-97/242-E

Abe, K. *et al.* (1995): The endcap Cherenkov Ring Imaging Detector at SLD. *IEEE Transactions on Nuclear Science* **NS-42**, 518

Abe, K. *et al.* (1996): Obtaining physics results from the SLD CRID. *Nuclear Instruments and Methods in Physics Research* **A371**, 195

Abramowicz, H., Caldwell, A. and Sinkus, R. (1995): Neural network based electron identification in the ZEUS calorimeter. *Nuclear Instruments and Methods* **A365**, 508

Abramowitz, M. and Stegun, I.A. (1970): *Handbook of Mathematical Functions*. Dover Publications Inc., New York

Abreu, P. *et al.* (1996): Performance of the DELPHI detector. *Nuclear Instruments and Methods in Physics Research* **A378**, 57

Abshire, G. *et al.* (1979): Measurement of Electron and Pion Cascades in a Lead-Acrylic Scintillator Shower Detector. *Nuclear Instruments and Methods* **164**, 67

Acosto, D., *et al.* (1992): Lateral shower profiles in a lead/scintillating fiber calorimeter. *Nuclear Instruments and Methods in Physics Research* **A316**, 184

Adloff, C. *et al.* (1997): Measurement of Event Shape Variables in Deep Inelastic *ep* Scattering. *Physics Letters B* **406**, 256

ADSP (1995): *ADSP-2106x SHARC User's Manual*. Analog Devices, First Edition 1995

Ahlen, S.P. (1980): Theoretical and Experimental Aspects of the Energy Loss of Relativistic Heavily Ionizing Particles. *Review of Modern Physics* **52**, 121

Akesson, T. *et al.* (1984): Properties of Jets in High-E_T Events Produced in pp Collisions at $\sqrt{s} = 63$ GeV. *Zeitschrift für Physik C* **25**, 13

Akesson, T. *et al.* (1985): Properties of a Fine-sampling Uranium–Copper Scintillator Hadron Calorimeter. *Nuclear Instruments and Methods in Physics Research* **A241**, 17

Akrawy, M. *et al.* (1990): Development studies for the OPAL end cap electromagnetic calorimeter using vacuum photo triode instrumented lead glass. *Nuclear Instruments and Methods in Physics Research* **A290**, 76

Albajar, C. *et al.* (1987): Events with Large Missing Transverse Energy at the CERN collider, I. W → τν Decay. *Physics Letters B* **185**, 233

Allen, A.O. (1978): *Probability, Statistics and Queuing Theory.* Academic Press, London

Allison, W. and Wright, P. (1987): The Physics of Charged Particle Identification: dE/dx, Cherenkov and Transition Radiation. In: *Experimental Techniques in High Energy Physics*, Addison-Wesley Publishing Company Inc., Menlo Park, California, Ed. T. Ferbel

Amaldi, U. (1981): Fluctuations in Calorimeter Measurements. *Physica Scripta* **23**, 409

Amsler, C. (1998): Proton–Antiproton Annihilation and Meson Spectroscopy with the Crystal Barrel. *Reviews of Modern Physics* **70**(4), 1293

Anderson, B. and Moore, J. (1979): *Optimal Filtering.* Prentice Hall, Englewood Cliffs, N.J.

Andrews, H.C. (1972): *Introduction to Mathematical Techniques in Pattern Recognition.* Wiley Interscience, New York

Antiero, D. *et al.* (1994): A high stability light emitting diode system for calibration and monitoring of the lead glass electromagnetic calorimeter in the NOMAD experiment at CERN. *Proceedings of the 1994 International Conference on Calorimetry in High Energy Physics at BNL*, World Scientific, Singapore, Eds. H.A. Gordon and D. Rueger

Aquilar-Benitez, M. *et al.* (1983): The European Hybrid Spectrometer – A Facility to Study Multihadron Events Produced in High Energy Interactions. *Nuclear Instruments and Methods* **205**, 79

Arnison, G. *et al.* (1983a): Experimental Observation of Isolated Large Transverse Energy Electrons with Associated Missing Energy at √s = 540 GeV. *Physics Letters B* **122**, 103

Arnison, G. *et al.* (1983b): Hadronic Jet Production at the CERN Proton–Antiproton Collider. *Physics Letters B* **132**, 214

Arnison, G. *et al.* (1986): Analysis of the Fragmentation Properties of Quark and Gluon Jets at the CERN SPS p̄p Collider. *Nuclear Physics* **B276**, 253

ATLAS (1994): ATLAS Collaboration, *ATLAS Technical Proposal.* CERN, Geneva, CERN/LHCC 94-13

ATLAS (1997): ATLAS Collaboration, *ATLAS Inner Detector Technical Design Report.* CERN, Geneva, CERN/LHCC 97-16

Aubert, J.J. and Broll, C. (1974): Track Parametrization in the Inhomogeneous Field of the CERN Split Field Magnet. *Nuclear Instruments and Methods* **120**, 137

Axelrod, T.S., Dubois, P.F. and Eltgroth, P.G. (1983): A Simulator for MIMD Performance Prediction – Application to the S-1 MKIIa Multiprocessor. *Parallel Computing* **1**, 237

Banner, M. *et al.* (1982): Observation of Very Large Transverse Momentum Jets at the CERN p̄p Collider. *Physics Letters B* **118**, 203

Barnett, R.M. *et al.* (1996): Review of Particle Properties. *Physics Letters D* **54**, 1

Barr, G.D. *et al.* (1996): Performance of an electromagnetic liquid krypton calorimeter based on a ribbon electrode tower structure. *Nuclear Instruments and Methods in Physics Research* **A370**, 413

Barry, D. (1994): ODBMS Benchmarks. *Object Magazine*, October 1994

Bartl, W. *et al.* (eds.) (1995): Proceedings of the Seventh International Wirechamber Conference, Vienna, Austria. *Nuclear Instruments and Methods in Physics Research* **A367**, Nos. 1–3

Baur, R., *et al.* (1994): The CERES RICH detector system. *Nuclear Instruments and Methods in Physics Research* **A343**, 87

Becker, J.J. *et al.* (1984): *A New Approach to Track Finding and Fitting in Vector Drift Chambers.* SLAC, Stanford, SLAC-PUB-3442

Beer, A., *et al.* (1984): The Central Calorimeter of the UA2 Experiment at the CERN p̄p Collider. *Nuclear Instruments and Methods in Physics Research* **A224**, 360

Behrens, U. *et al.* (1990): Test of the ZEUS forward calorimeter prototype. *Nuclear Instruments and Methods in Physics Research* **A289**, 115

Behrens, U. *et al.* (1992): Calibration of the forward and rear ZEUS calorimeters using cosmic ray muons. *Nuclear Instruments and Methods in Physics Research* **A339**, 498

Ben-Ari, M. (1982): *Principles of Concurrent Programming.* Prentice Hall International, London

Berkelman, K. (1981): *Track Finding and Reconstruction in the Cleo Drift Chamber.* Cornell University, Ithaca NY, Report CBX-81-6

Bethe, H.A. and Ashkin, J. (1953): Passage of radiations through matter. In: *Experimental Nuclear Physics*, Wiley, New York, Ed. E. Segre

Bichsel, H. (1970): Straggling of Heavy Charged Particles: Comparison of Born Hydrogenics Wave-Function Approximation with Free-Electron Approximation. *Physical Review B* **1**(7), 2854

Billoir, P. (1984): Track Fitting with Multiple Scattering: A New Method. *Nuclear Instruments and Methods in Physics Research* **225**, 352

Billoir, P. (1986): *Propagation of Transverse Errors for Charged Tracks*, CERN, Geneva, CERN-DELPHI, 86-66, PROG 52

Billoir, P. (1987a): *Precise Tracking in a Quasi-Homogeneous Magnetic Field*, CERN, Geneva, CERN-DELPHI, 87-6, PROG 65

Billoir, P. (1987b): *Error Propagation in the Helix Track Model*, CERN, Geneva, CERN-DELPHI, 87-4, PROG 63

Billoir, P. (1989): Progressive track recognition with a Kalman-like fitting procedure. *Computer Physics Communications* **57**, 390

Billoir, P., Frühwirth, R. and Regler, M. (1985): Track Element Merging Strategy and Vertex Fitting in Complex Modular Detectors. *Nuclear Instruments and Methods in Physics Research* **A241**, 115

Billoir, P. and Qian, S. (1990a): Simultaneous pattern recognition and track fitting by the Kalman filtering method. *Nuclear Instruments and Methods in Physics Research* **A294**, 219

Billoir, P. and Qian, S. (1990b): Further test for the simultaneous pattern recognition and track fitting by the Kalman filtering method. *Nuclear Instruments and Methods in Physics Research* **A295**, 492

Birsa R. *et al.* (1977): Reconstruction of the Momentum of a Particle Moving in an Axially Symmetric Magnetic Field. *Nuclear Instruments and Methods* **146**, 357

Bischof, H. and Frühwirth, R. (1998): Recent Developments in Pattern Recognition with Applications in High Energy Physics. *Nuclear Instruments and Methods in Physics Research* **A419**, 259

Bloom, E.D. and Peck, C.W. (1983): Physics with the Crystal Ball Detector. *Annual Review of Nuclear and Particle Science* **33**, 143

Blucher, E. *et al.* (1986): Tests of Cesium Iodide Crystals for an Electromagnetic Calorimeter. *Nuclear Instruments and Methods in Physics Research* **A249**, 201

Bock, R.K., Hansl-Kozanecka, T. and Shah, T.P. (1981): Parameterization of the Longitudinal Development of Hadronic Showers in Sampling Calorimeters. *Nuclear Instruments and Methods* **18**, 533

Bock, R.K. *et al.* (1984): Lorentz Force: Trajectory of a Charged Particle; Units. In: *Formulae and Methods in Experimental Data Evaluation.* Vol. 1. European Physical Society, Geneva

Bock, R.K. and Vasilescu, A. (1998): *The Particle Detector BriefBook.* Springer, Berlin

Booch, G. (1994): *Object-oriented analysis and design with applications.* Addison-Wesley, Menlo Park, California

Bouclier, R. *et al.* (1974): Proportional Chambers for a 50000 Wire Detector. *Nuclear Instruments and Methods* **115**, 235

Boudreau, J. (1997): Experiences with Open Inventor. *Proceedings HEPVIS 1996,* CERN, Geneva, CERN 97-01

Brammer, K. and Siffling, G. (1975): *Kalman-Bucy-Filter.* R. Oldenburg Verlag, München-Wien

Brandt, S. and Dahmen, H.D. (1979): Axes and Scalar Measures of Two-Jet and Three-Jet Events. *Zeitschrift für Physik C* **1**, 61

Breiman, N. *et al.* (1984): *Classification and Regression Trees.* Wadsworth, Belmont, California

Brinkmann, R. *et al.* (1997): *Conceptual Design of a 500 GeV e^+e^- Linear Collider with Integrated X-ray Laser Facility.* DESY, Hamburg, DESY 1997-048

Brückmann, H., Behrens, U. and Anders, B. (1988): Hadron Sampling Calorimetry, a Puzzle of Physics. *Nuclear Instruments and Methods in Physics Research* **A263**, 136

Brun, R. *et al.* (1979): *MUDIFI: Multidimensional Fit Program.* CERN, Geneva, Internal Note DD/US/69 (Revised 1979)

Brun, R. *et al.* (1985): *The GEANT3 Electromagnetic Shower Program and a Comparison with the EGS3 Code.* CERN, Geneva, Internal Note DD/85/1

Brun, R., Hansroul, M. and Kubler, J. (1980): *LINTRA – A Principal Component Analysis Program.* CERN, Geneva, CERN/DD/US/70

Bryant, P.J. and Johnsen, K. (1993): *The Principles of Circular Accelerators and Storage Rings.* Cambridge University Press, Cambridge

Buchanan, C.D. (1982): Proportional Mode Calorimeters of the TPC Facility. *Proceedings of the Gas Calorimeter Workshop,* Fermilab, October 1982

Budinich, M. and Esquivel, S. (1990): An efficient algorithm for charged tracks pattern recognition. *Computer Physics Communications* **58**, 83

Bugge, L. (1986): On the Determination of Shower Central Positions from Lateral Sampling. *Nuclear Instruments and Methods in Physics Research* **A242**, 228

Bugge, L. and Myrheim, J. (1981): Tracking and Track Fitting. *Nuclear Instruments and Methods* **179**, 365

Buontempo, S. *et al.* (1995): Calibration and performance of the CHORUS calorimeter. *Nuclear Physics B (Proc. Suppl.)* **44**, 45

Burke, S. *et al.* (1996): Track finding and fitting in the H1 Forward Track Detector. *Nuclear Instruments and Methods in Physics Research* **A373**, 227

Burrows P.N. (1987): *Multicluster Analysis of Hadronic Data at C.M.Energies between 12.0 and 46.8 GeV.* Contribution to the Lepton-Photon Conference, Hamburg 1987

Bursky, D. (1985): Digital GaAs ICs, Technology Report. *Electronic Design,* December 1985

Busi, C. *et al.* (1983): *Proposal to the CERN SPSC.* CERN, Geneva, CERN/SPSC/83-24 (P186)

Cashmore, R. *et al.* (1985): Monte Carlo Studies Towards the Design of Iron/Scintillator and Uranium/Scintillator Calorimeters. *Nuclear Instruments and Methods in Physics Research* **A242**, 42

Cassel, D.G. and Kowalski H. (1981): Pattern Recognition in Layered Track Chambers Using a Tree Algorithm. *Nuclear Instruments and Methods* **185**, 235

Cattell, R.G.G. (1991): *Object Data Management.* Addison-Wesley, Menlo Park, California

Ceccucci, A. (1996): The NA48 liquid krypton calorimeter project. *Nuclear Instruments and Methods in Physics Research* **A379**, 478

Charpak, G. *et al.* (1968): The Use of Multiwire Proportional Counters to Select and Localize Charged Particles. *Nuclear Instruments and Methods in Physics Research* **62**, 262

Charpak, G. (1978): Multiwire and Drift Proportional Chambers. *Physics Today* **31**, 23

Chen, P.P.-S. (1976): The Entity-Relationship Model – Toward a Unified View of Data. *ACM Transactions on Database Systems* **1**, 9

Chen, T.Y. and Yao, N.G. (1993): Track fitting in non-Gaussian coordinate errors. *Nuclear Instruments and Methods in Physics Research* **A329**, 479

Chernov, N. *et al.* (1993): Track and vertex reconstruction in discrete detectors using Chebyshev metrics. *Computer Physics Communications* **74**, 217

CMS (1994): CMS Collaboration, *CMS Technical Proposal.* CERN, Geneva, CERN/LHCC 94-38

CMS (1998): CMS Collaboration, *The Tracker Project Technical Design Report.* CERN, Geneva, CERN/LHCC 98-6

Coffman, D.M. (1987): *Properties of Semileptonic Decays of Charmed D Mesons.* Thesis, California Institute of Technology, Internal note CALT-68-1415

Conetti, S. (1984): A Review of Triggers and Special Computing Hardware for the Fermilab Fixed-Target Program. *Proceedings of the Symposium on Recent Development in Computing, Processor and Software Research for High Energy Physics*, Universidad Autonoma de Mexico, Guanajuato, Mexico

Cox, M.G. (1982): *Practical Spline Approximation.* National Physics Laboratory Report DITC 1/82, Taddington, Middlesex TW11 OLW, UK

Cushman, P. (1992): Electromagnetic and Hadronic Calorimeters. In: *Instrumentation in High-energy Physics*, World Scientific, Singapore, Ed. F. Sauli

Dahl-Jensen, E. (1979): *Track Finding in the R807 Detector.* CERN, Geneva, CERN/R807/8

Das, S.R. (1973): On a New Approach for Finding all the Modified Cut-sets in an Incompatibilty Graph. *IEEE Transactions on Computers* **C-22**, No. 2, 187

Dell'Orso, M. and Ristori, L. (1990): A highly parallel algorithm for track finding. *Nuclear Instruments and Methods in Physics Research* **A287**, 436

Denby, B. (1988): Neural networks and cellular automata in experimental high energy physics. *Computer Physics Communications* **49**, 429

Dewdney, A.K. (1985): Computer Recreations. *Scientific American*, April 1985

Diddens, A.N. *et al.* (1980): A Detector for Neutral-Current Interactions of High Energy Neutrinos. *Nuclear Instruments and Methods* **178**, 27

Diehl, M. *et al.* (1997): Global optimization for track finding. *Nuclear Instruments and Methods in Physics Research* **A389**, 180

Dieudonné, J. (1979): *Eléments d'analyse;* tome 1. Gauthiers-Villars, Paris

Dorenbosch, J. *et al.* (1987): Calibration of the CHARM Fine-grained Calorimeter. *Nuclear Instruments and Methods in Physics Research* **A253**, 203

Drevermann, H. and Grab, C. (1990): Graphical concepts for the representation of events in high-energy physics. *International Journal of Modern Physics C* **1**, 147

Drevermann, H., Kuhn, D. and Nilsson, B.S. (1995): Event Display: Can we see what we want to see? *Proceedings of the 1995 CERN School of Computing*, CERN, Geneva, CERN 95-05, p. 17

Drevermann, H., Kuhn, D. and Nilsson, B.S. (1997): Novel transformations for very high-multiplicity events. *Proceedings HEPVIS 1996*, CERN, Geneva, CERN 97-01, p. 55

Drijard, D., Ekelöf, T. and Grote, H. (1980): On the Reduction in Space Resolution of Track Detectors caused by Correlations in the Coordinate Quantization. *Nuclear Instruments and Methods* **176**, 389

Eadie, W.T., Drijard, D., James, F.E., Roos, M. and Sadoulet, B. (1971): *Statistical Methods in Experimental Physics.* North-Holland Publishing Company, Amsterdam

Eccles, J.C. (1994): *How the Self Controls Its Brain.* Springer, New York

Efron, B. (1982): *The Jackknife, the Bootstrap, and Other Resampling Plans.* SIAM, Bristol

Eichinger, H. (1980): Global Methods of Pattern Recognition. *Nuclear Instruments and Methods* **176**, 417

Eichinger, H. and Regler, M. (1981): *Review of Track Fitting Methods in Counter Experiments.* CERN, Geneva, CERN 81-06

Eisenhandler, E. *et al.* (1984): *Electron Shower Profiles in the Gondolas.* CERN, Geneva, Internal Note UA1/TN 84-64

Ellsworth, R.W. *et al.* (1982): A Study of Albedo from a Hadronic Calorimeter for Energies ∼ 100–2000 GeV. *Nuclear Instruments and Methods in Physics Research* **203**, 167

EUR 4100 (1972): *CAMAC, A Modular Instrumentation System for Data Handling*

EUR 4600 (1972): *CAMAC, Organization of Multi-Crate Systems*

Eyges, L. (1948): Multiple Scattering with Energy Loss. *Physical Review* **74**, 1534

Fabjan, C.W. (1985): Calorimetry in High Energy Physics. In: *Concepts and Techniques in High Energy Physics III*, p. 281, Plenum Press, New York, Ed. T. Ferbel

Fabjan, C.W. (1991): Calorimetry in High Energy Physics. In: *Experimental Techniques in Nuclear and Particle Physics*, World Scientific, Singapore, Ed. T. Ferbel

Fabjan, C.W. *et al.* (1977): Iron Liquid-Argon and Uranium Calorimeters for Hadron Energy Measurement. *Nuclear Instruments and Methods* **141**, 61

Fassò, A., *et al.* (1994): FLUKA: present status and future developments. *Proceedings of the 4th International Conference on Calorimetry in High Energy Physics, La Biodola, Italy*, World Scientific, Singapore, Eds. A. Menzione and A. Scribano

FASTBUS (1983): *A Modular High Speed Data Acquisition System for High Energy Physics and Other Applications.* Esone Committee, Esone/FB/01

FASTBUS (1985): *IEEE Standard FASTBUS Modular High-Speed Data Acquisition and Control System.* The Institute of Electrical and Electronic Engineers, Inc, ISBN 0-471-84472-1

Fernow, R.C. (1986): *Introduction to Experimental Particle Physics.* Cambridge University Press, Cambridge

Ferrari, A. and Sala, P.R. (1994): A new model for hadronic interactions at intermediate energies. *Proceedings of the MC93 International Conference on Monte Carlo Simulation in High Energy and Nuclear Physics, Tallahassee, Florida*, World Scientific, Singapore, Eds. P. Dragovitsch, S.L. Linn and M. Burbank

Ferrari, A. and Sala, P.R. (1998): Intermediate and High Energy Models in FLUKA. *Proceedings of the Conference on Nuclear Data in Science and Technology, Trieste*, The Italian Physical Society, Bologna

Fesefeldt, H. (1985): *The Simulation of Hadronic Showers, Physics and Applications.* RWTH Aachen Report PITHA 85/02, Aachen

Fesefeldt, H., Hamacher, T. and Schug, J. (1990): Tests of punch-through simulation. *Nuclear Instruments and Methods in Physics Research* **A292**, 279

Fesefeldt, H. (1990): GEANT 3.14 Benchmark Tests. *Proceedings of the International Conference on Calorimetry in High Energy Physics, Fermilab*, World Scientific, Singapore, Eds. D.F. Andersen *et al.*

Fidecaro, C. *et al.* (1980): Measurement of the Polarization Parameter in pp Elastic Scattering at 150 GeV/c. *Nuclear Physics* **B173**, 513

Forden, G.E. and Saxon, D.H. (1985): *Improving Vertex Position Determination by Using a Kinematic Fit.* RAL-85-037, Rutherford Appleton Laboratory, Chilton

Freytag, D.R. and Walker, J.T. (1985): Performance Report for the Stanford/SLAC Microstore Analog Memory Unit. *IEEE Transactions on Nuclear Science* **NS-32**, 622

Fricke, U. (1996): *Identification of neutral pions in the ZEUS Detector by feed-forward neural nets.* DESY, Hamburg, Diploma thesis, F35D-96-09

Friedman, J.H. (1977): A Recursive Partitioning Decision Rule for Nonparametric Classification. *IEEE Transactions on Computers* **C-26**, 404

Friedman, J.H. and Tukey, J.W. (1974): A Projection Pursuit Algorithm for Exploratory Data Analysis. *IEEE Transactions on Computers* **C-23**, 881

Frodesen, A.G., Skjeggestad, O. and Tøfte, H. (1979): *Probability and Statistics in Particle Physics.* Universitetsforlaget, Bergen, Oslo and Tromsø

Fröhlich, A., Grote, H., Onions, C. and Ranjard F. (1976): *MARC – Track Finding in the Split Field Magnet.* CERN, Geneva, CERN/DD/76/5

Frühwirth, R. (1986): Estimation of Variances in a Linear Model Applied for Measurements of Trajectories. *Nuclear Instruments and Methods in Physics Research* **A243**, 173

Frühwirth, R. (1987): Application of Kalman Filtering to Track and Vertex Fitting. *Nuclear Instruments and Methods in Physics Research* **A262**, 444

Frühwirth, R. (1988): *Application of Filter Methods for the Reconstruction of Tools and Vertices in Events in Experimental High Energy Physics.* Thesis, University of Technology, Vienna

Frühwirth, R. (1993): Selection of optimal subsets of tracks with a feed-back neural network. *Computer Physics Communications* **78**, 23

Frühwirth, R. (1995): Matching of broken random samples with a recurrent neural network. *Nuclear Instruments and Methods in Physics Research* **A356**, 493

Frühwirth, R. (1997): Track fitting with non-Gaussian noise. *Computer Physics Communications* **100**, 1

Frühwirth, R. and Frühwirth-Schnatter, S. (1998): On the treatment of energy loss in track fitting. *Computer Physics Communications* **110**, 80

Frühwirth, R., Kubinec, P., Mitaroff, W. and Regler, M. (1996): Vertex reconstruction and track bundling at the LEP collider using robust algorithms. *Computer Physics Communications* **96**, 189

Gabriel, T.A. and Bishop, B.L. (1978): Calculated Hadronic Transmission Through Iron Absorbers. *Nuclear Instruments and Methods* **155**, 81

Gaskins, T. (1992): *PHIGS Programming Manual.* O'Reilly, Sebastopol, California

Gelb, A., Ed. (1975): *Applied Optimal Estimation.* MIT Press, Cambridge, Mass.

Georgiopoulos, C.H., Goldman, J.H., Levinthal, D. and Hodous, M.F. (1986): A Non-numerical Method for Track Finding in Experimental High Energy Physics Using Vector Computers. *Nuclear Instruments and Methods in Physics Research* **A249**, 451

Gemmeke, H. (1997): *SAND/1, Simply-Applicable Neural Device.* Infosheet Forschungszentrum Karlsruhe, Postfach 3640, D 76021 Karlsruhe

Gingrich, D.M., *et al.*(1995): Performance of a large scale prototype of the ATLAS accordion electromagnetic calorimeter. *Nuclear Instruments and Methods in Physics Research* **A364**, 290

Gluckstern, R.L. (1963): Uncertainties in Track Momentum and Direction due to Multiple Scattering and Measurement Errors. *Nuclear Instruments and Methods* **24**, 381

Graf, N.A. (1990): The D0 Shower Library. In: *Proceedings of the International Conference on Calorimetry in High Energy Physics, Fermilab*, World Scientific, Singapore, Eds. D.F. Andersen *et al.*

Grant, A. (1975): A Monte Carlo Calculation of High Energy Hadronic Cascades in Matter. *Nuclear Instruments and Methods* **131**, 167

Grassmann, H. and Moser, H.G. (1985): Shower Shape Analysis in Longitudinally Sampled Electromagnetic Calorimeters. *Nuclear Instruments and Methods in Physics Research* **A237**, 486

Gratta, G., Newman, H. and Zhu, R.Y. (1994): Crystal Calorimeters in Particle Physics. *Annual Review of Nuclear and Particle Science* **44**, 453

Greenhalgh, J.F. (1984): A Trigger Processor for a Fermilab Di-Muon Experiment. *Proceedings of the Symposium on Recent Development in Computing, Processor and Software Research for High Energy Physics*, Universidad Autonoma de Mexico, Guanajuato, Mexico

Greville, T.N.E. (1969): *Theory and Applications of Spline Functions*. Academic Press, New York, London

Grote, H. (1981): *Data Analysis for Electronic Experiments*. CERN, Geneva, CERN 81-03 p. 136

Grote, H. and Zanella, P. (1980): Applied Software for Wire Chambers. *Nuclear Instruments and Methods* **176**, 29

Grote, H., Hansroul, M., Lassalle, J.C. and Zanella, P. (1973): Identification of Digitized Particle Trajectories. *Proceedings of the International Computing Symposion Davos*, North Holland, Amsterdam

Halpin, T. (1995): *Conceptual schema and relational database design*. Prentice Hall, Englewood Cliffs, N.J.

Hauptman, J. (1979): *Calorimeter Analysis*. Internal Note HEE-121, UCLA High Energy Group, July 1979

Hayakawa, S. (1969): *Cosmic Ray Physics*. J. Wiley and Sons, New York

Heller, D. and Ferguson, P.M. (1994): *Motif Programming Manual*. O'Reilly, Sebastopol, California

Hertzog, D. *et al.* (1990): A High-resolution Lead/scintillating fibre Electromagnetic Calorimeter. *Nuclear Instruments and Methods in Physics Research* **A294**, 446

Highland, V. (1975): Some Practical Remarks on Multiple Scattering. *Nuclear Instruments and Methods* **129**, 479

Holder, M. *et al.* (1978): Performance of a Magnetized Total Absorption Calorimeter between 15 GeV and 140 GeV. *Nuclear Instruments and Methods* **151**, 69

Hornik, K. (1991): Approximation Capabilities of Multilayer Feedforward Networks. *Neural Networks* **4**, 251

Huber, P.J. (1981): *Robust Statistics.* Wiley, New York

ICRU (1984): *Stopping Powers for Electrons and Positrons*, ICRU Report 37, Washington DC

IEC (1994): IEC 273-3 or HD 493.3 S2, 19 inch crate mechanics, CENELEC, B 1050 Brussels 1994

Iwata, S. (1980): *Calorimeters for High Energy Experiments at Accelerators*, Nagoya University Report DPNU-13-80

Jackson, J.D. (1998): *Classical Electrodynamics.* John Wiley and Sons, New York

Jain, A.K. (1989): *Fundamentals of Digital Image Processing.* Prentice Hall, Englewood Cliffs, N.J.

James, F. (1972): Function Minimization. *Proceedings of the 1972 Cern Computing and Data Processing School, CERN, Geneva, CERN 72-21*

James, F. (1981): Determining the Statistical Significance of Experimental Results. *Proceedings of the 1980 CERN School of Computing, CERN, Geneva, CERN 81-03*

James, F. (1983): Fitting Tracks in Wire Chambers Using the Chebyshev Norm instead of Least Squares. *Nuclear Instruments and Methods* **211**, 145

James, F. and Roos, M.(1986): *MINUIT – 'Function Minimization and Error Analysis'.* CERN, Geneva, CERN Computer Program Library

Jaroslawsky, S. (1977): Processor for the Proportional Chamber of the TASSO Experiment at DESY designed by S. Jaroslawsky, Imperial College, London

Jonker, M. *et al.* (1982): The Response and Resolution of a Fine-Grain Marble Calorimeter for Hadronic and Electromagnetic Showers. *Nuclear Instruments and Methods in Physics Research* **200**, 183

Kälviäinen, H. (1994): *Randomized Hough Transform: New Extensions.* Research Paper 35, Lappeenranta University of Technology, Finland

Kaplan, D.M. (1984): A Parallel, Pipelined Event Processor for Fermilab Experiment 605. *Proceedings of the Symposium on Recent Development in Computing, Processor and Software Research for High Energy Physics*, Universidad Autonoma de Mexico, Guanajuato, Mexico

Kendall, M.G. and Stuart, A. (1967): *The Advanced Theory of Statistics, Volume II* (Interference and Relationship, second edition). Charles Griffin and Company Limited, London

Kitagawa, G. (1989): Non-Gaussian seasonal adjustment. *Computers & Mathematics with Applications* **18**, 503

Kitagawa, G. (1994): The two-filter formula for smoothing and an inplementation of the Gaussian-sum smoother. *Annals of the Institute of Statistical Mathematics* **46**, 605

Kölbig, K.S. and Schorr, B. (1984a): A program package for the Landau distribution. *Computer Physics Communications* **31**, 97

Kölbig, K.S. and Schorr, B. (1984b): Asymptotic expansions for the Landau density and distribution functions. *Computer Physics Communications* **32**, 121

Kostarakis, P. *et al.* (1981): A Fast Processor for Di-Muon Trigger. *Topical Conference on the Application of Microprocessors to High Energy Physics Experiments,* CERN, Geneva, CERN 81-07

Kowalski, H., Moehring, H.-J. and Tymieniecka, T. (1987): *High Speed Monte Carlo with Neutron Component NEUKA.* DESY, Hamburg, DESY 87-170

Krammer, M. *et al.* (eds.) (1998): Proceedings of the Eighth International Wirechamber Conference, Vienna, Austria. *Nuclear Instruments and Methods in Physics Research* **A419**, Nos. 2, 3

Kunszt, Z. (1987): Large Cross Section Processes. *Proceedings of the Workshop on Physics and Future Accelerators,* CERN, Geneva, CERN 87-07

Lala, P.K. (1986): *Fault Tolerance and Fault Testable Hardware Design.* Prentice Hall International, London

Landua, R. (1996a): New Results in Spectroscopy. *Proceeedings of the 28th International Conference on High Energy Physics,* World Scientific, Singapore, Eds. Z. Ajduk and A.K. Wroblewski

Landua, R. (1996b): Meson Spectroscopy at LEAR. *Annual Review of Nuclear and Particle Science* **46**, 351

Lankford, A.J. (1984a): A Review of Trigger and Online Processors at SLAC. *Proceedings of the Symposium on Recent Development in Computing, Processor and Software Research for High Energy Physics,* Universidad Autonoma de Mexico, Guanajuato, Mexico

Lankford, A.J. (1984b): The ASP Energy Trigger, A Review of Trigger and Online Processors at SLAC. *Proceedings of the Symposium on Recent Development in Computing, Processor and Software Research for High Energy Physics,* Universidad Autonoma de Mexico, Guanajuato, Mexico

Lassalle, J.C., Carena, F. and Pensotti, S. (1980): TRIDENT: A Track and Vertex Identification Program for the CERN Omega Particle Detector System. *Nuclear Instruments and Methods* **176**, 371

Laurikainen, P. (1971a): On the Gemetrical Fit of Bubble Chamber Tracks by Treating Multiple Scattering as Measurement Error. *Commentationes Physico-Mathematicae* **41**, 131

Laurikainen, P. (1971b): *Multiple Scattering and Track Reconstruction,* Report Series in Physics 35, University of Helsinki

Laurikainen, P., Moorhead, W.G. and Matt, W. (1972): Least Squares Fit of Bubble Chamber Tracks Taking into Account Multiple Scattering. *Nuclear Instruments and Methods* **98**, 349

LeCroy (1985): *Catalog for fast electronics.* LeCroy Research Systems Corporation, New York

Leo, W.R. (1987): *Techniques for Nuclear and Particle Physics Experiments (A How-To-Approach),* Springer Verlag, Berlin

Levit, L.B. and Vincelli, M.L. (1985): A Modular Implementation of Sophisticated Triggers. *Nuclear Instruments and Methods in Physics Research* **A235**, 396

Lindström, M. (1995): Track reconstruction in the ATLAS detector using elastic arms. *Nuclear Instruments and Methods in Physics Research* **A357**, 129

Litchfield, P. (1984): Partial Wave Analysis. In: *Formulae and Methods in Experimental Data Evaluation*, Vol. 2, European Physical Society, Geneva

Longo, E. and Luminari, L. (1985): Fast Electromagnetic Shower Simulation. *Nuclear Instruments and Methods in Physics Research* **A239**, 506

Longo, E. and Sestili, I. (1975): Monte Carlo Calculation of Photon-initiated Electromagnetic Showers in Lead Glass. *Nuclear Instruments and Methods* **128**, 283

Lucha, W. and Regler, M. (1997): *Elementarteilchenphysik in Theorie und Experiment*. Verlag Paul Sappl, Kufstein, Austria

Lüdemann, J., Ressing, D. and Wurth, R. (1996): A Sharc DSP Cluster as HERA-B DAQ Building Block. *VITA Europe Congress VMEbus, October 7–9th 1996, Brussels, Belgium*

Lütjens, G. (1981): How Can Fast Programmable Devices Enhance the Quality of Particle Experiments. *Topical Conference on the Application of Microprocessors to High Energy Physics Experiments*, CERN, Geneva, CERN 81-07

Maples, C. (1984): Experience with Scientific Applications on the MIDAS Multiprocessor System. *Proceedings of the Symposium on Recent Development in Computing, Processor and Software Research for High Energy Physics*, Universidad Autonoma de Mexico, Guanajuato, Mexico

Margenau, H. and Murphey, G.M. (1964): *The Mathematics for Physics and Chemistry*. Van Nostrand, Toronto

Marx, J.N. and Nygren, D.R. (1978): The Time Projection Chamber. *Physics Today* **31**, 46

Matlab (1997): *MATLAB 5 Reference Guide*. The MathWorks, Inc., 24 Prime Park Way, Natick, MA

Mecking, B. (1982): On the Accuracy of Track Reconstruction with Inhomogeneous Magnetic Detectors. *Nuclear Instruments and Methods* **203**, 299

Mess, K.H., Metcalf, M. and Orr, R.S. (1980): Track Finding in a Fine-Grained Calorimeter. *Nuclear Instruments and Methods* **176**, 349

Messel, H. and Crawford, D.F. (1970): *Electron-Photon Shower Distribution Function*. Pergamon Press, Oxford

Metcalf, M. (1986): *Computers in High Energy Physics*. In: Advances in Computers, Vol. 25, Academic Press, New York, Ed. M. Yovitis, p. 277

Metcalf, M. and Regler, M. (1973): Solution of the Equation of Motion of a Charged Particle in a Stationary Magnetic Field. *Journal of Computational Physics* **11**, No. 2, 240

Metcalf, M., Regler, M. and Broll, C. (1973): *A Split Field Magnet Geometry Fit Program: NICOLE*. CERN, Geneva, CERN 73-2

Mitaroff, W. (1986): In *Report on Global Track and Vertex Fitting in the DELPHI Detector*. CERN, Geneva, Internal Note DELPHI 86-99 PROG-61

Mitaroff, W. (1987): *Status of Utility Library for Helix Tracking and Error Propagation*. CERN, Geneva, Internal Note DELPHI 87-51 PROG-85

Monolithic Memories (1985): *PAL Programmable Array Logic Handbook*

Moorhead, W.G. (1960): *A Program for the Geometrical Reconstruction of Curved Tracks in a Bubble Chamber*. CERN, Geneva, CERN 60-33

Moravec, H. (1988): *Mind Children: the Future of Robot and Human Intelligence.* Harvard University Press, Cambridge, Mass.

Morse, P.M. (1958): *Queues, Inventories and Maintenance.* John Wiley & Sons Inc., New York

Motorola (1977): *Bit Slice Processor 10800*

Mourou, G.A., Bloom, D.M. and Lee, C.H. (1986): Picosecond Electronics and Optoelectronics. *Proceedings of the Topical Meeting, Lake Tahoe, Nevada 1985,* Springer, Berlin

MULTIBUS II (1984): *Multibus II Bus Architecture Specification Handbook,* Intel Corp.

Muraki, Y., *et al.* (1985): Radial and Longitudinal Behaviour of Hadronic Cascade Showers Induced by 300 GeV Protons in Lead and Iron Absorbers. *Nuclear Instruments and Methods in Physics Research* **A236**, 47

Murzin, V.S. (1967): *Progress in Elementary Particle and Cosmic-Ray Physics,* Vol. IX, p. 247, North-Holland, Amsterdam, Ed. J.G. Wilson

Myrheim, J. and Bugge, L. (1979): A Fast Runge Kutta Method for Fitting Tracks in a Magnetic Field. *Nuclear Instruments and Methods* **160**, 43

Nagy, E. *et al.* (1978): Measurement of Elastic Proton–Proton Scattering at Large Momentum Transfer at the CERN Intersecting Storage Rings. *Nuclear Physics B* **150**, 221

Neider, J. *et al.* (1993): *Open GL Programming Guide.* Addison-Wesley, Menlo Park, California

Nelson, W.R., Hirayama, H. and Rogers, D.W.O. (1985): *The EGS4 Code System,* SLAC, Stanford, SLAC-265

Notz, D. (1981): Microprocessors at DESY. *Topical Conference on the Application of Microprocessors to High Energy Physics Experiments,* CERN, Geneva, CERN 81-07

Notz, D. (1984): A Review of Triggers and Special Computing Hardware at DESY. *Proceedings of the Symposium on Recent Development in Computing, Processor and Software Research for High Energy Physics,* Universidad Autonoma de Mexico, Guanajuato, Mexico

Nye, A. (1992): *Xlib Programming Manual,* Third Edition, O'Reilly, Sebastopol, California

Oed, A. (1995): Properties of micro-strip gas chambers (MSGC) and recent developments. *Nuclear Instruments and Methods in Physics Research* **A367**, 34

Ohlsson, M., Petersson, C., and Yuille, A. (1992): Track finding with deformable templates – the elastic arms approach. *Computer Physics Communications* **71**, 77

Olsson, J., Steffen, P., Goddard, M.C., Peace, G.F. and Nozaki, T. (1980): Pattern Recognition Programs for the JADE Jet-chamber. *Nuclear Instruments and Methods* **176**, 403

Papadimitriou, C. and Steiglitz, K. (1982): *Combinatorial Optimization: Algorithms and Complexity.* Prentice Hall, Englewood Cliffs, N.J.

Particle Data Group (1998): Review of Particle Properties. *The European Physical Journal* **C3**

Pavel, T.J. (1996): *Measurement of charged hadron spectra at the Z^0 with Cherenkov ring imaging.* Stanford University, PhD Thesis, August 1996

PAW (1994): *PAW Reference Manual.* CERN, Geneva, CERN Program Library Q121, Long Writeup

PCI (1993): *PCI Local BUS Specification,* Revision 2.0, issued by the PCI special interest group, Hillsboro, Oregon, USA, April 30, 1993

Penrose, R. (1994): *Shadows of the Mind.* Oxford University Press, Oxford

Pernicka, M., Regler, M. and Sychkov, S. (1978): Drift Time Relations in Rotated MWPCs and Their Advantage in Practice. *Nuclear Instruments and Methods* **156**, 147

Peterson, C. (1989): Track finding with neural networks. *Nuclear Instruments and Methods in Physics Research* **A279**, 537

Pimiä, M. (1985): *Track Finding in the UA1 Central Detector at the CERN $\bar{p}p$ Collider.* University of Helsinki, HU-P-D45

Platner, E. (1976): Paper presented at the 1976 Nuclear Science Symposium and Scintillation and Semiconductor Counter Symposium, October 20-22, 1976, New Orleans, Louisiana, *IEEE Transactions on Nuclear Sciences* **NS-24**, Feb. 1977

Poenaru, D. N. and Greiner, W. (1997): *Experimental Techniques in Nuclear Physics.* Walter de Gruyter, Berlin

Posse, C. (1990): An Effective Two-dimensional Projection Pursuit Algorithm. *Communications in Statistics, Simulation and Computation* **19**, 1243

Posse, C. (1995): Tools for Two-dimensional Exploratory Projection Pursuit. *Journal of Computational and Graphical Statistics* **4**, 83

Regler M. (1968): *EQUMOT – 'Equations of Motion of a Particle in a Static Magnetic Field'.* CERN, Geneva, CERN Computer Program Library

Regler, M. (1977): Vielfachstreuung in der Ausgleichsrechnung. *Acta Physica Austriaca* **49**, 37

Regler, M. (1981): Influence of Computation Algorithms on Experimental Design. *Computer Physics Communications* **22**, 167

Regler, M. and Frühwirth, R. (1990): Reconstruction of Charged Tracks. In: *Concepts and Techniques in High Energy Physics V*, Plenum Publishing Corporation, New York, Ed. T. Ferbel

Regler, M., Frühwirth, R. and Mitaroff, W. (1996): Filter Methods in Track and Vertex Reconstruction. *International Journal of Modern Physics C* **4**, 521

Rehlich, K. (1980): Processor for the Vertexdetector of the TASSO experiment at DESY designed by K. Rehlich, DESY, Hamburg

RICH (1996): Proceedings of the RICH 96 conference, Ed. T. Ekelöf. *Nuclear Instruments and Methods in Physics Research* **A371**

Richman, J. (1986): PhD Thesis, CALTECH Internal note CALT-68-1231

Rossi, B. (1965): *High Energy Particles.* Prentice Hall, Englewood Cliffs, N.J.

Rossi, B. and Greisen, K. (1941): Cosmic Ray Theory; §23. The Distribution Function. *Review of Modern Physics* **13**, 265

Rousseeuw, P.J. and Leroy, A.M. (1987): *Robust Regression and Outlier Detection*. Wiley-Interscience, New York

Rumbaugh, J. *et al.* (1991): *Object-oriented modeling and design*. Prentice Hall, Englewood Cliffs, N.J.

Sauli, F. (1978): Limiting Accuracies in Multiwire Proportional and Drift Chambers. *Nuclear Instruments and Methods* **156**, 147

Sauli, F. (1987): Principles of Operation of Multiwire Proportional and Drift Chambers. In: *Experimental Techniques in High Energy Physics*, Addison-Wesley Publishing Company Inc., Menlo Park, California, Ed. T. Ferbel, p. 79

Scherzer, J. (1997): *Global optimization for track finding in CMS*. Ph.D. thesis, Vienna University of Technology, Vienna

Schildt, P., Stuckenberg, H.-J. and Wermes N. (1980): An On-line Track Following Microprocessor for the Petra Experiment Tasso. *Nuclear Instruments and Methods* **178**, 571

Schmidt, B. (1998): Microstrip Gas Chambers: Recent Developments, Radiation Damage and Long Term Behaviour. *Nuclear Instruments and Methods in Physics Research* **A419**, 230

Schorr, B. (1974): *Introduction to Reliability Theory*. CERN, Geneva, CERN 74-16

Schorr, B. (1976): *Introduction to Cluster Analysis*. CERN, Geneva, Internal Note DD/76/3

Scott, W.M. (1963): The Theory of Small-Angle Multiple Scattering of Fast Charged Particles. *Review of Modern Physics* **35**, 231

Séguinot, J. and Ypsilantis, T. (1977): Photo-Ionization and Cherenkov Ring Imaging. *Nuclear Instruments and Methods* **142**, 377

SHARC (1997): *Analog Dialogue*, Volume 28, No. 3, 1994, published by Analog Devices

Sinkus, R. and Voss, T. (1997): Particle identification with neural networks using a rotational invariant moment representation. *Nuclear Instruments and Methods in Physics Research* **A391**, 360

Stampfer, D., Regler, M. and Frühwirth, R. (1994): Track fitting with energy loss. *Computer Physics Communications* **79**, 157

Sternheimer, R.M. (1952): The Density Effect for the Ionization Loss in Various Materials. *Physical Review* **88**, 851

Stimpfl-Abele, G. and Garrido, L. (1991): Fast track finding with neural networks. *Computer Physics Communications* **64**, 46

Stimpfl-Abele, G. (1991): Recognition of decays of charged tracks with neural network techniques. *Computer Physics Communications* **67**, 183

Strandlie, A. *et al.* (2000): Track fitting on the Riemann sphere. *Proceedings of the International Conference on Computing in High-Energy and Nuclear Physics (CHEP2000)*, Padova

Stuckenberg, H.-J. (1968): *Nukleare Elektronik I*. DESY, Hamburg, DESY F56-1

Stuckenberg, H.-J. (1981): Two Level Triggering in Storage Ring Experiments. *Topical Conference on the Application of Microprocessors to High Energy Physics Experiments*, CERN, Geneva, CERN 81-07

Sun, J. (1991): Significance Levels in Exploratory Projection Pursuit. *Biometrika* **78**, 759

Synertec (1976): The C10115 is produced by Synertec, 2050 Coronado Drive, Santa Clara, CA 95051, USA

Taylor, D. (1992): *Object-oriented Information Systems: Planning and Implementation.* John Wiley, New York

Toki, W. *et al.* (1984): The Barrel Shower Counter for the Mark III Detector at Spear. *Nuclear Instruments and Methods in Physics Research* **219**, 479

Trischuk, W. (1998): Semiconductor Trackers for Future Particle Physics Detectors. *Nuclear Instruments and Methods in Physics Research* **A419**, 251

Tysarczyk, G., Mättig, P. and Lohrmann, E. (1985): *Separation of π^0 and Single γ's at High Energies.* DESY, Hamburg, Tasso Note 351

VMEbus (1987): *The VMEbus Specification* Conforms to ANSI/IEEE STD1014-1978, IEC 821 and 297 VITA, Scottsdale, AZ 85253 USA Revision C, November 1986

Walenta, A.H., Heinze, J. and Schürlein, B. (1971): The Multiwire Drift Chamber, A New Type of Proportional Wire Chamber. *Nuclear Instruments and Methods* **92**, 373–380

Waloschek, P. (1984): *Fast Trigger Techniques.* DESY, Hamburg, DESY 80-114

Weilhammer, P. (1986): *Experience with Si Detectors in NA32.* CERN, Geneva, CERN-EP/86-54

Weinberg, G.M. (1972): *The Psychology of Computer Programming.* Computer Science Series, van Nostrand Reinhold Company, New York

Wells, L.K. and Trevis, J. (1996): *LabView for Engineers.* Prentice Hall, Englewood Cliffs, N.J.

Wernecke, J. (1993): *The Open Inventor Mentor.* Addison-Wesley, Menlo Park, California

Wigmans, R. (1991a): Advances in Hadron Calorimetry. *Annual Review of Nuclear and Particle Science* **41**, 133

Wigmans, R. (1991b): High-resolution Hadron Calorimetry. In: *Experimental Techniques in Nuclear and Particle Physics*, World Scientific, Singapore, Ed. T. Ferbel

Wimpenny, S.J. and Winer, B.L. (1996): The Top Quark. *Annual Review of Nuclear and Particle Science* **46**, 149

Wind, H. (1972): Function Parametrisation. *Proceedings of the 1972 CERN Computing and Data Processing School*, CERN, Geneva, CERN 72-21, p. 53

Wind, H. (1974): Momentum Analysis by Using a Quintic Spline Model for the Track. *Nuclear Instruments and Methods* **115**, 431

Wind, H. (1978): An Improvement to Iterative Tracking for Momentum Determination. *Nuclear Instruments and Methods* **153**, 195

Wind, H. (1979): The Use of a Non-diagonal Weight Matrix for Momentum Determination with Magnetic Spectrometers. *Nuclear Instruments and Methods* **161**, 327

Wolfram, S. (1991): *Mathematica.* Addison-Wesley, Menlo Park, California

Wu, S. L. (1984): e^+e^- Physics at Petra – The First Five Years. *Physics Reports* **107**, 95

Young, T.Y. and Calvert, T.W. (1974): *Classification, Estimation, and Pattern Recognition.* Elsevier Publishing Company, Amsterdam

Ypsilantis, T. (1981): Cherenkov Ring Imaging. *Physica Scripta* **23**, 371

Ypsilantis, T. and Seguinot, J. (1994): Theory of ring imaging Cherenkov counters. *Nuclear Instruments and Methods in Physics Research* **A343**, 30

Zacharov, V. (1982): Parallelism and Array Processing. *Proceedings of the 1982 Cern School of Computing*, CERN, Geneva, CERN 83-03

Zahn, C.T. (1973): Using the Minimum Spanning Tree to Recognize Dotted and Dashed Curves. *Proceedings of the International Computing Symposion Davos*, North Holland, Amsterdam (1974) p. 381

ZEUS (1986): The ZEUS Detector; Technical Proposal; DESY, Hamburg, March 1986. The ZEUS Detector; Status Report; DESY, Hamburg, February 1993

ZEUS (1989): *ZEUS Calorimeter First Level Trigger.* DESY, Hamburg, ZEUS Note 89-85

Zhu, R., *et al.* (1995): An RFQ calibration system for the L3 BGO calorimeter. *Nuclear Physics B (Proc. Suppl.)* **44**, 109

Zupančič, Č. (1986): A Simplified Algorithm for Momentum Estimation in Magnetic Spectrometers. *Nuclear Instruments and Methods in Physics Research* **A248**, 461

Index

Printed in the United States
By Bookmasters